GENERAL BOTANY

By

E.N. Transeau

DPH

DISCOVERY PUBLISHING HOUSE

NEW DELHI-110002

Reprinted - 2023

ISBN: 978-81-7141-252-5

General Botany

Published by:

DISCOVERY PUBLISHING HOUSE

4383/4B, Ansari Road, Darya Ganj

New Delhi-110 002 (India)

Phone: +91-11-23279245; 23253475; 43596065

+91 9811179893 / +91 9871656464

E-mail: discoverybooksindia@gmail.com

orderdphbooks@gmail.com

namitwasan9@gmail.com

web: www.discoverypublishinggroup.com

Printed at:

Infinity Imaging Systems

Delhi

PREFACE

THIS book is a by-product of a rather extended experience in the teaching of botany and especially of seven years of effort to develop an introductory course that would give students a broad view of the subject and enable them to see its problems and appreciate the importance of the solution of these problems.

The first step in developing the course was the arranging of laboratory and field work. The second was the working out of the classroom discussions so that the class work and the actual work with the plant materials would run parallel and supplement each other. The laboratory directions have been used and revised at intervals during the past five years, and the textbook has been tested and revised and used in mimeographed form for the past two years. During this period the work has been presented to upward of three thousand students and the course has been improved in every detail through the experience and constructive criticisms of the eight instructors who have tried out various methods and devices and different arrangements of the material with their classes at The Ohio State University.

Suggestions and ideas for guidance in the selection of subject matter have been derived from four sources. The first of these is the traditional course in general botany. This embodies the facts and principles which most botanists agree are essential for a foundation in the subject, and these are to be retained unless there is definite reason why they should be set aside.

The second source from which suggestions as to subject matter have come is the large body of men engaged in the teaching or practice of horticulture, agriculture, and forestry. These workers in the applied fields of botany are the men who more than any others use information concerning plants, and certainly a course in general botany should afford a foundation for their courses and their practice.

A third source of suggestions and criticism is the students to whom the course has been given. The questions they have asked, the relative interest they have manifested in different kinds of subject matter, and their responses to various methods of presentation have furnished a valuable basis for evaluating ideas and suggestions from the first two sources.

A fourth set of suggestions has come from the questions asked by the public. These inquiries are usually very practical. Often they are unanswerable, but an introductory course should enable a student either to answer many of them or to find the information that is called for.

Suggestions from the above four sources have been consciously sought during the working out of the course. We have attempted to avoid trespassing on allied fields, but we have not hesitated to point out many important uses and applications of botanical principles.

The selection of subject matter, however, is only one of two important elements in presenting any science course. Equally important is the efficient use of the time given to field and laboratory exercises, which after all are the heart of the course. Textbooks may furnish a fund of information that will conserve time in the classroom, but it is the work in the field and the laboratory that tests the ability and insight of the student, and makes real (or sometimes unreal) classroom and textbook statements.

The use of the field and laboratory time for the answering of questions and the solving of problems, rather than for the making of detailed drawings, has changed the attitude of our students toward laboratory work. It has also made it possible to cover a far greater range of materials and principles than formerly, and to give the student with a scientific mind as good a chance as the student with an artistic hand. We prefer to use the laboratory and field periods not for drawing exercises but for study and recitation in the presence of the materials.

Although the laboratory and field work were developed first, the textbook is being published in advance of the laboratory outline. This is done because having an illustrated textbook in the students' hands renders certain changes in the outline advisable. These changes are now being made and the outline will be published in the near future. In it a fuller explanation of how the laboratory and field work has been conducted will be given.

Throughout the book the author has tried to avoid purposeful explanations and words implying such explanations. Teleology answers all questions by the easiest method, and closes the mind of the student to the means by which scientific explanations may be discovered. It is an inheritance from the dark ages and should be eradicated from the laboratory and classroom. Students should learn at the very beginning that plant phenomena, so far as we now know, take place in the plant in accord with the laws of physics and chemistry; that they do not happen because of some alleged purpose any more than hydrochloric acid unites with soda in order to form table salt. If certain phenomena and structures eventuate to the advantage of the plant, well and good. There are many that do not! And neither the advantageous nor the disadvantageous should be cited as a cause.

E. N. T.

ACKNOWLEDGMENTS

THE author is deeply indebted to his associates at The Ohio State University. Professor H. C. Sampson has had direct charge of this general course since it was first given and has conducted various experiments in teaching methods that have greatly improved the efficiency of the course. He has read and criticized the various revisions of the manuscript, has suggested numerous changes in the text, and has contributed much both to the form and substance of the book. Professor A. E. Waller, Professor W. G. Stover, Dr. L. H. Tiffany, Dr. E. L. Stover, and Dr. J. D. Sayre have read a part or all of the manuscript and its revision. Without their coöperation and constructive ideas neither the course nor the book would have had its present form.

Professor H. N. Whitford, and Professor G. E. Nichols of Yale University and Professor W. S. Cooper of the University of Minnesota, made helpful suggestions regarding the chapters on plant distribution. Dr. Cooper also generously supplied a number of photographs. The author's thanks are also due to various men in the United States Department of Agriculture who have furnished photographs of plants used in their investigations. These are credited below the illustrations in the book.

CONTENTS

Contents

GENERAL BOTANY

CHAPTER ONE

PLANTS FROM OUR STANDPOINT

One of the most pressing economic problems of the world today is the securing of an adequate food supply. In the older and more densely populated parts of Asia an unfavorable growing season has for centuries meant famine and death for thousands of persons, and as a result of the world war nearly every nation on the globe has recently experienced inconvenience or suffering because of limited food resources. The fact that the population of the earth is increasing far more rapidly than the food supply should give us an increased interest in plants, the primary source of all foods. When we realize further that our resources of lumber, fuel, fibers, paper pulp, oils, resin, rubber, and numerous other products come from plants, our absolute dependence on plant life for our necessities and comforts is apparent.

How plants affect our own lives. Before taking up the study of plant life let us enumerate briefly some of the more important ways in which plants contribute to our welfare or detract from it.

1. *Foods derived from plants.* The principal foods of all nations are derived directly from plants. Animal foods are of secondary importance, and all animals live directly or indirectly upon plants. Agriculture is, therefore, the most fundamental of human occupations, for of all living beings green plants alone are able to organize the simple materials found in the air, water, and soil into the complex substances which all plants and animals must have for food.

2. *Fuel a plant product.* A second necessity of all nations is a fuel supply. Like food, fuel is primarily a plant product. Wood is the most universal source of heat and light energy. Coal,

petroleum, and natural gas, although obtained from the earth, are the products of plants which lived in former geological times. When wood is burned, the great store of energy which the tree has accumulated from sunlight is released in the form of heat. When coal, petroleum, or natural gas is burned, the energy stored by plants from sunlight of former geological ages is liberated.

3. *Plant fibers.* Of almost equal importance is the production of fibers for clothing and many other articles. Such plants as cotton, flax, and hemp supply the bulk of these fibers, and the fibers not directly derived from plants come from animals that feed on plants. Artificial silk and vegetable wool are recent substitutes for animal fibers that are being manufactured from plant fibers.

4. *Wood products.* Lumber is a primary necessity in the construction of houses and buildings of all kinds. It is also the chief material for furniture and countless other articles of household use and ornament. Paper, the principal medium of communication and commercial exchange, is also essentially a wood product.

5. *Oils, resins, and drugs.* Plants have played an important part in the progress of civilization by supplying oils and fats, gums, resins, dyes, rubber, drugs, alcohols, cork, the materials for explosives, and many other basic substances for the arts and industries.

6. *Other uses of vegetation.* Trees and grasses are the great stabilizers of the soil on mountains and in valleys. They help to retain flood water and prevent destructive erosion. They provide food and shelter for numerous wild animals that are of great economic importance. The plants of our lakes, ponds, and streams are the primary sources of food and shelter for fishes, ducks, and other water animals.

7. *Importance of bacteria and fungi.* Certain small plants, the bacteria and fungi, are beneficial in bringing about the decay of the bodies of plants and animals. This results in the production of substances that can be used again by green plants in the mak-

ing of foods and the construction of their bodies. Some bacteria increase the fertility of the soil by building nitrates and other

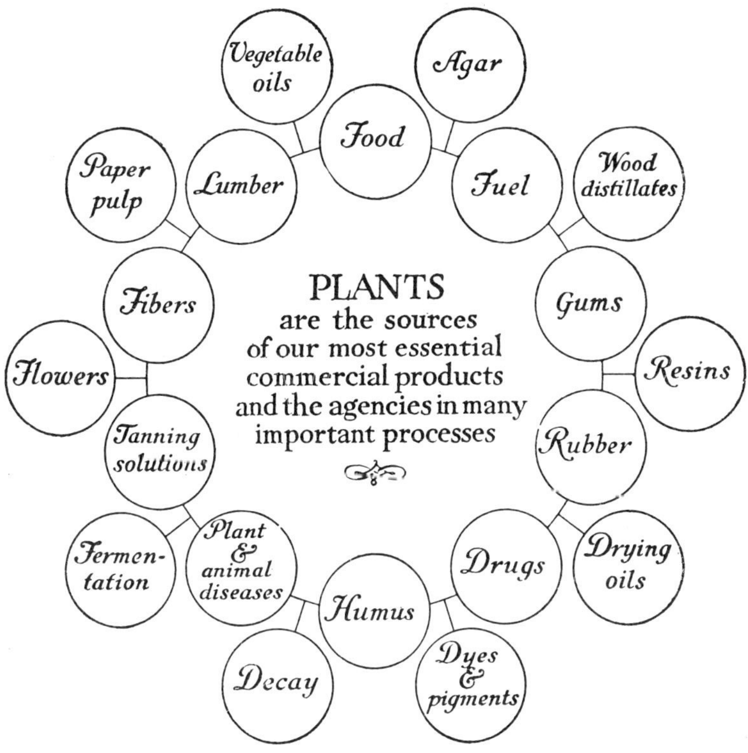

FIG. 2.

nitrogen compounds. Other bacteria and fungi cause most of the diseases to which plants and animals are subject, and they are being studied extensively in order that people may know how to avoid, control, or destroy them.

8. *Plants a source of pleasure.* Aside from all these great economic considerations plants afford us an æsthetic pleasure that cannot be measured in money but which is nevertheless real

FIG. 3. Primitive peoples in the tropics are dependent upon plants even more than are civilized races. Most of their foods are from plant sources, and their houses, mats, cloth, boats, ropes, and household utensils are made almost wholly of plant materials.

and important. Cities, towns, and individuals are expending millions of dollars every year in beautifying parks, boulevards, and residences with artistic groupings of trees, shrubs, and flowering plants.

Similarity of plants and animals. We have a further interest in plants in that they form one of the two great divisions of living things. They differ from animals in many particulars and to such an extent superficially that the largest and more complex forms are not only readily distinguished but are commonly thought of as being quite unrelated. It is difficult for those who have not been students of plant life to realize how similar the life processes of plants and animals are.

Nevertheless, they have many points of similarity. Both plants and animals use food as a source of building material and energy, and the foods are essentially the same. Both plants and animals use oxygen in respiration, and give off **carbon dioxide.**

Digestion, assimilation, growth, and reproduction are carried on by similar chemical and physical processes. As will be brought out more fully later, all the processes of both plants and animals depend upon the properties and activities of cells which are very much alike in the two groups.

Need for scientific study of plants. The profitable cultivation of plants for food, fiber, timber, and ornamental purposes, and the control of plant and animal diseases, depend primarily on our knowledge of the structures, products, and processes of the particular plants involved. Only a beginning has been made in the application of science to the industry of plant production.

For centuries agricultural practices have depended almost entirely on observation, experience, and tradition. Only recently has it been possible to explain on a scientific basis many of the principles underlying agricultural practice, and many problems of plant growing still await solution. The production of increased yields of crops per acre, the improvement in the quality and variety of the products, and the prevention of the ever increasing losses from diseases — in fact, the future development of agriculture and of all industries dependent upon plant products — will be based on scientific experiments with plants. For this development a better understanding of the laws of inheritance is fundamental. In addition, we must have clear insight into the effects of the environment on plant processes and structures.

The need for conserving plant resources. With the growth of population the conservation of our natural plant resources and the proper utilization of our lands for forestry and agricultural purposes becomes increasingly important. The United States started with a huge bank account of natural forest resources, much of which has now been dissipated. Every year we are using timber at a rate greatly in excess of the annual growth of all trees on our forest land. The future outcome of this system of timber destruction is clear. To formulate wise plans for the better use of forest lands requires a complete understanding of the relation of forest growth to climate and soil.

These are a few outstanding products and problems of plant life which have been enumerated in order to emphasize the fundamental importance of botany, which is the *science of plant life.*

PROBLEMS

1. Make a list of the uses you are making of plant products today.
2. What percentage of your diet is derived directly from plants?
3. How many acres of forest must be destroyed each week to furnish paper for our weekly magazine with the largest circulation?

REFERENCES

SMITH, J. RUSSELL. *The World's Food Resources.* Henry Holt & Co.
EAST, E. M. "Population in Relation to Agriculture," *Eugenics in Race and State,* Vol. II, page 215. Williams and Wilkins Company.
PEARL, RAYMOND. *The Nation's Food.* W. B. Saunders Company.
BERRY, JAMES B. *Farm Woodlands.* World Book Company.

CHAPTER TWO

PLANTS AS LIVING THINGS

THUS far plants have been discussed in their relation to men; they have been considered as objects of interest, and as a part of man's environment that may promote or interfere with his welfare. But, of course, plants do not grow, or flower, or fruit for the sake of animals or man. Their various organs grow and their structures develop as a result of their own life processes. A plant is successful in nature when it secures nourishment for its complete development, and when it produces offspring and thus insures the perpetuation of its kind.

It is important for the beginner in the study of botany to realize that plants are living things. We are accustomed to think of movement as a necessary evidence of life, and to one who has given no thought to the subject a tree may seem more akin to the stones among which it is rooted than to the animals that live about it. But when we study living beings, we find there are activities more fundamental than movement that are regularly associated with all life. As we shall see later, these more fundamental vital processes — such as respiration, growth, and reproduction — take place in plants just as they do in animals, and plants may therefore be considered as truly alive as are animals.

Parts of a plant. The vegetative body of the ordinary flowering plant is made up of three parts: root, stem, and leaves. The root anchors the plant and absorbs water and mineral salts from the soil. The leaf carries on a remarkable process in which water and carbon dioxide are united by the energy of the sunlight to form sugar, thus providing food for the plant. The stem supports the leaves and conducts water and mineral salts from the roots to the leaves, and foods from the leaves to the roots and other organs. The chief advantage of an erect stem is that it can display a large number of leaves to the light. In the roots, stems and leaves, all those processes which are related to the nourishment of the plant are carried on.

In the course of the plant's life three other organs are developed. These are the flower, fruit, and seed, which are the reproductive parts. The flower is usually very different in structure, texture, and color from the vegetative parts and also much shorter lived. The fruit follows the flower and is usually developed by the continued growth of one or more of the parts of the flower. Within the fruit are the seeds. They are commonly small bodies containing within them a young undeveloped plant (embryo) and a food supply. Seeds can withstand cold and drying; so, in addition to multiplying the plant, they carry the species through winters and periods of drought. Under favorable conditions they germinate and reproduce the plant from which they sprang. Flowers, fruits, and seeds are reproductive organs.

Interdependence of the parts of a plant. The roots, stems, and leaves make up the plant's machinery of nutrition, and the nourishment of the entire plant depends upon each part doing its work. If the roots are broken, the water supply is cut off and the leaves wither. If the leaves are removed, food manufacture stops and all the parts die for lack of nourishment. If the conducting vessels in the stem are cut, the water supply to the leaves fails and the roots have no food. The farmer destroys bushes by keeping them cut down so closely that they cannot expose leaves to the light, and he knows how to kill trees by cutting a ring about them through the bark, so that food cannot pass from the leaves to the roots.

It will thus be seen that when we discuss the relation of any particular part of a plant to the energy-supplying and nutritive processes, we must ever keep in mind the interrelation and interdependence of all parts of the plant. Just as no part of the human body lives an independent life but is dependent for its welfare on the activities of all the other parts, so the life of each part of a plant is bound up with the life of the plant as a whole.

Reproduction an essential process in plant life. To be successful, plants must not only maintain themselves, but in addition they must provide for the continuation of similar plants in the future; for plants maintain their kind from year to year and from one century to the next by producing new plants like themselves. Reproduction is sometimes accomplished by the separation and further development of a part of the parent body, as the tuber of a potato or the runner of a strawberry plant. In most plants, however, it takes place also through the development of seeds, and it is upon the growth of the young plant within the seed that the production of another generation of that particular kind of plant depends. A sunflower may develop a tall stem and a large leaf area, but unless it flowers and produces seed, no young plants will be grown from it. If it were the only sunflower in existence, there could be no more sunflowers after its death. Reproduction in plants must, therefore, be considered as an essential process, for without it plant life would soon disappear from the earth.

Summary. Plants are not nearly so complicated as the higher animals. Nevertheless, in the course of their long history on the earth, the plant body has become differentiated into several rather distinct organs which differ from each other in structure, and each of which carries on quite a different group of physiological processes. Roots, stems, and leaves are the chief organs of nutrition. Flowers, fruits, and seeds are the organs concerned with the reproduction of the plant. Each organ of a plant is more or less dependent upon the other organs, and the plant attains its greatest development only when all the organs are working in harmony together. This is possible only when each of the organs is placed in favorable conditions.

CHAPTER THREE

THE PLANT AND ITS ENVIRONMENT

The seasonal changes in plants and landscapes are so marked that they have always attracted attention. Numerous explanations as to why and how these changes are produced have been given. You yourself have probably accounted for the opening of buds in the spring, the blooming of certain flowers during certain seasons, the autumn colorations, and the fall of leaves as the effects of temperature, light, and moisture conditions.

Any one who travels extensively will also be impressed by the striking changes in the vegetation as he passes from one region to another. These differences, too, are to be accounted for by the differences in the conditions that surround the plants; the various types of forests, grasslands, and deserts are the results of climates and soils. Even locally one notices how different the plants are that grow in ponds and swamps, or on cliffs, from those that occur in valley bottoms and on gentle slopes. The plants differ not only in kind, but in size, form, and structure. The plant is profoundly affected by its *environment.*

So firmly established is this fact in our minds that when attention is directed to a familiar plant we at once call to mind the situation in which it grows and perhaps some of the conditions surrounding it.

Definition of environment. *By the environment of a plant is meant the complex of all those influences outside the plant which directly or indirectly affect its physiological processes, its structures, and its development and propagation.* These influences are numerous and are usually spoken of as *environmental factors.* The factors of the environment include the physical and chemical properties of the soil and the air surrounding the plant; also light, gravity, and the influence of other plants and of animals (Fig. 4).

Development of plants influenced by the environment. Each factor of the environment affects the growth of every plant. We

FIG. 4. Results of an experiment to show effects of environmental factors (light and moisture) on the growth of potato shoots: *A*, light but no water; *B*, light and water; *C*, water but no light.

are all familiar with the fact that light may determine the position of leaves and stems; that drought may reduce the size of a plant; that gravity has something to do with the upward growth of stems and the downward growth of roots; and that insect injuries and plant diseases may reduce the vitality of plants so that they produce neither flowers nor fruits. So the texture and the fertility of the soil, the temperature of the soil and air, and all the other environmental factors influence the plant's development and its growth. Successful farmers know that they cannot secure vigorous plants and profitable crops except under favorable conditions of light, temperature, moisture, and soil.

Limiting factors of plant growth. Wherever plants grow the several environmental factors are not all equally favorable. One or more conditions may be somewhat unfavorable, and when this is the case these unfavorable factors interfere with certain physiological processes, and the final form of the plants is greatly modified. Just as the strength of a chain is determined by the weakest link, so the development of a plant is limited by the factor which is least favorable.

Fig. 5. Differences in form of the water smartweed (*Polygonum amphibium*) due to the environment: *A*, grown on moist soil, stem erect, covered with hairs; *B*, grown in water, upper leaves floating, smooth throughout; *C*, grown on dry soil, stems and leaves rough hairy, and prostrate. (*After Massart.*)

If soil conditions prevent the entrance of an adequate supply of water into the roots, the stem and leaves will be dwarfed, no matter how rich the supply of minerals may be. If the leaves are exposed to unfavorable conditions of light or temperature, their work will be retarded and the nourishment of the whole plant curtailed even though water in abundance be supplied. Corn grown in sand under the most favorable conditions of light, temperature, and moisture is small and may fail entirely to produce seed, because sand does not supply the needed minerals. Or, if during the winter months corn is grown in a greenhouse in the richest of soil, it attains only a small size, because the intensity, or the duration, of light is insufficient for normal development.

The practical farmer knows that the yield of a crop plant is limited by the least favorable conditions of the environment.

He does not, therefore, try to improve all the environmental factors that affect his plants, but gives his attention to those that require it most. He irrigates if the soil lacks water. He adds fertilizer if the soil is deficient in certain minerals. Insect injuries he tries to prevent by spraying the plants with substances that kill the insects. These are all efforts directed toward removing or taking care of the limiting factors. It is often difficult to determine, in a particular case, what the limiting factor may be. But the ever increasing study of effects of environmental factors on plants is every year making the discovery, and the correction, of the limiting factors in crop production easier and more efficient.

Distribution of plants determined by the environment. Every plant has a somewhat definite form and structure. In order that it may live, it must carry on certain chemical and physical processes. Because of these processes and its particular structure and composition, a plant also has certain indispensable requirements as to its environment. The requirements of some plants are very definite, of others more or less indefinite; and the requirements of different plants vary greatly. If we survey any extensive tract of land, we see that its surface is more or less broken into elevations and depressions containing streams, lakes, or ponds. Slopes extend in various directions and are thus differently exposed to sunlight. The soil may be shallow on some slopes, and deep on others; it may vary in texture and mineral components. Some areas may be wet or moist, and others dry. In some places the soil is fertile, in others more or less sterile. Each of these smaller divisions of the land affords different opportunities and conditions for plant growth, and under natural conditions we find very different plants growing on them.

Habitats. The smaller areas into which every large land surface is broken are characterized by various groups and combinations of environmental factors. These areas from the standpoint of plant distribution are called *habitats*. Lakes, ponds, swamps, bogs, cliffs, bottom lands, sand plains, and clay slopes are examples.

Every habitat affords a particular complex of environmental factors. In a particular habitat we shall find only those plants growing whose requirements are satisfied by the factors of that habitat. In similar habitats, therefore, we expect the same or similar plants; in different habitats, different kinds of plants having different requirements. Plants whose requisites are very definite may occur only in a single habitat, while those whose requirements are rather indefinite may live in a variety of habitats.

Summary. It is quite impossible to understand the life of a plant without having constantly in mind the environment in which the plant lives. The environment is made up of several, or many, factors, among which are light, temperature, water, gravity, and the various properties of the air and soil. Every individual plant is modified in its development by these external conditions. Plants growing wild, or as crops, are limited in their development by the least favorable factors of the habitat. These are called *limiting factors*. In nature, plants are not distributed haphazard, but each occurs only in those habitats that afford the conditions which are necessary for its development. Similar habitats have the same or similar plants living in them. Habitats that differ in character are occupied by dissimilar plants.

CHAPTER FOUR

THE CELLULAR STRUCTURE OF PLANTS

THE body of every plant either is a single cell or is made up of a mass of cells. Many one-celled plants are so small that they can be seen only with a microscope. A large plant like a tree is composed of so many cells that their number can scarcely be conceived. A cubic inch of a potato, for example, contains at least 600 million cells, and a cubic inch of pine wood more than a billion.

The cellular structure of plants was first noticed about the middle of the seventeenth century, soon after the invention of the microscope. Cells were first seen in examining thin sections of cork, which is composed of layers of minute rectangular box-like structures. Observations were extended to other parts of plants, and to other plants, and it was finally recognized that *cells are the units of which all plant structures are built.*

The plant cell. Nearly two hundred years passed, however, after the discovery of cells before it was known that the cell walls which had been previously studied constitute merely the framework of the plant; that the most important part of the cell is a transparent, jelly-like, living substance inclosed by the walls. This living matter is called *protoplasm* (Greek: *protos*, first, and *plasma*, form). When we now speak of cells, we usually have in mind both the protoplasm and the wall around it. *A cell may be defined as a unit mass of protoplasm, capable of exhibiting all the phenomena usually associated with life*, such as growth and respiration.

We have omitted the cell wall from the definition because some cells are without walls for a part or the whole of their existence. It is through the activity of the protoplasm that the walls and other structures that are found in different types of cells are formed; so it seems best to define the cell in terms of the fundamental material of all cells.

In young plant cells the protoplasm occupies all the space within the walls. As cells grow older and enlarge, small cavities containing water appear in the protoplasm. As the cell takes up more water, these cavities, or *vacuoles*, expand and unite, until finally there is a single large vacuole within the protoplasm, containing the *cell-sap*. There are, then, three primary divisions of the plant cell: the protoplasm, the vacuole, and the cell wall (Fig. 6).

Protoplasm. The living matter of the cell when active has about the same consistency as the white of egg. Active protoplasm contains a large amount of water, while in dry seeds the protoplasm contains less water and may be quite rigid. Like gelatine, protoplasm is more or less liquid when it contains a high percentage of water; when it contains smaller amounts of water it becomes more nearly solid. The ability to absorb and hold large amounts of water is one of its most important qualities.

In composition protoplasm is very complex and may vary considerably, not only in different plants and in different parts of a plant, but also in the same cell, as a result of a change of environment or of increasing age. Analyses show that aside from water about one half of the protoplasm consists of protein. The remainder is made up of sugars and other carbohydrates, fats, and smaller amounts of salts and other substances. In some manner that is not fully understood at present, the components of protoplasm maintain a continuous group of activities which result in the phenomena known as life. Its most remarkable property is its ability to take up food and to construct from it more protoplasm like itself.

Protoplasm can also use food substances as a *source of energy*, and it carries on physical and chemical processes with the energy thus obtained. These processes are regulated in one way or another by the protoplasm itself. We may, therefore, look upon protoplasm as a body of matter that can absorb food materials,

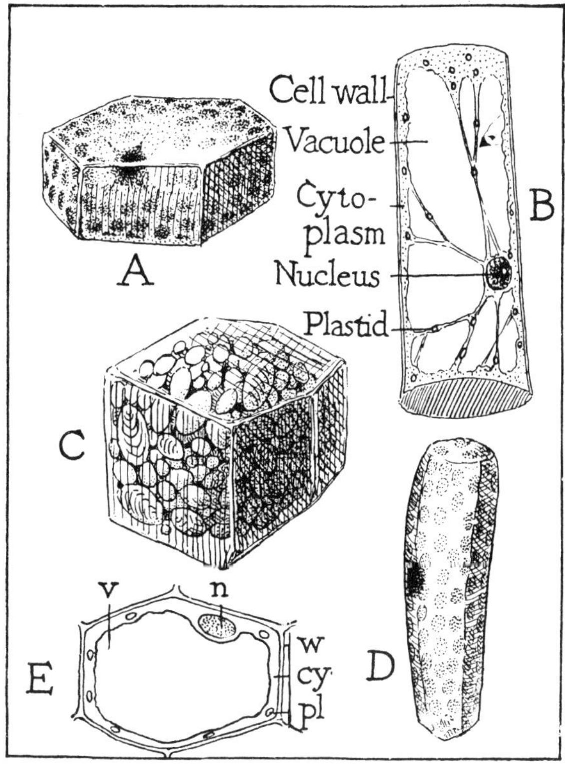

FIG. 6. Plant cells: *A* is from a moss leaf; *B* is from a squash-vine hair; *C* is a starch-filled cell from a potato tuber; and *D* is a cell from the palisade layer of a leaf. *E* shows a cell in cross-section.

that can liberate energy from a part of its food and with the remainder construct more matter like itself, and that regulates its own activities. It is self-constructing and self-regulating, the only truly automatic mechanism known.

Differentiation of the protoplasm. The living matter of the cell is primarily differentiated into two parts, the *cytoplasm* and the *nucleus*. The cytoplasm constitutes the bulk of the protoplasm and forms a definite layer within the cell wall and surrounding the central vacuole.

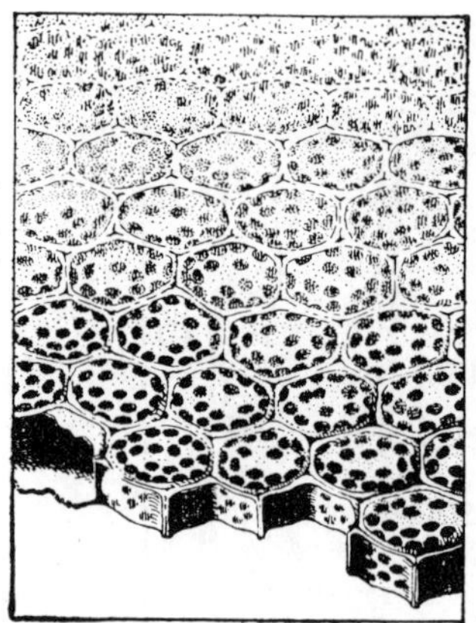

FIG. 7. Part of a moss leaf that is composed of a single layer of cells. The dark bodies shown in the cells are the plastids which contain the green coloring matter.

Portions of the cytoplasm may be organized into definite structures called *plastids*, in which food substances or coloring matters accumulate. Starch is formed in plastids, as is also the green pigment of leaves, and some plastids accumulate fats and oils (Fig. 7).

The nucleus is a small, round body of greater density than the cytoplasm. It occupies the center of the young cell, but as the vacuole enlarges it is carried with the cytoplasm close to the cell wall. The nucleus seems to be the starting point of many of the activities of the cell. It is believed to control and determine the course of development of the cell and the formation of new cells in growth and reproduction.

Vacuoles. The cell sap inclosed by the cytoplasm is a small drop of water containing sugars, salts, acids, and other soluble substances. As will be seen later, the cell sap influences many of the cell processes, especially those concerned with absorption and growth. The vacuole is the reservoir of the cell into which dissolved substances may pass from the cytoplasm and from which substances may again move into the cytoplasm as they are being utilized in cell activities.

The cell wall. The walls of plant cells are composed of nonliving materials deposited by the cytoplasm. They support the soft protoplasm of the cells in somewhat the same way that the wax of the honeycomb supports the honey. They also give stiffness to all parts of the plants. Most cell walls are composed in part of cellulose, a substance closely related to starch and sugar. You have seen pure cellulose in the form of cotton fiber. Filter paper and most book papers are made of cellulose derived from wood cells. The walls of cells may be modified in various ways by changes in composition, and thickened by the deposit of additional layers. Other substances may be added to the cellulose which render the wall hard and rigid, as is found in the shells of nuts, or impervious to water, as in the outer cells of leaves.

Animals, as well as plants, are composed of cells; but the animal cell, instead of having a stiff cellulose wall like a plant cell, has a soft wall, or it may lack a wall entirely, as in nerve cells and white blood corpuscles. Consequently, the tissues of animals (except the skeletal tissues) are usually softer and more pliable than plant tissues. This makes it easy for an animal to bend and to move about. The difference in cell walls and in the pliability of tissues is so general throughout the plant and animal kingdoms that it is one of the important distinctions between plants and animals.

Cell division and enlargement. Among simple one-celled plants, new individuals are formed by the division of the cell into two. The cell first enlarges; then the nucleus divides and the two newly formed nuclei separate. The cytoplasm then divides, the division beginning at the outside and gradually extending to the middle of the cell. As the cytoplasm divides, a new division wall is formed between the two daughter cells. In one-celled plants this wall splits and the two cells separate. In the more complex plants the same kind of cell division takes place when growth occurs and when new parts are formed, but the cells remain together.

Cell division is accompanied, or immediately followed, by an increase in the amount of protoplasm and the taking in of additional water. These two processes lead to the enlargement of the newly formed cells. Cell division and cell enlargement are first steps in the *growth* of all plants, whether the plants be small and simple or large and complex.

Cell differentiation. If we could trace the development of one of the more complex plants, we should find that it begins as a

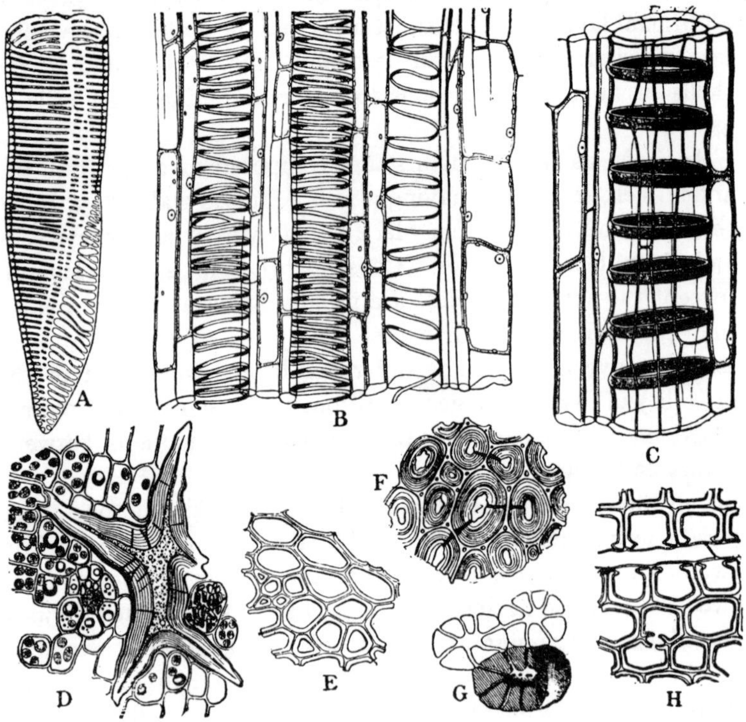

FIG. 8. Types of plant cells and tissues, resulting from the differentiation of cells: *A*, a supporting and water-conducting cell (tracheid) from fern stem; *B*, water-conducting vessels with spiral thickenings on the walls and soft, thin-walled cells (parenchyma) between them, from sunflower stem; *C*, water-conducting vessel with ring-like thickenings; *D*, giant stone cell in parenchyma of camellia leaf; *E*, wood cells from sunflower stem; *F*, stone cells, with greatly thickened walls, from stem of club moss; *G*, stone cells from shell of pecan; *H*, wood cells from pine. *D*, *E*, *F*, and *G* are cross-sections. (*After Sachs.*)

single cell. This cell divides, forming two cells, each of which divides again, forming four. Cell division continues until an embryo composed of hundreds of cells, all very much alike, is formed.

But as the embryo grows farther, some rows or groups of cells begin to differentiate, and they quickly change into the different types of cells found in the mature plants. Some of them enlarge but remain round; others greatly elongate; still others become flattened, disk-shaped bodies. In some the walls remain thin, while in others the walls are greatly thickened or are marked by lines and pits and irregular thickenings. They may differ in size, form, thickness of cell wall, content, or color, as will be made evident by our later studies of different plant parts.

Tissues and organs. Cells of a given kind are usually concerned with the carrying on of some particular process, and they tend to be grouped together. Such a group of cells is called a *tissue.* The epidermis of a leaf is a tissue covering the leaf; the soft green inner part of a leaf is a tissue that is concerned with the manufacture of food; the shell of a nut is a hard tissue inclosing the kernel.

An actively working tissue needs supplies and a means of disposing of its products. Hence, tissues are grouped together, and by their coöperation carry on some general function of the plant more efficiently. A number of tissues grouped and working together in this way form an *organ.* The leaf, for example, is an organ especially concerned with the manufacture of foods. It is made up commonly of five different tissues, each composed of thousands of cells.

Summary. Before proceeding to the study of structures and processes of plants it is important that we understand (1) that all plant structures are either single cells or masses of cells, (2) that the protoplasm is the active living part of the cells, and (3) that the processes carried on by any plant organs are the combined results of processes going on in the cells of which it is composed. Further details of cell structures and activities will be given as they are needed to understand the processes of particular parts of plants.

CHAPTER FIVE

LEAVES AND THEIR STRUCTURES

THE leaves of plants are generally their most conspicuous part. The prominence of leaves is the natural result of their relation to light. Leaves manufacture food, and sunlight is necessary in this process. In this chapter we shall study the structure of a leaf, and in subsequent chapters we shall discuss the work of the leaves and the processes that take place within these organs of the plant.

The parts of a leaf. If we examine a leaf closely, we see that it consists of a broad, thin *blade*, marked into small divisions by veins. The vein near the middle of the blade is commonly larger than the others and is called the *midrib*. In some forms of leaves there are several prominent veins, which we may call the *principal veins*. In general, the smaller veins form a network uniting with the larger ones, and these in turn connect with the midrib or with the principal veins. These large veins are smallest at

FIG. 9. Leaves with prominent stipules: pea, black willow, red clover, Japanese quince, rose.

the apex or outer end of the leaf and gradually become larger toward the base of the blade. They continue down through the *petiole*, or leafstalk, into the interior of the stem.

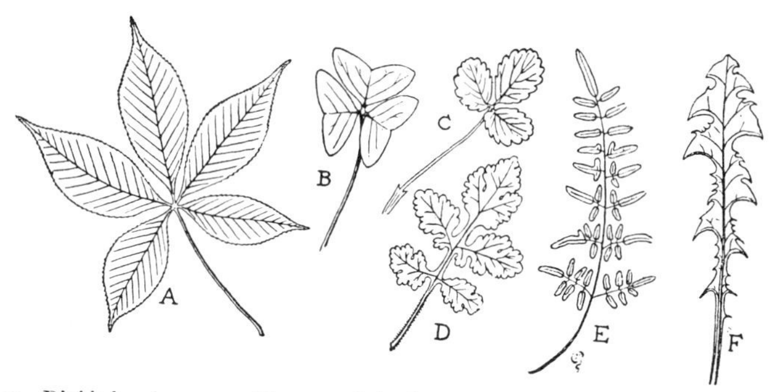

FIG. 10. Divided and compound leaves: *A*, buckeye; *B*, oxalis; *C*, avens; *D*, celandine; *E*, cliff fern; *F*, dandelion.

At the base of the petiole there is in many leaves a pair of sma appendages, the *stipules* (Fig. 9). These are usually unimportant structures, but occasionally, as in the pansy and garden pea, they are large and blade-like. These enlarged stipules supplement the blade, or in some plants may even take its place in food manufacture. *The primary divisions of the leaf are the blade, the petiole, and the stipules.*

The leaves of many grass-like plants have no petioles or stipules. In such plants the blade is attached to the stem by a *sheath*, which may be long or short. At the top of the sheath is a short, collar-like extension called the *ligule*. In the bamboo the ligule consists of several long bristles.

When the blade of a leaf is attached directly to the stem without an intervening petiole, it is said to be *sessile* (Latin: *sedere*, to sit).

Compound leaves. When several blades are attached to a single petiole, as in clover, buckeye, walnut, ash, and hickory, the leaf is called a *compound* leaf. The blades of the compound leaf are called *leaflets*. There is usually a distinct joint between the leaflet and the petiole. The leaf of the orange may be said to be compound because it has such a joint. The fact that some species of oranges have three leaflets gives support to this view. If the leaflets are joined to the end of the petiole, like the fingers of the hands, the leaf is described as *palmately compound;* if joined to the sides of the petiole, it is termed *pinnately compound* (Fig. 10).

The leaf made up of tissues. The soft green tissue essential to food production is found chiefly in the blade of the leaf. This may be shown by dissecting a fleshy leaf like that of the common houseleek or live-for-ever. Cutting across the blade of such a leaf, we find that there is a skin covering it above and below. The skin is readily stripped off, leaving the interior of the leaf as a green, granular mass of cells with veins running through it in all directions. The skin is called the *epidermis* or epidermal tissue (Greek: *epi*, upon, and *derma*, skin). The soft tissue

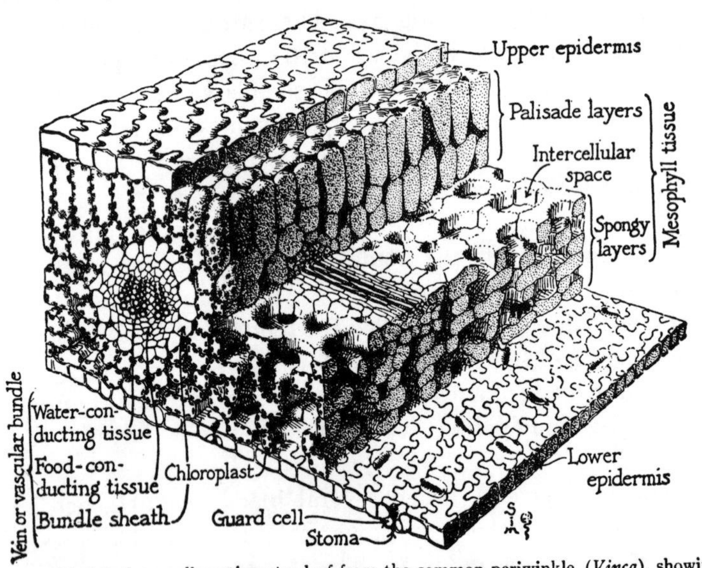

FIG. 11. Model of a small portion of a leaf from the common periwinkle (*Vinca*), showing cells and tissues.

between the upper and lower epidermis is the mesophyll tissue (Greek: *meso*, middle, and *phyll*, leaf). This tissue is green in

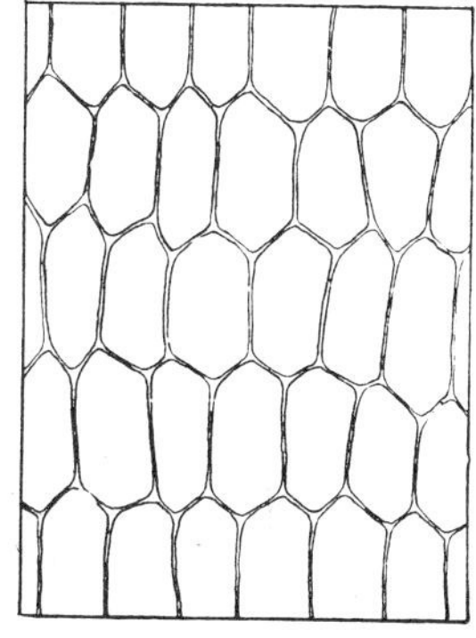

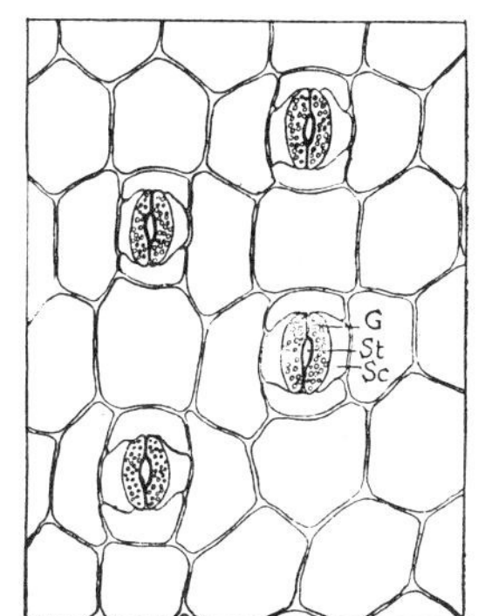

FIGS. 12 and 13. Upper epidermis of "Wandering Jew" (*Zebrina*) leaf, on the left, and lower epidermis, on the right. *St* is a stoma, *G* a guard cell, and *Sc* a subsidiary cell. The stomata are found only on the lower surface of this leaf.

color and may also be called chlorenchyma (Greek: *chlor*, green, and *enchyma*, tissue).

The veins consist of three tissues: the water-conducting, food-conducting, and mechanical tissues. The blade, therefore, commonly contains five tissues: the epidermis, the mesophyll, and three tissues of the veins (Fig. 11).

How cells are held together. The cells which form the tissues of plants are held together by a layer known as the *middle lamella*. This layer binds the cells together much as cement binds the bricks in a wall. The middle lamella is usually composed of calcium pectate. If it is dissolved out or changed by chemical action, the cells fall apart, just as bricks fall apart when the mortar between them is dissolved and removed by weathering. In boiling, the pectate between the cells is dissolved, and it is this action that causes fruits and vegetables to break up into a soft mass when cooked. The cells of ripe fruits are also easily pulled apart because of changes in the pectic compounds of the cell wall during the process of ripening.

The epidermis and the stomata. The cells of the epidermis are flat (Fig. 14), irregularly shaped, closely united, and in most

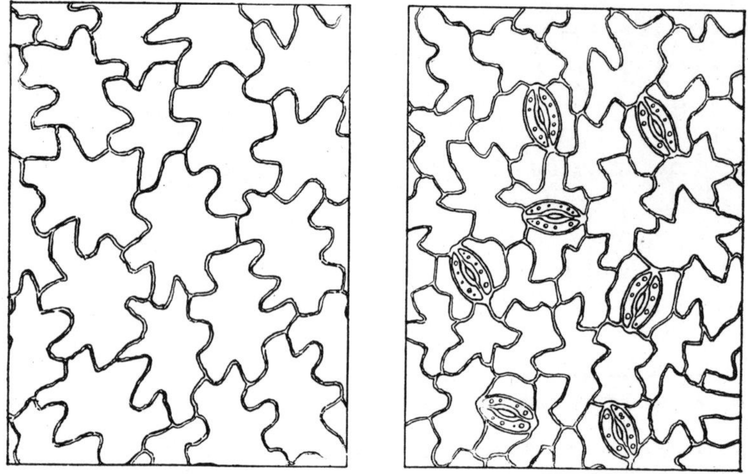

FIGS. 14 and 15. Upper and lower epidermis of leaf of common periwinkle (*Vinca*). Lobed, interlocking epidermal cells are strikingly different from the regular type shown on the preceding page.

plants colorless. The cell walls on the side of the epidermis which is exposed to the air become thickened with a wax-like material called *cutin*, which forms a layer over the surface of the leaf. This layer is called the *cuticle*. It is useful to the plant because water does not pass through it readily, and it reduces the amount of water that would otherwise evaporate from the epidermal cells. It may be compared to the enamel covering of

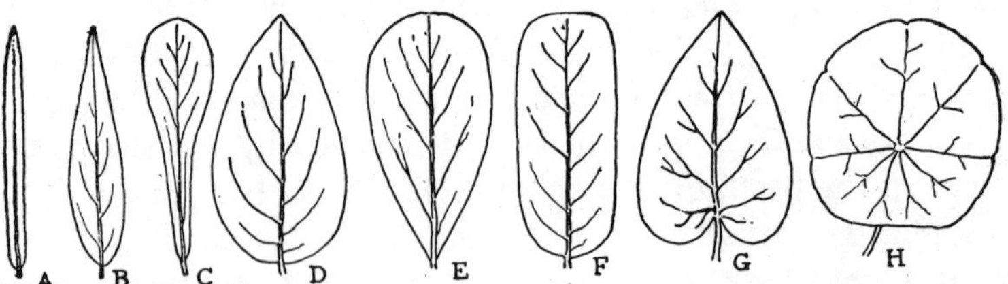

FIG. 16. Illustrating terms used in describing the shapes of leaves: *A*, linear; *B*, lanceolate; *C*, spatulate; *D*, ovate; *E*, obovate; *F*, oblong; *G*, cordate; *H*, peltate.

oilcloth, and when thick it is quite as impervious to water. The cuticle is useful to the plant also because it serves as a first line of defense against disease germs. The importance of the epidermis as a protective covering for the delicate inner tissues of the plant may be judged from the drying and decay that follow the breaking of the thin epidermal coat of an apple or a pear.

Scattered among the colorless cells of the epidermis are pairs of small, crescent-shaped green cells, the guard cells. Each pair of these surrounds a small opening or pore, the *stoma* (Greek: *stoma*, mouth; plural, *stomata*),[1] which is opened or closed by the expansion or contraction of the guard cells. The stomata are very important, for they connect the air spaces among the cells inside the leaf with the external atmosphere. When open, they allow the exchange of water vapor and other gases through the epidermis; and when closed, they complete the barrier to gas movements in either direction (Fig. 11).

In most of our trees and in many other plants the stomata occur only in the epidermis on the lower surfaces of the leaves. In some plants, especially in those growing in shaded situations, they are found in the epidermis on both the upper and lower leaf surfaces. In such leaves the number of stomata is always greater on the lower side.

[1] Stomata are so small that 2500 of them have an area about equivalent to that of an ordinary pinhole. They are so numerous, however, that they occupy about $\frac{1}{100}$ of the area of the average leaf. On a square centimeter of the lower surface of a sunflower leaf there are about 15,000 of them.

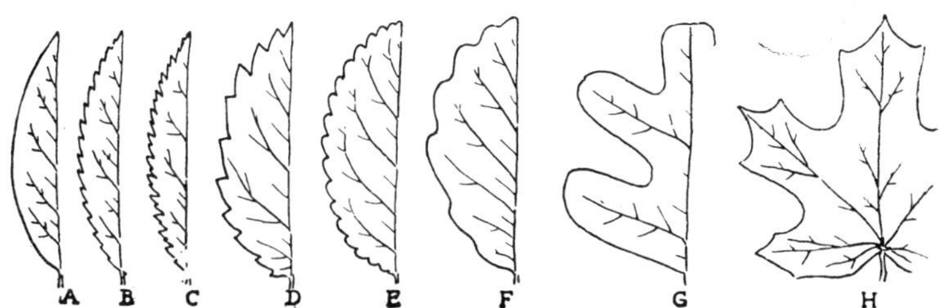

FIG. 17. Illustrating terms used in describing leaf margins: *A*, entire; *B*, serrate; *C*, doubly serrate; *D*, dentate; *E*, crenate; *F*, undulate; *G*, pinnately lobed; *H*, palmately lobed.

The mesophyll tissue. The mesophyll tissue is composed of the soft, thin-walled cells that lie among the veins in the interior of the leaf. In most leaves there are beneath the upper epidermis one or more palisade layers, which are composed of elongated cells standing close together, as is shown in Figure 11. The remainder of the mesophyll tissue is made up of ovoid or irregularly shaped cells joined quite loosely, so that air spaces are left between them. In fact, a much larger part of the surfaces of these cells than of other cells is in contact with air spaces. The air spaces within the leaf are continuous, and through them the oxygen and carbon dioxide of the atmosphere can reach every cell in the leaf. We shall see later that the differences in the epidermal and mesophyll cells, and in the way they are arranged, are definitely related to the different processes carried on by each of them.

The mesophyll in many plants contains other cells in addition to chlorenchyma. These additional cells are filled with water, and sometimes form a compact layer between the epidermis and the chlorenchyma, as in the begonia; they may also occur in long lines along the veins, as in the corn plant. It is the loss of water from these colorless water-storage cells that causes leaves

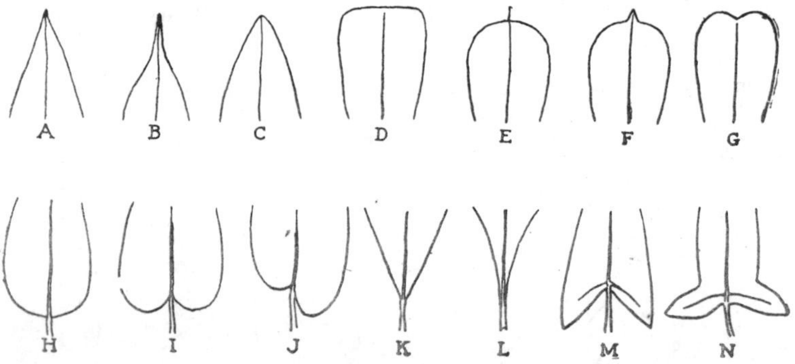

FIG. 18. Illustrating terms used in describing the apexes and bases of leaves: *A*, acute; *B*, acuminate; *C*, obtuse; *D*, truncate; *E*, aristate; *F*, mucronate; *G*, emarginate; *H*, rounded; *I*, cordate; *J*, obliquely cordate; *K*, acute; *L*, acuminate; *M*, sagittate; *N*, hastate

of many grasses to curl up during a period of drought. Sometimes these colorless mesophyll cells have heavy walls and contribute to the stiffening of the leaf.

The chloroplasts. Of the several structures found within the mesophyll cells, the most important in the primary process of food manufacture are the chloroplasts. These are round or lens-shaped plastids which contain a green coloring matter called *chlorophyll*. They are living organized bodies of protoplasm and multiply by division as the cells grow or new cells are formed. Cells may contain many or only a few chloroplasts, and these may be located deep within the leaf or near its surface. Since the chloroplasts are the special apparatus which manufactures food, the amount of food produced by a plant under any given conditions is roughly proportional to their number.

The chlorophyll. Chlorophyll is formed in the chloroplasts and colors them green. It can be taken out of the chloroplasts by putting the leaf in alcohol. After the chlorophyll is removed, the chloroplasts remain in the cell, but they are then colorless and the leaf is white or yellowish instead of green.

For the development of chlorophyll, light is usually necessary. The white sprouts on potatoes in a dark cellar, the blanching of celery when the lower part of the leaves is covered, and the whitening of young growing grass under a board are familiar evidences of this fact. In the inner tissues of plants and in the underground parts the plastids are usually colorless, but in many plants these

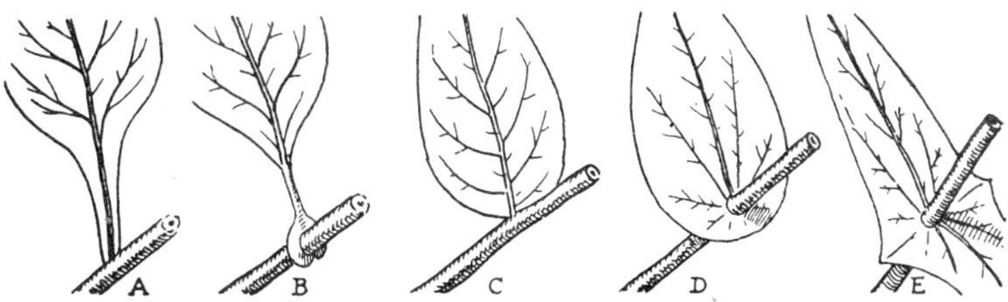

FIG. 19. Illustrating terms used in describing the attachment of leaves to stems: *A*, with margined petiole; *B*, petiole clasping; *C*, sessile; *D*, perfoliate; *E*, connate perfoliate.

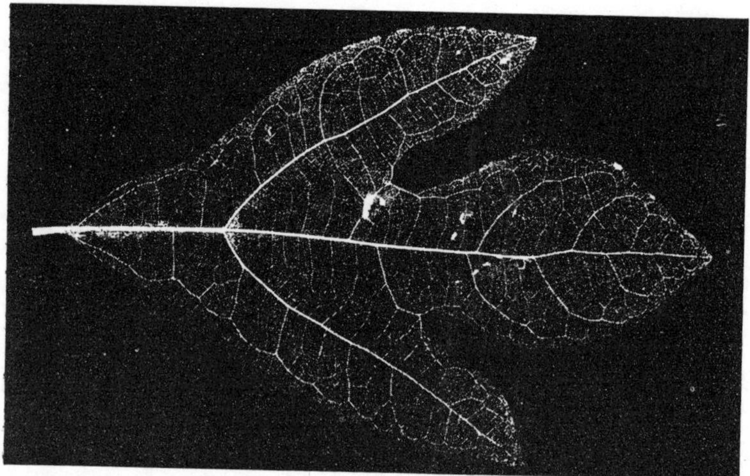

FIG. 20. The vein system of a skeletonized sassafras leaf. The leaf was prepared by placing it in water and allowing bacteria to digest the epidermis and mesophyll.

parts become green if they are exposed to the light. This is why potatoes that grow at the surface of the soil are likely to be green.

Seedlings of pine, spruce, lemon, and lotus develop green color even in the dark. One occasionally finds in lemons, for example, green sprouted seeds that have developed inside the fruit in complete absence of light. Evidently there are substances in some plants which make possible the formation of chlorophyll without light.

The veins. The veins in a leaf branch again and again, forming a fine meshwork through all its parts.[1] Each vein is composed of a bundle of water-conducting and food-conducting tissues surrounded by a bundle sheath. The water-conducting tissues are located in the upper side of the vein. These tissues are made up

[1] The *venation*, or arrangement of the veins of leaves, is of three general types: *parallel*, extending more or less parallel from the base to the apex; *dichotomous* or *forked*, when the veins divide at intervals into two smaller veins; and *net veined*, when the veins form an irregular network throughout the blade. The principal veins may be arranged either *palmately*, as in maple leaves, or *pinnately*, as in oak leaves

of long, cylindrical cells placed end to end. Usually the inner walls of these cells have spiral thickenings, and sometimes the end walls of the cells are absorbed, leaving continuous tubes, or vessels, several cells in length. After the growth of the cells is completed, the living protoplasm within them dies, and the dead cases of the cells, with their porous walls, lie like bundles of very fine pipes within the leaf. Through these vessels the water and mineral salts that are absorbed by the roots pass into the leaf and supply its living cells. The supplies of water and mineral salts pass out through the walls of the water-conducting vessels into the cells that adjoin them, and then from these they pass to other cells of the leaf.

The food-conducting tissues or vessels lie below the water-conducting vessels within the leaf veins. They provide an elaborate system of channels by which the surplus foods manufactured in the leaf are distributed throughout the plant. The foods pass from the mesophyll cells into these food-conducting tissues, and then down through the petiole of the leaf to the living cells of the stem and roots. The conductive tissues, or bundles, may be readily studied in the petioles of celery leaves.

In the smaller veins the bundle sheath is a layer of mesophyll cells. In the larger veins it contains one or more layers of thick-walled elongated cells, which act as a mechanical or supporting tissue. The mechanical tissue is rigid and gives stiffness to the leaf.

CHAPTER SIX

THE MANUFACTURE OF FOOD

You will probably remember from your study of physiology that the principal foods used by animals belong to three classes of chemical substances: carbohydrates, fats, and proteins. These same classes of substances constitute the food of plants. A grain of corn contains a supply of starch, oil, and protein which is used by the young plant, and these same foods that are used by animals are accumulated in many plants. The difference in the nutrition of plants and animals lies, then, not in any differences in the foods used, but in the way their foods are secured. In this chapter the manner in which plants obtain their foods will be discussed.

Plants the source of all food. Mineral soils and the air do not contain any of the substances that we class as foods. Yet green plants may grow luxuriantly on mineral soils. It follows, therefore, that *green plants are able to manufacture their own foods.* They can synthesize, or build together, simple substances that they obtain from the soil and air into the complex foods that they require. Animals lack this power. They must have foods that have already been built up, rather than the simple materials of which foods are made. These foods they secure either directly or indirectly from plants. The ability of plants to manufacture complex foods from simple substances brings up several questions:

What is the method by which plants produce food? Just what parts of the plants do the work? What constitutes the machinery? Out of what materials is the food manufactured? How is the energy supplied? And what are the conditions under which the process goes on?

Photosynthesis. The primary step in the making of food is the building of simple carbohydrates through the process called *photosynthesis* (Greek: *photos*, light, and *synthesis*, putting together). In this process carbon dioxide from the air and water from the soil are brought together in the chloroplasts and united to form carbohydrates. Sugar is the first abundant product, but being soluble in the water of the cell, it is quite invisible. In most plants a large part of the sugar is rapidly changed to starch, and as the starch is insoluble in water, it accumulates temporarily in the chloroplasts in the form of little grains which may be readily seen with a microscope. There is a very simple test for the presence of starch. A solution of iodine stains most substances yellow or brown, but it colors starch blue or purple. So any object that contains starch — a cell, a leaf, or a piece of cloth — will be colored purple if iodine is applied to it.

Light and photosynthesis. If we take a leaf from a plant that has been in the dark for two days, place the leaf in warm alcohol to remove the chlorophyll, and then put it in a solution of iodine, it is stained yellow. This proves the absence of starch. If the plant is then put in the light for an hour, a leaf tested in the same way will be colored purple, showing that starch is present. *Evidently light is necessary for photosynthesis.*

It is not surprising to find that light is so effective in building up compounds in the green parts of plants, for it is a powerful agent in causing chemical change. You may be familiar with its use in photography. The film and the printing paper have on them a layer of gelatin containing certain chemicals. Exposure to the light for even a fraction of a second effects changes in these which may be seen when the film or paper is developed. Many chemical substances kept in drug stores must be protected from the light; otherwise they soon change their composition and become different substances.

The amount of light required for photosynthesis varies in different plants. Among trees, for example, the beech, sugar maple, and hemlock do not require as much light as the willow, cottonwood, pine, and aspen. Usually a reduction in intensity to one fifth of full sunlight does not decrease the rate of photosynthesis. In some shade plants the rate does not fall off until the light is reduced to one twelfth. This is still several times the intensity of light in an oak, maple, or spruce forest, where one finds herbs on the forest floor that must be able slowly to manufacture sufficient food with a fiftieth or a hundredth of full sunlight.

Chlorophyll necessary for photosynthesis. By using a plant with variegated leaves, the iodine test will show that the white parts form no starch. Since starch is formed only in the green part of the blades, it is evident that *chlorophyll is necessary for photosynthesis.* Any green part of a plant can carry on photosynthesis, but the principal food factories are the leaves.

Effects of temperature on photosynthesis. The effects of temperature on photosynthesis may be demonstrated by taking plants that have been in the dark long enough for the starch to be removed from the leaves, placing them in the light under different conditions, and noting the time that it takes for starch to form. Such tests show that the *ordinary summer temperature is most favorable for photosynthesis.* When the temperature falls nearly to the freezing point, photosynthesis slows up and finally ceases entirely; and, on the other hand, when it rises above 100° F., the process is slowed down rapidly.

Materials and products. Experiments have shown that the materials used in photosynthesis are carbon dioxide and water. Carbon dioxide is a gas that makes up from three to four out of every 10,000 parts of the air. Its molecule contains one atom of carbon and two atoms of oxygen (CO_2). Water, which the plant gets from the soil, has two atoms of hydrogen and one atom of oxygen in every molecule (H_2O). The simple sugars made in photosynthesis from the carbon dioxide and water contain these same elements (Fig. 23).

Carbohydrates include many substances commonly classified as sugars, starches, and celluloses. The simple sugars, glucose and fructose, have a formula $C_6H_{12}O_6$. The double sugars like sucrose (cane and beet sugar) and maltose (malt sugar) may be built up by combining two simple sugars,

$$C_6H_{12}O_6 + C_6H_{12}O_6 \longrightarrow C_{12}H_{22}O_{11} + H_2O,$$
$$\text{glucose} + \text{fructose} \longrightarrow \text{cane sugar} + \text{water},$$

one molecule of water being lost in the process. Cane sugar may be split into glucose and fructose by heating it in dilute sulfuric acid for a few minutes. This brings about the addition of a molecule of water (by hydrolysis) and the subsequent splitting:

$$C_{12}H_{22}O_{11} + H_2O \longrightarrow C_6H_{12}O_6 + C_6H_{12}O_6$$
$$\text{cane sugar} + \text{water} \longrightarrow \text{glucose} + \text{fructose}$$

The starches and celluloses are formed by combining many molecules of the simple sugars and removing a molecule of water for each molecule that enters into the combination:

$$n(C_6H_{12}O_6) \longrightarrow (C_6H_{10}O_5)_n + n(H_2O)$$
$$\text{glucose} \qquad \text{starch} + \text{water}$$

Consequently their formulas are $(C_6H_{10}O_5)_n$, in which n represents a rather large number. The starches and celluloses may also be split up into simple sugars by adding the required number of molecules of water. This last process is the one by which corn sirup (glucose) is made from corn starch. Corn starch is hydrolyzed in the same way as cane sugar, mentioned above, with the result that it breaks down into glucose. The process may be represented by the equation,

$$(C_6H_{10}O_5)_n + n(H_2O) \longrightarrow n(C_6H_{12}O_6)$$
$$\text{starch} \qquad \text{water} \qquad \text{glucose}$$

Those sugars like glucose, which are the first abundant products of photosynthesis, contain six atoms of carbon, twelve atoms of hydrogen, and six atoms of oxygen in each molecule. For every molecule of glucose manufactured, therefore, it would require six molecules of carbon dioxide to furnish the carbon and six molecules of water to provide the hydrogen. These amounts of water and carbon dioxide, however, contain eighteen atoms of oxygen, twelve more than are needed for the making of glucose,

$$6\,CO_2 + 6\,H_2O \longrightarrow C_6H_{12}O_6 + 6\,O_2$$

We should, therefore, expect oxygen to be given off from leaves during photosynthesis. That this actually happens may easily be shown by inverting under water a bundle of the branches of some water plants, like *Elodea*, with the cut ends placed under the mouth of a test tube that is filled with water. When exposed to the light for a day, the tube will be partly filled with gas. By testing with a glowing match or splinter (Fig. 21), the gas may be shown to be mostly oxygen.[1]

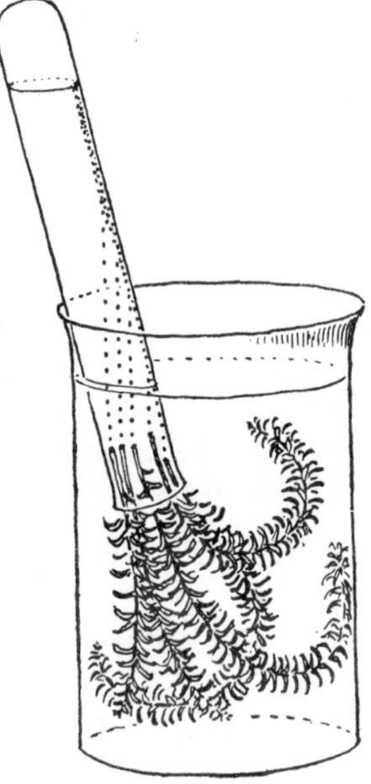

FIG. 21. Experiment to show the giving off of oxygen from a water plant (*Elodea*) during photosynthesis.

How the supplies are obtained. Every industrial workshop must constantly be provided with the raw materials needed in the manufacture of its product. Likewise the leaf must be supplied with the substances that it uses in the making of food. These necessary supplies come to the leaf through the veins and the stomata. The water passes into the leaf through the water-conducting tissue of the veins. The supply of carbon dioxide reaches the cells of the mesophyll through the stomata and the intercellular spaces. When the stomata are closed, little or no carbon dioxide can enter, and at such times the process of photosynthesis is of necessity greatly retarded or completely stopped.

That the carbon found in a plant does not come from the soil was shown 300 years ago by one of the earliest students[2] of plant

[1] Water containing a considerable amount of dissolved carbon dioxide should be used in this experiment so that photosynthesis may go on rapidly. Pond water is better than tap water.

[2] Van Helmont (1577–1644) grew the branch of a willow tree for 5 years. At the beginning it weighed 5 pounds, at the end 164 pounds. The loss in weight of the soil was 2 ounces.

physiology. He grew a plant for several years in an accurately weighed body of soil. He then carefully removed the plant, dried it, weighed it, and also reweighed the soil. He found that the increase in dry weight of the plant was more than a thousand times the loss in weight of the soil. This proved that the plant must have obtained most of its materials from some other source than the soil. The plant, of course, used vast quantities of water during its growth, but since water contains no carbon, the only other source of this material must therefore have been the carbon dioxide of the air.

How the products and wastes are removed. The manufacture of carbohydrates in the leaf goes on only during the hours of sunlight; the removal of food goes on at all times. The food-conducting tissue of the veins furnishes the outlet for the product, which is transferred in the form of sugar. During the day the rate of manufacture is so much greater than the rate of removal of food that starch and sugar accumulate. During the night the movement of food into the stem nearly empties the chlorenchyma. The waste product, oxygen, passes from the cells to the intercellular spaces and out through the stomata to the atmosphere.

A leaf, then, is carrying on photosynthesis at its full capacity only when there are sunlight, a favorable temperature, and an abundant supply of water, and when the stomata are open. Even under these conditions the work may be interfered with if more than a certain amount of the products accumulates in the cells.

The amount of the product. The amount of carbohydrates produced in photosynthesis varies so greatly in different plants and under dissimilar conditions that it is very difficult to make a general estimate of it. The result of many experiments shows that under favorable conditions a square meter of leaf surface makes on an average about 1 gram of carbohydrates per hour. At this rate a square meter of leaf surface in midsummer would require 2 months to produce food equivalent to that consumed by the average man in a day. This average rate of carbohydrate manufacture may also be expressed by saying that the leaf makes enough sugar in a summer to cover it with a layer 1 millimeter thick. Because as a whole the factors involved in photosynthesis are most favorable during the morning hours, the greater part of food manufacture occurs before noon.

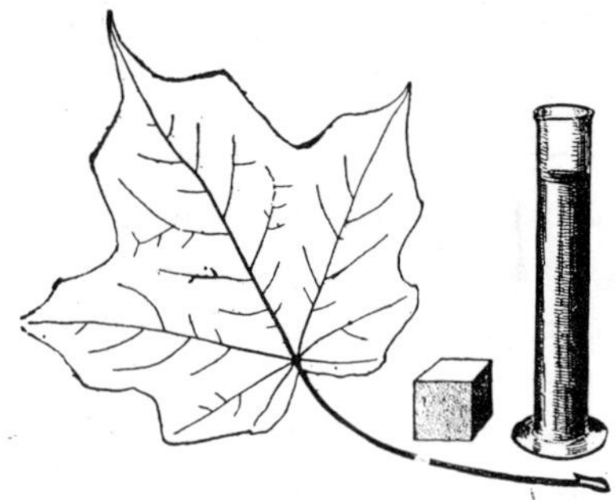

FIG. 22. A maple leaf and the sugar and maple sirup equivalent to the amount it could manufacture in a season. All drawn to the same scale.

An acre of corn exposes about 2 acres of leaf surface to the light. The total weight of organic material in an acre of mature corn plants having a yield of 100 bushels of corn is about 7 tons. Of this amount, about 3 tons is carbon. To secure such a large quantity of carbon, not less than 11 tons of carbon dioxide were taken in by the plants. Furthermore, as we shall see in connection with respiration, not all the carbon taken in and built into organic compounds remains in the mature plants. It is estimated that the plants of the United States manufacture nearly a cubic mile of sugar each year.

Hindrances to photosynthesis. Aside from the lack of light, water, and carbon dioxide, the process of photosynthesis may be interfered with in several ways. In cities where there are much dust and smoke, plants do not grow well because (1) the amount of sunlight is greatly reduced; (2) the dust forms a layer on the upper surface of the leaf and reduces still further the amount of light that actually reaches the chlorenchyma; and (3) soot and dust collect in the stomata and interfere with the entrance of carbon dioxide. If the dust and smoke are very abundant, the stomata may even become completely blocked and photosynthesis stopped altogether.

Insects and plant diseases are often serious hindrances to photosynthesis. When insects eat the leaves of plants, they decrease the supply of carbohydrates in proportion to the amount of leaves they destroy. If the plant happens to be a crop plant, the injury done by insects may result in the failure of the plant to manufacture sufficient food for filling out the fruit, grain, or seed for which it was grown. Diseases of plants caused by fungi or bacteria also greatly interfere with the power of the plant to manufacture carbohydrates.

Carbohydrates as storehouses for energy. When the carbon dioxide and water are converted into carbohydrates by photosynthesis, the energy supplied by the sunlight in doing this work is stored as potential energy in the new substances formed. Then, when these carbohydrates are oxidized or burned (or in other words, when they are changed back into carbon dioxide and water), the exact amount of energy that was stored is set free. *Thus the plant acts as a storehouse from which we can draw energy at any time.*

The importance of photosynthesis as a life process. Photosynthesis is not only important to the plant itself, but, broadly speaking, it is the most important of all life processes. The sun pours a constant flood of energy on the earth, and this energy warms the earth, causes the winds and rains, and in general furnishes the power for the work that we see going on in nature about us. From running water, winds, and direct sunlight man obtains a certain amount of energy for his own use, but the great source of the energy that we use for heating purposes and for power is wood, coal, petroleum, or gas. The energy stored in these was accumulated through photosynthesis. It came originally from the sun, and but for the plants would have radiated off into space as heat waves from the earth. But through the work of green plants it was locked up in the molecules of the wood and coal, and by burning these fuels man can release the energy that is stored in them and use it for his own purposes. We may, therefore, say that most of the work of the world, including that done by men and animals, is accomplished by the use of energy accumulated by plants in photosynthesis.

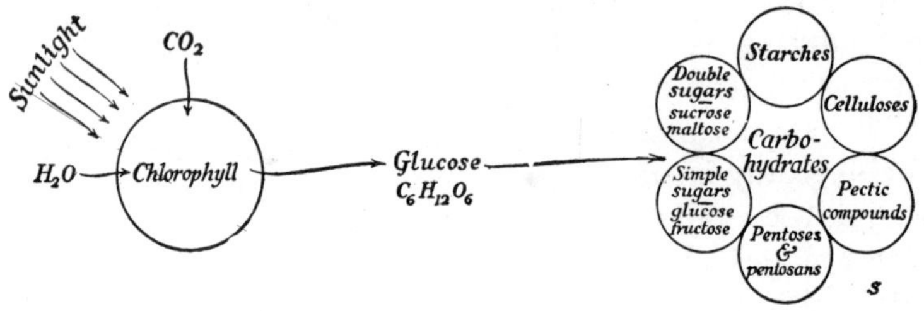

FIG. 23. Diagram of the process of photosynthesis, showing primary and secondary products.

The importance of photosynthesis as a source of wealth. The value of all the plant products of field and forest for one year is many times as great as the value of all the minerals dug from the earth during the same amount of time. Furthermore, minerals are limited in amount and are gradually being exhausted; while on the other hand the products of photosynthesis are being constantly renewed, and we may continue to collect them indefinitely.

Summary of photosynthesis. We may summarize the facts we have learned regarding the process of photosynthesis by likening it to a manufacturing process of human invention:

The factory	is the green tissue, especially that of the leaves.
The workrooms	are the cells.
The machinery	is the chloroplasts and the chlorophyll.
The energy	is the sunlight.
The raw materials	are the carbon dioxide and water (CO_2 and H_2O).
The supply department	is the stomata and intercellular spaces, and the water-conducting tissue.
The products	are carbohydrates: sugars ($C_6H_{12}O_6$) and starches $(C_6H_{10}O_5)_n$.
The forwarding department	is the food-conducting tissues, and it works both day and night.
The waste material	is oxygen, which escapes through the intercellular spaces and the stomata.
The working hours	are all the hours of sunlight.

The production of fats. In addition to carbohydrates, plants make and use two other important classes of foods: fats and protein. The fats are quite similar to the carbohydrates in composition. They contain the same chemical elements: carbon, hydrogen, and oxygen. The proportion of the oxygen to carbon, however, is smaller. At ordinary temperatures fats occur in plants both as solids and liquids. The liquid fats are commonly called oils. They are probably made directly from the carbohydrates, for the plant has no special fat-producing apparatus comparable with the carbohydrate-producing chloroplasts of the leaves. The chemical changes are probably effected by the protoplasm; therefore fat can be formed in any living part of the plant.

In some plants belonging to the lily family (the onion, for example) small drops of oil appear in the cells of the leaf as the first *visible* product of food manufacture. The primary product of photosynthesis (probably glucose) is changed directly into oil when it accumulates, instead of into starch as it is in the leaves of most plants. Starch does not form in these leaves at any time, but when the materials of which the fats are composed are transferred and accumulate in the underground bulbs of these plants, they then assume the form of starch. This emphasizes the close relationship existing between starch, glucose, and fat.

Fats and oils, like starch, are inactive storage substances; that is, before being used or transferred they must be converted into substances soluble in water. Although fats are widely distributed in the plant body, they are especially abundant in seeds and fruits. Some of the commonest fats and oils of commerce derived from plants are corn, coconut, cottonseed, linseed, castor, pea, peanut, and olive oils, and cocoa butter.

Fats are formed from carbohydrates by two different series of chemical changes. In the first of these series the carbohydrates are changed to fatty acids, the more important of which are oleic, palmitic, and stearic. In the second series of chemical changes the carbohydrates are transformed into glycerin. This unites with the fatty acids and forms fats and oils.

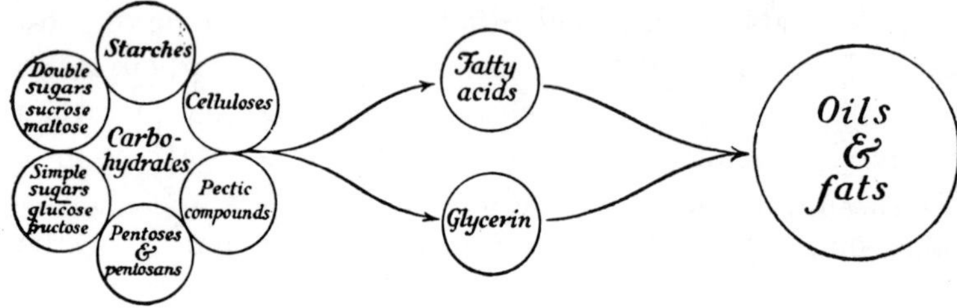

FIG. 24. Diagram of fat synthesis.

carbohydrate ⟶ glycerin
carbohydrate ⟶ fatty acid
⟩ fats and oils

When fats are broken down or digested, they are changed back again to glycerin and fatty acid, and may be finally altered to glucose. The digestion of fats consists in forcing water into the oil molecule, thus breaking it up into two or more molecules.

$$\underset{\text{fat}}{C_3H_5(C_{18}H_{35}O_2)_3} + \underset{\text{water}}{3\ H_2O} \longrightarrow \underset{\text{glycerin}}{C_3H_5(OH)_3} + \underset{\text{stearic acid}}{3\ HC_{18}H_{35}O_2}$$

In plant cells the glycerin and stearic acid may be further transformed into sugar before leaving the cell. Compare the formula of a fat with the formula of a sugar. Does it contain a larger or smaller proportion of oxygen? In changing from sugar to fat is oxygen added or removed?

Sunlight is the direct source of energy used in photosynthesis. The energy used in the transformation of the simple sugars into fats and many allied compounds is derived from the oxidation of a part of the sugar formed, not directly from sunlight. This is discussed more fully under respiration.

The making and use of proteins. The proteins are the third class of foods. They too are constructed in large part from the carbohydrates; but their molecules are vastly more complex than are the molecules of carbohydrates and fats, and they all contain the elements nitrogen and sulfur and some of them contain phosphorus, in addition to carbon, hydrogen, and oxygen. In protein synthesis the amount of sulfur and phosphorus consumed is small, but a very large amount of nitrogen is required. Furthermore, nitrogen in the gaseous condition in which it occurs in the air does not readily unite with other substances; so, although it makes up four fifths of the atmosphere, green plants cannot take it directly from the air. For the nitrogen needed for protein making, plants must depend, therefore, on the supply which comes from the soil in the form of nitrates. This is carried to the cells with the water that is absorbed by the roots.

Protein synthesis, like the synthesis of fats, is probably effected by the protoplasm. It may occur in nearly all parts of a plant, but it takes place for the most part in the leaves where the carbohydrates are being made and where their constituent parts are in a condition to unite with the nitrogen, sulfur, and phosphorus compounds. Light may be a factor in the process when it takes place in the leaves, but it has been definitely proved that it may also take place in the absence of light. Proteins like fats and starch, are mostly inert storage substances, and many of them are insoluble in water. Because of their chemical composition they are especially used in building up protoplasm (Fig. 25).

Steps in protein synthesis. The various steps in the building of proteins are not fully known. It is probable, however, that the nitrates derived from the soil are transformed into ammonia (NH_3) within the plant and that this unites with certain acids, derived from the carbohydrates, forming amino acids. They are called amino acids because the amino group (NH_2) forms a part of the molecule.

The amino acids are comparatively simple substances, but, like the simple sugars, they may be built together to form large and very complex molecules. Just as many glucose molecules may be joined together in the formation of starch, so amino acids may be joined together to form protein. In fact, there is good evidence that in some proteins the molecules are formed by the union of a hundred or more amino-acid molecules. Just as starch yields many molecules of glucose when it is digested, or broken down, so when proteins are digested they yield many amino-acid molecules.

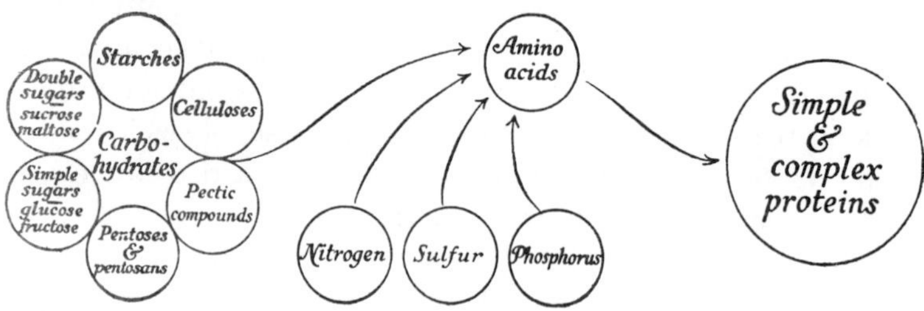

FIG. 25. Diagram of protein synthesis.

The proteins are transported from the leaves in the food-conducting tissue of the bundles, usually after they have been broken down into simpler and more soluble substances (amino acids and amides).

Importance of nitrogen in soil. Since the proteins make up more than half of the living protoplasm, and since all of them contain a considerable percentage of nitrogen, the need for abundant nitrates in the soil is evident. Any kind of moist land would furnish the raw materials for making carbohydrates and fats, but to supply the necessary materials for protein manufacture, the land must contain nitrogen, sulfur, and phosphorus. It is the varying amounts of these three substances in the soil that make the difference in agricultural land values when other conditions are equally favorable.

Sources of protein in the human diet. The most expensive portion of the diet of human beings is the proteins. Figure 26 shows that in soy beans we possess the richest source of protein. It also shows why the soy bean is one of the most important of foods in the Asiatic nations, where animal foods are very limited. One dollar will buy several times as much protein in soy beans as it will in any other plant or animal food. However, recent experiments in animal feeding have shown that for maintenance and growth some proteins are more

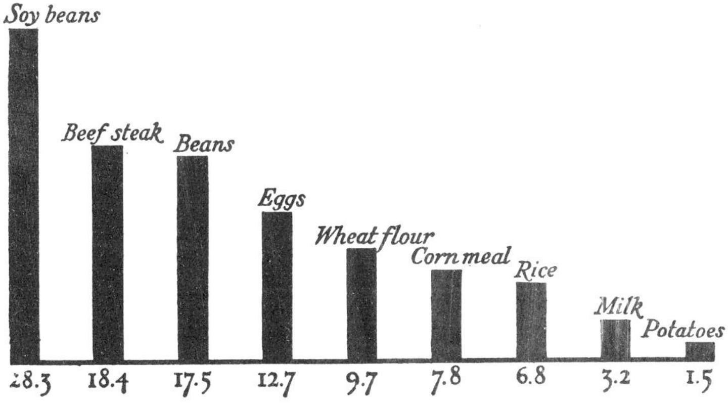

FIG. 26. Percentage of protein in various foods.

valuable pound for pound than others. Curiously enough, the protein of the soy bean is not only furnished in large amounts, but in its ability to be digested and assimilated it stands at the very top of the vegetable proteins. It is of interest to know that even in the United States, where meat is consumed in comparatively large quantities, the principal source of protein in our diet is wheat.

Importance of understanding the food-making processes. A knowledge of the essential facts of food manufacture by plants lies at the foundation of all agricultural, horticultural, and silvicultural practices.

We have gone far enough now to be perfectly sure that *plants do not get their food from the soil* any more than animals do. Both plants and animals require various salts. Plants get these salts from the soil, but they constitute only from 1 to 3 per cent of the plant body. Both plants and animals require *carbohydrates*, *fats*, and *proteins* as their principal food. Animals can obtain these from plants, but green plants must manufacture them.

It is therefore evident that to obtain the best crop yields it is not only essential to have sufficient nutrient salts in the soil but that the temperature and light conditions be favorable for photosynthesis. Water must always be available. It is possible to increase crop yields by increasing the supply of carbon dioxide and water, as well as by adding more mineral salts (fertilizers) to the soil.

Amount of food produced per acre. Since the food supply of all living beings depends primarily upon these synthetic processes that are carried on in plants, it is of interest to inquire how much food may be derived from an average acre of land when planted to different crops. It must be remembered that the plants that produce this food take a considerable part for their own maintenance, and that the part which the farmer harvests is the plant's surplus. The following table shows the average yield per acre, its food value calculated in Calories, and the number of men that 1 acre planted to different crops might feed for 1 day, assuming that each man required 3000 Calories per day:

Food Products	Yield per Acre		Millions of Calories[1] Equivalent	No. of Men That Might Be Fed for One Day
	Bu.	Lbs.		
Corn	35	1960	3.1	1000
Sweet potatoes	110	5940	2.8	900
Irish potatoes	100	6000	1.9	600
Wheat	20	1200	1.8	600
Rice	40	1154	1.7	560
Soy beans	16	960	1.5	500
Beans	14	840	1.1	375

If the plant products of an average acre are fed to cattle, the dressed beef produced amounts to only 125 pounds, yielding an energy equivalent to the food of 43 men for 1 day. If transformed into pork, the yield is 273 pounds, or sufficient food for

[1] A Calorie is the amount of heat necessary to raise the temperature of 1 kilo of water to 1 degree Centigrade.

220 men for 1 day.[1] This shows the great loss of energy that results when plant foods are converted into meat before they reach the human consumer. It is evident that as the human family becomes larger and food becomes scarcer, we shall have to take more and more of our foods directly from plants.

There are, however, certain animals that feed, either directly or indirectly, on plants that cannot be used for human food. All of our sea-food animals, such as fish, clams, and oysters, are able to convert otherwise unusable food into food that can be used, thus adding much to our diet. Sheep and cattle grazing on the open range and forest reserves in the Western states and on the pampas of Argentina may be looked upon as gatherers and converters into available forms of food not directly usable by man.

PROBLEMS

1. How do the white parts of a variegated leaf get food?
2. Occasionally in a field of young corn a stalk that lacks chlorophyll will be found. How long will it live?
3. Geraniums with variegated leaves occasionally produce branches that are entirely white. A noted horticultural firm offered $1000 to any one of its gardeners who would root one of these branches and thus produce a white-leafed geranium. What was the chance for success? Why?
4. Why do trees in the open retain their lower as well as their upper branches, while the same trees grown in a dense forest retain only their uppermost branches?
5. Why are there comparatively few weeds in a cornfield in the autumn as compared with an adjoining field in which wheat has been grown?
6. Bushbeans cannot be grown profitably between rows of corn in a cornfield, but polebeans, if properly spaced in the field, will yield abundantly and not interfere with the corn. Explain.
7. Why is it best to wait until celery is well grown before tying it up with paper, or covering it with boards to blanch it?
8. In how many ways could you cause a plant to starve to death? Are any of the methods used in controlling weeds?

[1] United States Department of Agriculture, Farmers' Bulletin No. 877. The table does not take into account the necessity for using a variety of food substances in our diet. Milk cattle return a larger proportion of food for human consumption than the above statistics indicate.

CHAPTER SEVEN

THE RELEASE OF ENERGY

In order to do work, every machine in a manufacturing establishment must be supplied with energy, and every living cell in a plant requires energy for carrying on its work of repair, growth, and movement. In manufacturing establishments the energy is usually generated at one place and is then transmitted by means of shafts and belts or by wires and motors to all parts of the factory. It has already been shown that the plant obtains energy from sunlight during photosynthesis, and that this energy is stored as potential energy in the food. Since the food passes from cell to cell, some of the stored or potential energy finally reaches every living cell of the plant. Here the energy that is in the food may be liberated, or changed to free energy, and used in the life processes of the cell, such as the synthesis of fats, proteins, and other compounds.

Respiration. A steam engine is supplied with energy by the oxidation of fuel beneath the boiler that is connected with it. A cell is supplied with energy by the oxidation of food within it. *The process by which the cells obtain energy through the oxidation of foods is called respiration.* In the process oxygen is absorbed and carbon dioxide is given off. Respiration takes place in all living cells, and to carry on this necessary process all living parts of the plant must have access to oxygen.

The substance most commonly oxidized in plants is glucose. Other carbohydrates like starch are changed to glucose before oxidation takes place. Fats occurring in seeds are first oxidized to sugars, and the sugars may be used in building tissue or they may be further oxidized to carbon dioxide and water in respiration. Protein may be oxidized in respiration, but this does not usually occur unless sugar is scarce or lacking entirely. The leaves and stems of land plants obtain their oxygen from the atmosphere; the roots obtain theirs from the air that is in the soil. Wet soils are unsuited to the growth of many plants, not because of the water present, but because of the lack of a sufficient oxygen supply for the roots. Drainage is a valuable agricultural practice, not only because it removes excess water, but also because it draws air (oxygen) into the soil. When the farmer breaks the crust on the surface, he is making it possible for more oxygen to reach the roots of his crop.

The plant and the process of respiration may be compared to a manufacturing establishment and the work that goes on in it.

The power stations	are every living cell of root, stem, and leaf.
The machinery	is the protoplasm and enzymes.
The fuel	is foods, especially carbohydrates.
The process	is the combining of food and oxygen.
The product	is energy.
The waste	is carbon dioxide and water.
The working hours	are twenty-four hours a day.

Respiration and photosynthesis contrasted. Respiration is the reverse of photosynthesis. In photosynthesis, carbon dioxide and water are combined, complex molecules of carbohydrates are formed, and a large number of oxygen atoms are set free in the process. In respiration, the complex carbohydrate molecules are broken up, oxygen is again combined with them, and simple molecules of carbon dioxide and water are formed. In photosynthesis, the energy of the sunlight used in building up the carbohydrates is stored in them. In respiration, this energy is released when carbohydrates are oxidized and changed back to the simple substances out of which they were made.

When we wind up a clock spring, we put energy into the tightened coil. When the spring is allowed to uncoil, this energy is released and turns the wheels of the clock. So in photosynthesis the energy is stored in the carbohydrates, and in the process of respiration this energy is released and used in the life processes of the cell.

In photosynthesis	*In respiration*
Oxygen is released.	Oxygen is consumed.
Energy is accumulated.	Energy is released.
Simple molecules are built up into complex ones.	Complex molecules are broken down into simple ones.
Plants accumulate food and increase in weight.	Plants consume food and decrease in weight.

Comparative rates of respiration. The rate of respiration is greatest where there is rapid growth, as in germinating seeds, opening flowers, and ripening fruits. In some of these it is much more rapid, bulk for bulk, than in animals. A man gives off in respiration about 2.5 per cent of his dry body weight of carbon dioxide every twenty-four hours. Actively growing parts of plants, like opening flower clusters, may give off 10 per cent of their dry weight in the same time. Some kinds of germinating seeds give off carbon dioxide equivalent to 30 per cent of their dry weight in a day. The average growing herbaceous plant, like corn, loses carbon dioxide at a rate not far from 1 per cent of its dry weight per day. About one fourth of the food manufactured by an acre of corn is used in respiration. Thus a mature plant contains only about three fourths of the carbon that was absorbed in photosynthesis. Since photosynthesis takes place only during sunlight, the average rate of photosynthesis in a corn plant is how many times the rate of respiration?

The lowest rates of respiration occur in dry seeds and other dormant structures, and there is comparatively little respiration in woody stems and other hard parts in which there are only a few living cells.

Respiration of fruits and vegetables. How important is the recognition of the respiratory requirement of living cells may be illustrated by the difficulties that have been met with in storing and shipping fruits and bulbs. Peaches, during shipment, sometimes develop brownish spots where they touch each other. These spots were formerly thought to be due to jarring in transportation, but they are now known to be caused by packing the peaches so closely that the air does not have full access to all the fruit. The respiration of the cells at the points of contact is in consequence interfered with, and these cells are suffocated and die.

One sometimes finds large potatoes that are hollow in the center, the cells lining the interior colored brown or black. Otherwise the potatoes are sound. This also is a respiration injury. While the tuber was in the soil or after it had been placed in storage, the outer layers of tissue used all the available oxygen, and the innermost tissue died, leaving a hollow. Cellars, pits, and storage houses for fruits and vegetables must be carefully ventilated.

Ships with specially ventilated holds are used in importing bulbs from Holland and fruits from the tropics. The building of ventilated holds came as a result of the death through suffocation of several men who attempted to unload a cargo of bulbs from an unventilated ship bottom.

CHAPTER EIGHT

SUBSTANCES MADE FROM FOODS

ALL plants contain a variety of substances made from foods that cannot properly be classed as foods. Some of them are of great importance in plant processes; others form the constituents of cell structures. Some may be changed again into foods, and others seem to be waste, or by-products, of cell activities. The most important of these substances will be briefly described in this chapter.

Colorless plant tissues. One occasionally finds on plants leaves that are wholly, or partly, white. This is simply the natural color of living plant tissues that lack chlorophyll or other pigments. The protoplasm, cell sap, and cell walls are transparent and colorless. The presence of air spaces among the cells makes these tissues appear white, just as ice is white when it is filled with minute air bubbles. White leaves and flowers merely show the natural appearance of plant tissues in the absence of chlorophyll and other pigments.

The pigments in green leaves. We can best approach the matter of plant colors by inquiring into the composition of the pigments that color the leaves of deciduous trees in summer and the leaves of evergreen trees throughout the year. The most abundant of these pigments is chlorophyll (Greek: *chloros*, green, and *phyll*, leaf), which is bright green in color. In addition to chlorophyll, two other pigments, one yellow and one orange, are found in a green leaf. These three pigments may exist quite independently of one another.[1]

[1] The coloring matter in a green leaf is composed of about 66 per cent green pigment (chlorophyll); 23 per cent yellow pigment (xanthophyll); and 10 per cent orange pigment (carotin; so named because of its abundance in the carrot). The green pigment is not a simple substance, however, but a mixture of two kinds of chlorophyll, one of which is blue-green and the other yellow-green. The depth of the green color in a leaf depends in part on the proportions in which these various pigments are combined. Chlorophyll contains carbon, hydrogen, nitrogen, oxygen, and magnesium. Carotin contains only carbon and hydrogen, while xanthophyll contains in addition a small amount of oxygen.

In the chloroplasts all three are present at the same time, so that we cannot distinguish them under the microscope. As the three are soluble in alcohol, the presence of the yellow and orange pigment does not become apparent when the coloring matter is extracted from leaves by means of alcohol. The chlorophyll within a leaf is constantly breaking down, and new chlorophyll is being formed constantly in the chloroplast exposed to light. Since it is the chlorophyll in the chloroplast that effects the union of carbon dioxide and water in photosynthesis, it is scarcely an exaggeration to say that chlorophyll is the most important pigment in the world.

Conditions affecting the development of chlorophyll. Chlorophyll is produced only in the presence of light, but the yellow and orange pigments are developed in the dark as well as in the light. When we lay a board on grass or shut out the light to blanch the leaves of celery, the green color disappears, exposing the yellow or orange. Likewise seedlings grown in the dark and the inner leaves of head lettuce show a yellow but not a green hue. These facts make it clear that the yellow pigments do not require light to develop, while the green pigment does.

There are a number of conditions besides absence of light that result in the partial, or complete, disappearance of the green pigment, but these affect various plants quite differently. Low temperature, drought, injuries, and diseases of various kinds may interfere with the nutrition of the leaf; even a slight decrease in light may do so. All these factors tend to affect the green pigment more than the yellow and orange. Although these same influences — low temperature, drought, reduced light, injuries, and diseases — may be effective at other seasons, they become generally operative in late summer and autumn. Hence it is at this time of the year that the green pigment disappears from the leaves of most deciduous plants and unmasks the yellow pigments in the chloroplasts. There is every gradation in the readiness with which the green pigment disappears from the leaves of different species of deciduous trees, from the cottonwood, in which the leaves become yellow during a midsummer drought, to the peach, in which they may still be vivid green when shed. In evergreens the chlorophyll is less sensitive, and external conditions are not so effective in causing changes in the color of the leaves.

Red pigment in plants. The red colors of autumn leaves are not due to changes in the content of the chloroplasts, but to the formation in the cell sap of a red pigment called *anthocyan*. This pigment is present in the cells of many young leaves in early spring. It occurs also in beets, in red cabbage, in the petioles and veins of many different kinds of leaves, in the coleus and other foliage plants, and in many flowers. The presence of anthocyan in the cell sap makes the whole cell red, and any or all of the cells may develop the pigment. The anthocyans are soluble in water, as is shown by the red color of water in which beets have been cooked.

Autumn colors of leaves. In spring and summer the most prominent feature of the landscape is the green color of the vegetation. The most striking feature in autumn is the varied colors of the foliage on the trees and shrubs. In the northern provinces of Canada most of the trees are evergreen, and the most abundant deciduous trees, like the aspen, birch, and tamarack, merely turn yellow. But in our Northern states the vivid greens of the sugar maple, white oak, gum, and sumac disappear in a blaze of red that contrasts strongly with the greens of the hemlock, spruce, and pine. Every one who has seen the colors of autumn woods and the annual falling of the leaves must have wondered what processes go on within the leaves to bring about these changes.

The development of the most brilliant red coloring of autumn is commonly ascribed to the action of frost. This explanation is probably incorrect, for careful observation indicates that the color is most intense when a moderately low temperature is accompanied by bright sunshine. In warm, cloudy autumns the colors are more likely to be dull, with the yellows predominant. In other seasons, when cold weather is delayed, autumn coloration may be brilliant and near its climax before the first frost occurs. That sunlight is important in the development of the red pigment in many plants may be shown also by an examination of a leaf that has been closely shaded by another. The pigment stops so abruptly where the shade begins that a perfect print of the uppermost leaf results. An abundance of nitrogen in the soil prevents anthocyan formation in some plants. This fact may explain in part the greater brilliancy of colors seen on hillsides and river bluffs than on adjoining floodplains.

Among different plants there is much variation in the amount of light that is required for the development of anthocyan colors. This accounts for the great variation in the brilliancy of autumn coloration in different years. One autumn affords light conditions which promote the formation of anthocyan in only a few trees and shrubs; another autumn furnishes conditions so favorable that many plants become brilliant.

Colors of fruits and flowers. The red colors of the fruits of peaches, apples, and pears likewise are due to anthocyan. Here again we may see the effects of sunlight on the intensity of color by comparing fruits from the brightly illuminated top of the tree with others from the shaded under parts. Certain varieties of apples grown in the Northwestern states are more brilliant in color than the same varieties grown in the Eastern states, and this higher coloration is probably due to exposure to more intense light.

The red, blue, and purple colors of many flowers are due to anthocyans, which are red when acid, purple when neutral, and blue when alkaline. The anthocyan pigment that occurs in some vegetables like beets, radishes, and purple cabbage bears no relation to light.

Other pigments. A number of other pigments occur widely distributed among plants, particularly the yellow pigments of many flowers, the yellow bark of some trees, and certain yellow fruits. Some of these pigments have a commercial value as dyes. Indigo and litmus are blue dyes of vegetable origin. Madder is one of a group of red dyes used in making artists' colors.

Cell-wall constituents. Cellulose is the best known of the substances found in cell walls. It belongs to the more complex of the carbohydrates and is a strong, white, insoluble substance. It forms the framework of most plants and is the important constituent of all textile fibers, like cotton, hemp, flax, and jute. It is also the basis of a large number of manufactured products, such as paper, celluloid, acetic acid, artificial rubber, lamp black, charcoal, vegetable silk, and numerous explosives.

Pectic compounds, which closely resemble cellulose in chemical composition, occur in most cell walls. The *middle lamella*, which holds together the cells of the higher plants, is made of pectose or of calcium pectate. It is this layer which breaks down in the boiling of fruits and vegetables and allows them to soften and separate. Pectic compounds occur in many fruits, and when these fruits are boiled with sugar, jelly is formed. In the living plant pectic compounds aid in holding water in the cell.

Lignin, suberin, cutin, and wax. Closely associated with cellulose is a group of substances which modify the cell walls of certain tissues as they increase in age. These substances form mixtures, or chemical combinations, with the cellulose already present. Lignin increases the hardness and rigidity of cellulose walls and is present in most woody tissues. Suberin is the important constituent of the walls of cork tissue. Cutin and wax are usually present in the outer walls of the epidermis of land plants. Suberin, cutin, and wax are all related chemically to the fats, and when present in, or on, cellulose walls render them less permeable to water.

Resins, gums, and mucilages. Resins and gums are products frequently formed in all parts of plants. Resins are insoluble in water and render walls impervious. They occur usually in

FIG. 27. Tapping the Pará rubber trees, in the Malay States, to get the latex from which crude rubber is made.

definite glandular structures, or in tubes extending throughout the plant. Gums are soluble in water, forming a sticky solution. Gums and resins form the bases of a variety of varnishes and paints. Mucilages frequently occur in plant cells. Like gums, they are soluble in water and are often useful in holding water in plant tissues. The drought resistance of some plants is due to the presence of mucilages.

Latex. Many plants like the milkweeds, euphorbias, figs, and rubber plants have a milky juice, called *latex*. This is a mixture of resins, gums, fats, and food substances. It is the source of commercial rubber. Whether it has a definite function in the living plant is unknown.

Alkaloids. Under this general name may be grouped a large variety of chemical substances that seem to be of little importance in the economy of the plant, but which have been of great importance in medicine. They are nitrogen compounds, are generally odorless, and have a bitter taste and marked physiological effects upon animals. They are extensively used as stimulants and narcotics. The best known are nicotine, from tobacco; atropin, from nightshade; strychnine, from strychnos; cocaine, from coca leaves; quinine, from cinchona bark; morphine and codeine, from the poppy; and caffeine from coffee, tea, and cacao seeds.

Essential oils. The odors of flowers and the taste of many fruits and vegetables are due to minute quantities of these substances. Because of their medicinal uses their composition is well known. You are familiar with menthol, the characteristic oil of mint, camphor, oils of lavender, bergamot, bitter almonds and vanilla. Some of the oils contain sulfur. These produce the odor and taste of onions, garlic, water cress, radishes, and many kinds of mustard.

Vitamins. These substances, which are essential in the nutrition of animals, occur only in minute quantities. We cannot test for them or find them in the cell, and we know of their occurrence only through feeding experiments with animals. If they are destroyed by prolonged boiling before the food containing them is fed to animals, the animals fail to grow properly and gradually weaken and die. Scurvy, beriberi, and rickets are diseases produced by lack of the essential vitamins. Vitamins are manufactured mostly by plants, and accumulate in certain animal tissues and in milk, from the plants eaten by the animals.

Tannins. The bark of many trees, the galls occurring on oaks, and certain unripe fruits like the persimmon, contain bitter astringent substances known as *tannins*. These substances react with gelatin or raw hides, forming insoluble compounds, and this reaction is the basic one in the tanning of leather. With iron salts, tannins produce black or green colors. Ink was

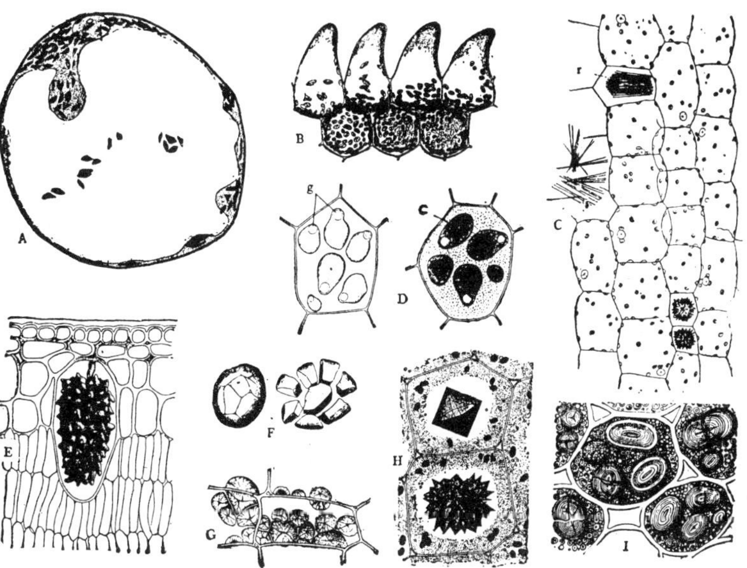

FIG. 28. Structures and substances occurring in plant cells: *A*, cell from pulp of ripe tomato, showing chromoplasts in which the red pigment arises; *B*, vertical section of upper cells of petal of yellow lupine — a yellow pigment forms in the chromoplasts; *C*, cells from green bark of grapevine, some of which contain needle crystals (*r*) and other crystal aggregates of calcium oxalate; *D*, cells from castor bean containing aleurone (protein) grains; *E*, part of a vertical section of leaf of rubber plant showing crystal aggregate of calcium carbonate; *F*, compound starch grains of oats; *G*, sphæro-crystals of inulin in cells of dahlia roots after storage in alcohol: *H*, calcium oxalate crystals in cells of spiderwort; *I*, cells from seed of pea containing starch grains and protein granules. (*After Frank.*)

formerly made in this way. Many fruits are discolored when cut with a steel knife, because of black compounds formed by tannic acid and the iron in the knife.

Enzymes. The enzymes make up another group of very important substances found in all living cells. Their composition is unknown, and we know of their presence only through the effects that they produce. They are usually soluble in water, in dilute salt solutions, or in glycerin, and are insoluble in alcohol.

They are important in speeding up all chemical reactions in cells; without them the chemical changes would be so slow that life could not continue.

Chemists have known for a long time that many reactions can be accelerated by the addition of small quantities of certain substances which do not appear to take any immediate part in the reactions. For example, if we boil cane sugar in pure water, glucose and fructose are formed very slowly.

$$\underset{\text{cane sugar}}{C_{12}H_{22}O_{11}} + \underset{\text{water}}{H_2O} \longrightarrow \underset{\text{glucose}}{C_6H_{12}O_6} + \underset{\text{fructose}}{C_6H_{12}O_6}$$

If a very small amount of a mineral acid is added, the reaction takes place very rapidly. Substances which accelerate chemical reactions are called *catalysts*, or catalytic agents.

All cell processes, including oxidation, take place at ordinary temperatures, often indeed at very low temperatures; and it would be quite impossible for them to take place so rapidly in the absence of catalytic agents. *Enzymes are the catalytic agents of the cell.* Many enzymes have been isolated from plant tissues, the best known of which are diastase, used in digesting starch; lipase, employed in breaking down fats; and papain, used in digesting proteins.

Enzymes will be more fully discussed later in connection with digestion, but they are mentioned at this time because of their occurrence in all cells and because they are concerned in all chemical processes that occur in cells. Enzymes not only aid in breaking down complex substances in cells, but under slightly different conditions bring about the reverse process, the building up of complex substances. They are concerned in photosynthesis, fat synthesis, protein synthesis, and the formation of the many substances described in this chapter.

Protoplasm. The living substance of the cell is the most important product made from food. Carbohydrates, fats, and proteins are in some way by the aid of enzymes built into protoplasm. This process can only be carried on by previously existing protoplasm. How non-living materials are transformed into living protoplasm is one of the greatest problems in biology.

Assimilation. *Assimilation may be defined as the process by which protoplasm, cell walls, and other essential constituents of cells are made from foods.* Protoplasm must be made in the formation of new cells, and it must be constantly renewed in cells already formed. Of all the foods the proteins most nearly approach protoplasm in composition and are most used in the building of the living matter. However, carbohydrates and fats also enter into the construction of both protoplasm and cell walls. Before being assimilated, the complex foods are broken up into simpler and more active compounds. Assimilation takes place in all living cells, but it is most active in growing parts. Respiration is also most active in these parts, and some of the energy liberated by respiration is used in forming other compounds.

Summary. The many substances of which plants are composed are made from foods. Some of these substances, like chlorophyll and enzymes, are of vital importance; others, like tannins, alkaloids, and essential oils, may be merely by-products of the nutritive processes. Protoplasm is an organization of many substances possessing various properties. Cell walls are composed primarily of cellulose, which is modified by the addition of other substances. Vitamins, which are formed mostly by plants, are essential additions to the food of animals, and there can be little doubt that the presence of enzymes in the vegetable food of animals aids in digestion. The building of new tissues is known as assimilation, and is considered the culmination of all the chemical processes occurring in cells. Of the sugar made in photosynthesis by a corn plant, about one fourth is used in respiration, about one half is assimilated in the construction of the plant, and the remaining one fourth is accumulated in various forms of food within the plant.

REFERENCES

THATCHER, R. W. *The Chemistry of Plant Life.* McGraw-Hill Book Co.
HAAS AND HILL. *Chemistry of Plant Products.* Longmans, Green & Co.

CHAPTER NINE

LEAVES IN RELATION TO LIGHT

LEAVES grow from points variously arranged on stems that have all sorts of positions. If these leaves grew out in random directions, many of them would receive little light. But an examination of a plant shows its leaves arranged in positions which display them advantageously to the light. The raised leaves of the pumpkin, the mosaics formed on the sides of buildings by the leaves of vines, and the successive tiers of leaves on a beech, maple, or dogwood illustrate different arrangements by which large numbers of leaves are efficiently displayed to light. Evidently something controls the positions of leaves on a plant.

FIG. 29. Vertical branch of magnolia. Note the alternate arrangement of the leaves.

Growth influenced by light. Light affects growth in all organs of the plant, including the leaf. The amount of light received by a leaf blade not only affects the growth of the blade, but also the petiole, and in some plants the adjoining stem. The influence of the light, during the growth of leaves, is such that when they are mature most leaves are favorably placed with respect to light.

The arrangement of leaves on stems. Leaves develop from somewhat thickened places on the stems, called the *nodes*. Each node may bear one, two, or several leaves. According to the number of

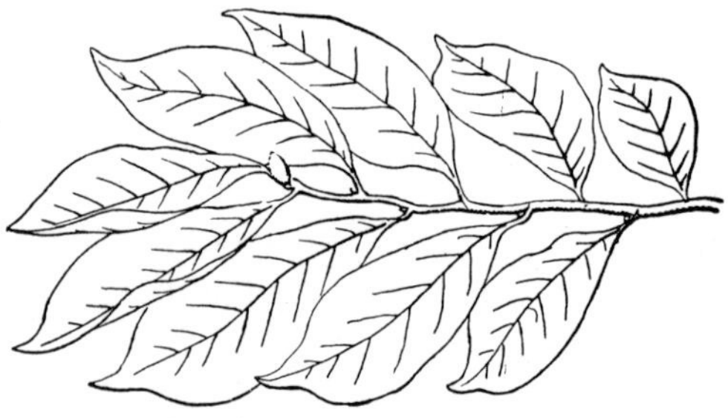

FIG. 30. Horizontal branch of magnolia. Compare leaf positions with those of Figure 29.

leaves that the node bears, the leaf arrangement is designated as *alternate*, *opposite*, or *whorled*.

In the alternate arrangement each node bears one leaf. This is also called the *spiral arrangement*, because a line drawn through successive leaf bases forms a spiral about the stem. Sometimes, as in the corn plant, the spiral passes half around the stem in going from one node to the next. In other plants, like the sedges, the spiral passes but a third around the stem between nodes. In several of our common fruit trees, as, for example, the apple and the peach, the spiral between nodes passes two fifths around the stem. These variations of the spiral arrangement of leaves on stems are called the two-ranked (Fig. 37), three-ranked (Fig. 31), and five-ranked arrangements (Fig. 29).

FIG. 31. A sedge (*Dulichium*), showing three-ranked arrangement of the leaves.

In the *opposite arrangement* two leaves occur at each node (Fig. 33). The leaves at successive nodes, however, are at right angles to each other, giving four ranks of leaves. The maple, ash, dogwood, and lilac furnish examples of the opposite arrangement. In the whorled arrangement the leaves are in a circle about the node (Fig. 32). The Indian cucumber root (*Medeola*) and the wood lily furnish excellent examples of the whorled arrangement.

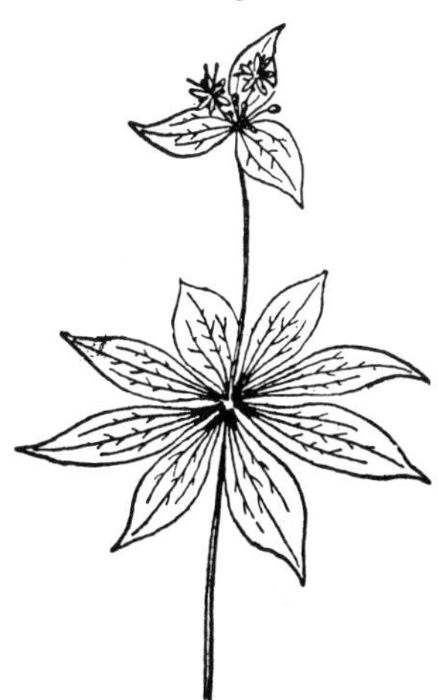

FIG. 32. Indian cucumber root, showing the whorled arrangement of the leaves.

However, it is only on upright stems which receive the light equally on all sides that the blades take their normal positions directly out from the nodes. If an erect shoot be placed in an inclined position, it is easy to see that the leaves are no longer well displayed to the light. As may be readily seen by examining the branches of trees and the stems of trailing plants, horizontal or inclined stems become twisted during development through the influence of unequal illumination upon the relative growth of different sectors of the stem. The twisting of the stems brings the leaves into better-illuminated positions, but it often obscures the normal arrangement of the leaves.

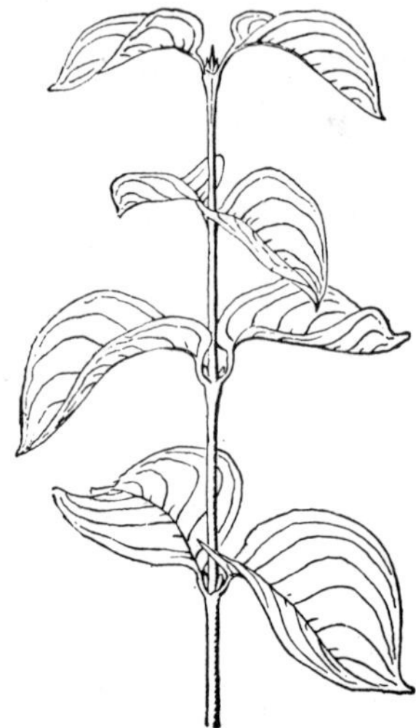

FIG. 33. Vertical branch of dogwood, showing the opposite arrangement of the leaves.

The positions of leaves with reference to light. If leaves are moderately sensitive to light, their positions when mature are approximately at right angles to the line along which the greatest amount of light reaches them. Consequently the leaves on most of our common trees, shrubs, and herbs have an approximately horizontal position (Figs. 33, 34). The sugar maple and the magnolia are examples of trees whose leaves are displayed in this manner (Figs. 30, 35). In the cottonwood

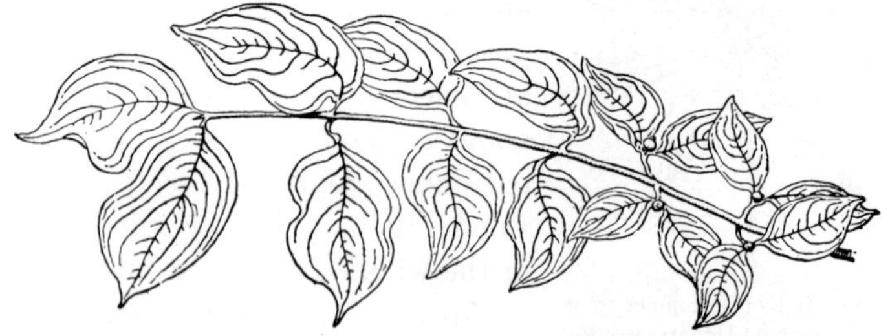

FIG. 34. Horizontal branch of dogwood. Compare with Figure 33.

W. S. Cooper

FIG. 35. Leaf mosaics formed by maple leaves (*Acer macrophyllum* and *Acer circinatum*), Olympic Mountains, Washington. The light affects the growth of the petioles and branches in such a manner that the leaves are fitted together side by side.

and tulip trees the leaves are less sensitive to light, and the result is that their leaves assume a great variety of positions. If leaves are extremely sensitive to light, the blades may turn toward the sun in the early morning and follow it throughout the day, always keeping the broad face of the leaf to the light. The leaves of the common mallow move in this way.

Leaf mosaics The leaves of many plants, like the Boston ivy, sugar maple, and beech, are so arranged that if we look at them from the direction from which they receive the most light they seem to be fitted together like the stones in a mosaic. In this way each leaf is exposed to the most possible light.

Many small herbs, like the dandelion, moth mullein, common plantain, and evening primrose, form rosettes of leaves near the soil. An examination of these rosettes will show that each leaf is arranged so that it fills a space in the circle. Further examination will show that the leaves that would otherwise be more or less shaded have changed their positions and occupy the spaces between the leaves above them.

Compass plants. There is another class of plants which are sensitive to light, but which respond to it in a very different manner. These are the so-called compass plants, of which the wild prickly lettuce is a widely distributed example. In sunny situations the leaves of these plants tend to take positions edgewise to the direction of the most intense light. As the sunlight is most intense at noon, it is only in the morning and late afternoon that the flat sides of the leaves are perpendicular to the sun's rays. This response to the light also places most of the leaves in a vertical north-and-south plane and suggests the name "compass plant." When grown in partial shade, the leaves of these same plants are horizontal. Hence it is clear that the position of their leaves in sunny situations is the result of light conditions (Fig. 36).

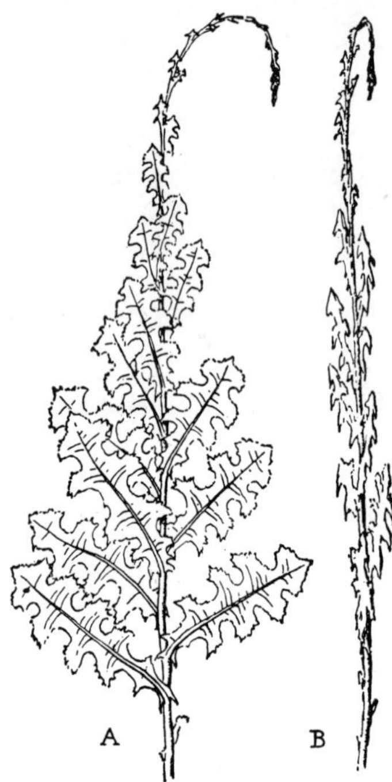

FIG. 36. Prickly lettuce, which is called "compass plant" because its leaves stand in a north-south plane: *A*, viewed from west; *B*, viewed from south. Drawn from a specimen grown under exposure to bright sunlight.

How the blade attains its position with reference to the light. The position of the leaf blade is partly attained, as has been noted, by the bending and twisting of the plant stem during its development. To a much greater extent the blade owes its position to the bending, twisting, and elongating of the petiole. Indeed, its ability to place the leaf in an advantageous position toward the light is the particular advantage of the petiole. Its length and direction of growth are for the most part determined by the way in which the light falls on the blade during the period of development.

Strong light retards growth in length of the petiole. If the blade is shaded, the petiole elongates more than usual; if shaded on one side, the petiole grows unequally on its two sides until the blade is about equally illuminated. The position of the leaf, when it has stopped growing, is usually fixed, and shading will no longer affect the growth of the petiole. When a leaf that has attained its full growth is overshadowed, it loses its chlorophyll, turns yellow, and dies. You can find examples of such leaves under the green leaves of rosettes, or on the lower branches of trees that form mosaics.

Vertical leaves. In a number of common plants, including the iris, cat-tail, calamus, and many grasses, the leaves are vertical because they are held in this position by their sheathing bases rather than because of a response to light. These plants usually occur in dense growths (Fig. 38), and the vertical position of the leaves permits the light to penetrate to their bases. This has the advantage of allowing photosynthesis to go on throughout the entire length of the leaves.

Differences in vertical and horizontal leaves. The structure of vertical leaves differs from that of horizontal leaves in several particulars:

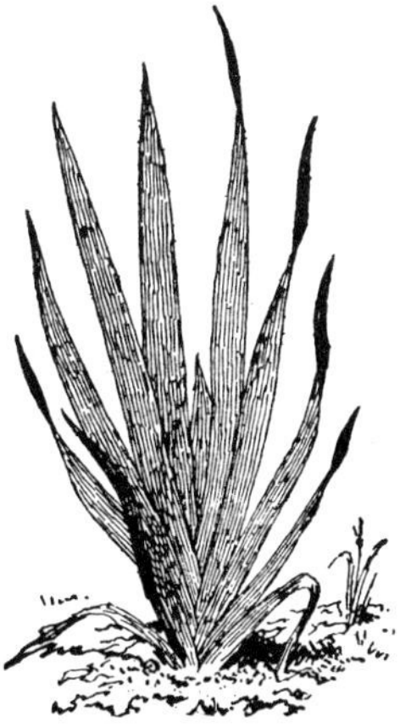

In vertical leaves the mesophyll may be composed of spongy tissue, or it may be composed entirely of palisade cells. More rarely there are palisade layers on both sides, with a spongy layer between. In contrast, a horizontal leaf usually has a palisade layer beneath the upper epidermis, and the lower portion of the mesophyll is composed of loosely arranged cells. In vertical leaves stomata usually

FIG. 37. Two-ranked vertical leaves of iris, held in position by their sheathing bases.

Fig. 38. Guinea grass, a plant grown in the tropics for fodder. Vertical leaves expose a large surface to the sunlight in spite of the crowding.

occur on both surfaces, while in most horizontal leaves the stomata are confined to the lower surface (page 27). Vertical leaves are likely to be of the same color on both surfaces, while horizontal leaves are generally of a darker green on the upper surface.

The difference in the color of the two sides of a horizontal leaf is due in part to the presence of a larger amount of chlorophyll in the compact palisade layers of the mesophyll than in the loose spongy layer beneath. In vertical leaves the similarity of structure in the mesophyll on each side, and the fact that both surfaces of the leaf are equally illuminated, account for the sameness of color of the two surfaces. The color of leaves is sometimes modified by the presence of hairs, wax, or drops of resin.

Motile leaves. The leaves of which we have been speaking have their positions rather definitely fixed when they reach maturity. There is another class of leaves, however, in which the positions of the blades are not fixed, but are changed according to the intensity and direction of light. A familiar example is the roadside sweet clover. At night the three leaflets of the compound leaf droop downward from the petiole; in the medium light of a cloudy day they are held perpendicular to the light; in the most intense sunlight the blades are raised above the petiole until they are edgewise and point toward the light. Some observation of lima bean seedlings (Fig. 41), which may readily be grown in the laboratory, will be instructive in this connection. Other examples of motile leaves may be seen in the honey locust, the leaflets of which fold upward at night, and in white clover, oxalis, and the red-bud tree. The leaflets of the sensitive plant vary their positions according to light intensity, and also when touched or injured in any way (Fig. 39).

The changes of position in motile leaves is brought about by changes in the water content of the cells on opposite sides of a special organ called the *pulvinus* (Fig. 40), which is located at the base of the leaves and the leaflets. This device may readily be studied in the leaf of the bean.

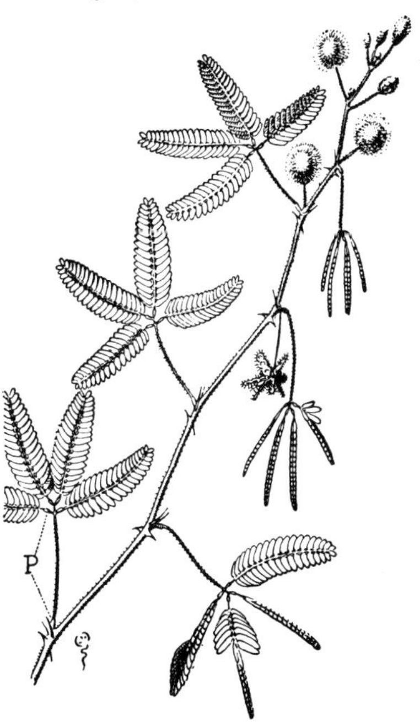

Fig. 39. Sensitive plant. The leaves on the left side are in normal positions; those on the right side have been touched and the leaflets have folded together wholly or in part, and the petioles have folded toward the stem. *P* is the pulvinus.

The leaves of shade plants. As may be observed by a trip to the woods, the leaves of plants growing in the shade are usually darker and more bluish-green than the leaves of plants growing in full sunlight. This difference in color is accounted for in part by the amount of chlorophyll near the surface and in part by a slight difference in the color of the green pigment in the chloroplasts. In a few shade plants the depth of the green color is increased by the presence of chloroplasts in the epidermal cells. Shade plants are not subjected to drying, as are plants growing in exposed situations, and, generally speaking, their leaves are broad and thin. The leaves of these plants differ further from the ordinary leaf in that the cuticle is less developed, the mesophyll is composed almost entirely of spongy tissue, and usually stomata are present on both surfaces of the leaf.

Submerged leaves. Every one who has gone fishing or rowing knows that a great deal of sunlight is reflected from the surface of water. A smooth water surface reflects about one fourth, and a rough water surface about one half, of the light that falls on it. This means that the amount of light that passes into the water is reduced by the amount that is reflected. The penetration of the water by the sun's rays is further interfered with by the fine sediment that clouds our ponds and lakes. Every one who has dived and opened his eyes under water knows that it is dark at a comparatively slight depth. Measurements have shown that one half to three fourths of all the light that enters the water may be stopped in the first three feet, depending upon the amount of suspended particles present. Hence submerged plants always grow in

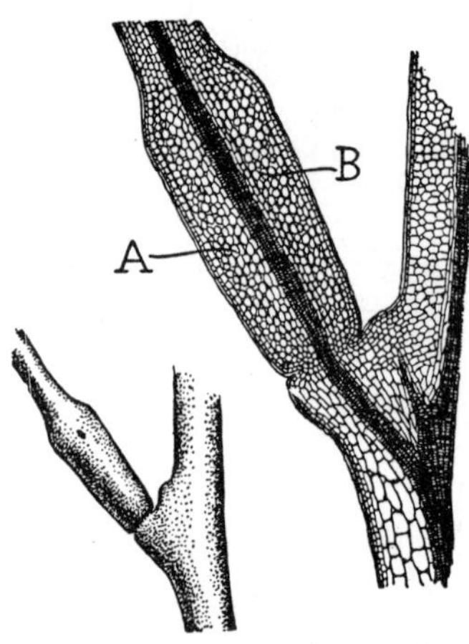

FIG. 40. Pulvinus and section of pulvinus from leaf of sensitive plant, both enlarged. When the leaf is touched, the water in the cells on the side *A* passes outward into the intercellular spaces, causing the cells partially to collapse. The pressure of the cells on the side *B* then forces the leaf downward.

light of reduced intensity. They receive an amount of light comparable to that received by the shade plants found in

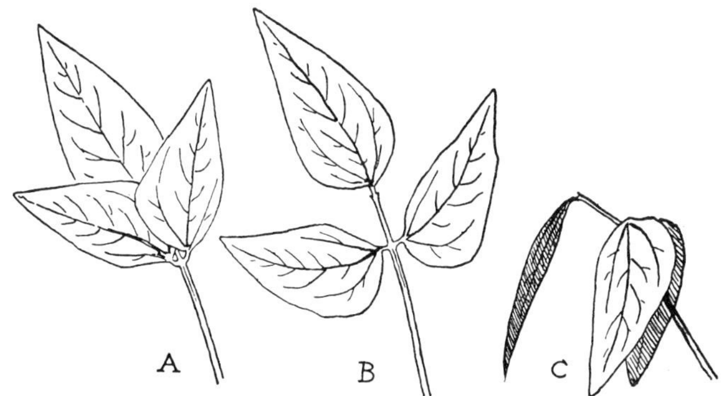

FIG. 41. Various positions taken by leaflets of lima bean: *A*, position in intense light; *B*, position in diffuse light; *C*, position in darkness.

forested ravines. Submerged leaves, too, are of very soft texture, and are quite without mechanical tissue in the veins, so that they are unable to support themselves when lifted from the water. They are kept upright in the water by their buoyancy.

Summary. Light has marked effects upon the positions, the color, and the structure of leaves. Leaves tend to be placed directly outward from the nodes to which they are attached, but light affects them during their development, and most leaves come to occupy positions that have more relation to the light than to the stem which bears them. The position of leaves and the movements of leaves are determined by differences in water content and in the rate of growth on opposite sides of the stems and petioles that support them.

PROBLEMS

1. Why do house plants flourish best at south windows in the winter time?
2. What part of full sunlight is received by a plant that stands near a window?
3. Why do gardeners shade lettuce plants in midsummer?
4. What other condition, besides light intensity, is affected by shading?
5. Why is tobacco that is intended to be used for cigar wrappers usually grown under canvas or beneath lattice frames?

CHAPTER TEN

THE WATER RELATIONS OF LEAVES

DURING a prolonged drought in Illinois, in 1914, oats in some places failed to attain a height of more than 4 inches and produced practically no grain. Corn which should have averaged 10 feet in height reached only 5 feet in many fields, and yielded only half the normal amount of grain. In the four great corn-growing states there must be 3 inches of rainfall in July for the best yield of corn; and if the rainfall during July is $2\frac{1}{2}$ inches instead of 3, it is estimated that at normal prices there is an average loss of $5 an acre, or a total loss of $150,000,000. Those who cultivate plants know from experience the importance of a sufficient water supply in the production of crops. The reason why the water supply is important will be apparent when we understand the uses made of water by the plant.

Why water is necessary to a plant. The active protoplasm of all plant cells is in a semiliquid condition. More than 90 per cent of its weight is made up of water, and in consistency it closely resembles white of egg. The several parts of the protoplasm — the cytoplasm, the nucleus, and the plastids — differ somewhat in their water content, but all of them must be nearly saturated with water to carry on the life processes. When the amount of water in the cell falls much below this point, the protoplasm becomes rigid and all its activities are retarded. The curled-up leaves of corn during a summer drought illustrate this effect. The corn manufactures little food, and consequently growth is retarded. In many plants the protoplasm may even die if the water content is greatly reduced. For example, the seeds of the soft or silver maple which are shed in late spring and germinate soon afterward die if the water content is reduced below 30 per cent. *Water is necessary for the life of the protoplasm of plant* *cells.* We have previously shown that water enters into the composition of all carbohydrates; therefore *water is necessary for photosynthesis.*

Substances can enter plants only when they are in solution. Both the gases and the mineral compounds that are used by the plant in its various processes must be in solution in water before they can be absorbed or pass from one cell to another within the plant. Indirectly as well as directly, water is necessary to photosynthesis; for water keeps the mesophyll cells wet and thus makes it possible for the carbon dioxide to enter the cells. *Water is necessary for the absorption of minerals and gases and for the transfer of materials within the plant.*

Growth and reproduction result from a series of many physical and chemical changes within the cells. These changes can take place only in the presence of water. *Water is necessary for all physical and chemical changes within the plant* and consequently for all plant activities

Transpiration. If we expose a wet cloth to the air, the water evaporates; that is, it changes from a liquid to a vapor and passes off into the atmosphere. The same thing happens when a plant is exposed to the air. The mesophyll cells of the leaf are continually losing water vapor to the intercellular spaces, from which, if the stomata are open, this vapor passes out into the atmosphere. The epidermis of the leaf also allows some water to pass through it, but in land plants this is a relatively small amount, because the cuticle hinders the process. The loss of water vapor from plants is called *transpiration.*

The loss of water in the form of vapor is a process that takes place in animals as well as in plants. If you hold your hand near a window-pane on a cool day, a halo of minute water drops condenses on the glass. These water particles come from the moist cells of your skin. If you blow on a glass, water collects even more abundantly. The vapor in the breath is water that has evaporated from the moist cells of the lungs.

Importance of transpiration. Just how important transpiration is to the plant may be easily seen by a study of the energy changes that take place in a leaf. First of all, we must understand that light energy is very readily changed to heat energy, and that when heat energy accumulates in a body it raises its temperature. When a body loses heat energy, it is cooled. When the sun shines on a leaf, it is estimated that about 10 per cent of the light is reflected by the leaf surface and about 25 per cent goes through the leaf. Sixty-five per cent is retained by the leaf. This is sufficient energy to raise in a few minutes the temperature of the leaf from air temperature to the danger point for protoplasm.

As soon as the temperature of a leaf rises through exposure to sunlight, the water particles become more active, and as they leave the surface and fly off into the air the excess heat energy is reduced. In this way the leaf is kept at, or within a few degrees of, the air temperature. Transpiration cools the leaf just as water evaporation from your hand cools the skin. *Transpiration is important to plants because it helps to regulate the temperature and prevent overheating. As will be shown later, it is also an important factor in raising water and mineral salts from the roots to the leaves.*

It is estimated that nearly one half the energy of sunlight that falls on a cornfield in Illinois is used in transpiration.

Transpiration and stomata. Most of the water vapor that leaves the plant in transpiration (80–97 per cent) is derived from the mesophyll cells and passes through the stomata from the intercellular spaces. Comparatively little (3–20 per cent) is lost through the epidermis directly into the air. It is evident, then, that the movements of the guard cells, as they result in opening or closing the stomata, modify the rate of transpiration.

In most plants the stomata are closed at night, and as there is little or no heat energy added to the leaves, the rate of transpiration is very low. The stomata open slowly after sunrise, but as soon as the light strikes the leaf its temperature rises. Transpiration increases rapidly. Toward the middle of the afternoon the stomata begin to close, and before sundown are completely closed. The rate of transpiration begins to decrease about 2 o'clock and reaches the slow night rate before, or soon after, sunset. The stomata, therefore, modify the rate of transpiration greatly.

It must not be supposed, however, that they act as safety mechanisms to conserve water in the plant. They may open in the light, whether the plant has an adequate water supply or not. Likewise they may close when the plant has an abundance of water. Usually the stomata close when the leaves wilt, but there are exceptions even to this rule. The opening of the stomata in light not only allows the outward diffusion of water vapor, but also the inward passage of carbon dioxide used in photosynthesis and the escape of oxygen liberated in this process.

The amount of water transpired by plants. The amount of water lost in transpiration is surprisingly large. During its lifetime, a well-watered corn plant may lose 40 gallons of water. The water lost by a field of wheat during its entire period of development would cover the field to a depth of 4 or 5 inches. A medium-sized date palm growing in the Sahara Desert under irrigation is estimated to require from 100 to 190 gallons of water per day during at least four months of the year. For the best growth of plants, therefore, there must be available in the soil enough water to replace all that is lost by transpiration and the smaller amount used in the growth of new parts.

When we consider that the quantity of water transpired by wheat in cultivation is one fifth to one eighth of the rainfall of the central United States, we begin to realize how large a fraction of all the water that falls on the soil is actually used by the plants. In all rainfall, some water runs off the soil without penetrating the surface, some evaporates from the soil surface itself, and some sinks below the level of the plant roots. Consequently, it is only when there are abundant rains, distributed throughout the growing season, that the amount of water needed by the plants for transpiration and their best development is available in the upper layers of the soil. It has been shown by experiment that for production of every pound of solid matter in the above-ground parts of crop plants, from 300 to 500 pounds of water are required in the central United States, and that from 400 to 1000 pounds are needed on the plains of Colorado. The amount of water used in transpiration is, therefore, many times the amount used in the manufacture of food. It is estimated that an acre of corn uses 1700 tons of water in transpiration and $4\frac{1}{2}$ tons in photosynthesis.

Substances and structures modify transpiration. Most leaves possess certain structures that reduce the rate of transpiration. The possession of these structures enables the plants to live under somewhat drier conditions than they otherwise could.

(1) *Thickened cuticle and "bloom."* The cuticle of a leaf checks transpiration, and in plants of dry climates the cuticle may be so thick as to reduce transpiration through the epidermis to almost nothing. There are many plants which secrete, in addition to the cuticle, particles of wax on their leaves or other parts. This is the so-called "bloom" which may be seen on the leaves of the houseleek and cabbage and on the fruits of the grape, plum, and blueberry. The bloom consists of a layer of wax particles scattered thickly over the surface of a leaf or fruit. It forms a layer that is nearly impervious to water and helps to reduce water loss through the epidermis.

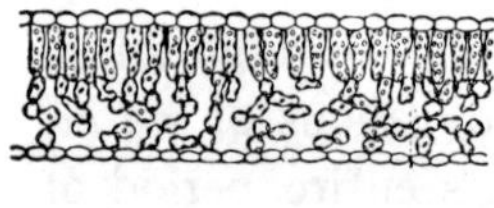

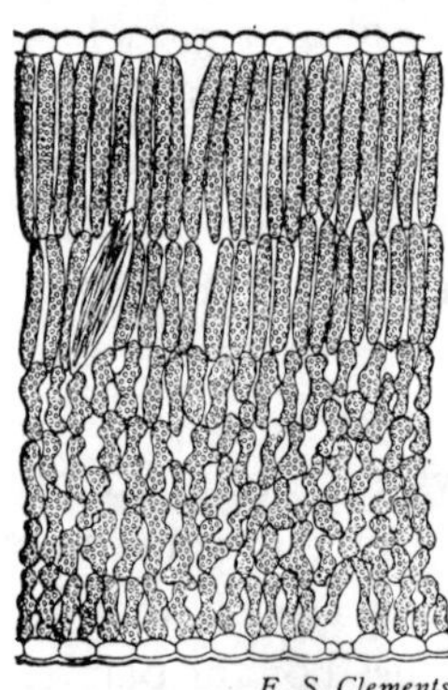

E. S. Clements

FIG. 42. Vertical sections of leaves of *Mertensia*, showing differences in structure when growing in moist, shaded situation (above), and when growing in dry, intensely lighted situation (below).

In transpiration, however, most of the water is lost through the stomata. So we may have a heavy cuticle and still have a

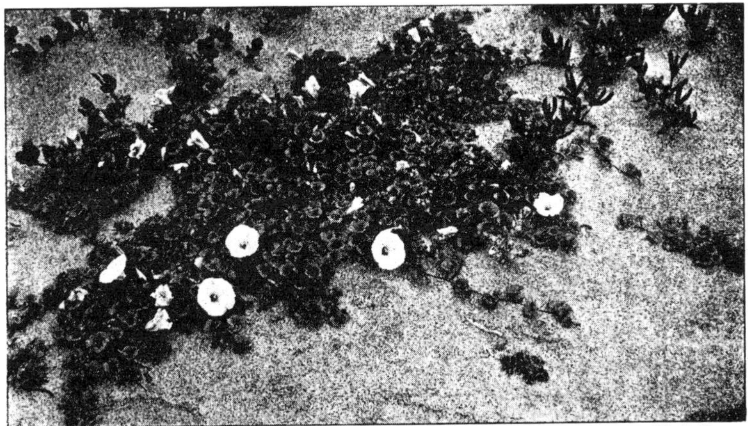

W. S. Cooper

FIG. 43. A xerophytic morning-glory and a succulent-leafed *Mesembryanthemum* (above at right) growing on the dunes near Coronado, California. The thick cuticle of the morning-glory leaf reduces the transpiration rate. The other plant has a relatively small leaf area and holds its water tenaciously because of substances within its cells.

high transpiration rate if the stomata are open. In leaves with a heavy cuticle the stomata are usually small and do not open so freely as in leaves with a very thin cuticle; consequently transpiration is generally less from hard, thick, and heavily cutinized leaves, even though the cuticle prevents evaporation only from the outer leaf surfaces.

(2) *Compact leaves.* A plant may *become* adjusted to an inadequate water supply by the development of leaves with compact tissues, as a result of exposure to drought or bright sunshine. In such leaves the intercellular spaces are much reduced, and evaporation from the mesophyll cells is greatly lessened. In extreme cases the mesophyll cells are all of the compact palisade type, which leaves the minimum of air space within the leaf. Compact tissues reduce the rate of transpiration through the stomata. The tissues are compact under these circumstances simply because drought prevented the further expansion of the leaf, leaving the cells close together (Fig. 42).

(3) *Small leaf area.* A third way in which plants become ad-

justed to dry conditions is by a decrease in the total leaf area. When a plant is brought into the house in autumn, some of its leaves usually fall off. The air inside houses being much drier than the air outside, transpiration is greatly increased. As the water supply remains about the same, the loss of a few leaves restores the water balance of the plant. Some trees, like the cottonwood, lose part of their leaves during a summer drought. If a wet period follows, more leaves may be added, and in this way a nearly uniform water balance is maintained.

(4) *Hairs.* Hairs are frequently described as structures that greatly reduce transpiration. Some of the very broad scale-hairs may reduce the rate slightly but experiments show that the hairy covering of the common mullein, which is exceedingly dense, has little or no effect on its rate of transpiration either in still air or in wind.

(5) *Resin, wax, and mucilage.* Some plants produce resin, wax, or mucilage, which retard transpiration. For example, the horsechestnut has a coating of resin on its buds. The bayberry has wax covering its fruits. The tissues of cactus contain mucilaginous substances that have a great water-holding capacity.

External factors that modify transpiration. That plants growing under moist conditions have larger leaves and more leaves than the same kinds of plants growing under dry conditions has been noted by every one. Experiments show that their rates of transpiration are far greater than when grown under dry conditions. Humidity of the air directly affects transpiration, because when the air is nearly saturated the difference between the humidity inside the leaf and outside is so small that water vapor passes out through the stomata into the air very slowly.

The amount of available water in the soil and the humidity of the air are important because they determine the amount of water in the cells of the plant. The condition of the mesophyll cells in turn regulates the rate of water loss to the intercellular spaces. A dearth of water in the plant may also prevent the opening of the stomata. Light affects the opening of stomata and raises the temperature of the leaf, and consequently increases the rate of transpiration.

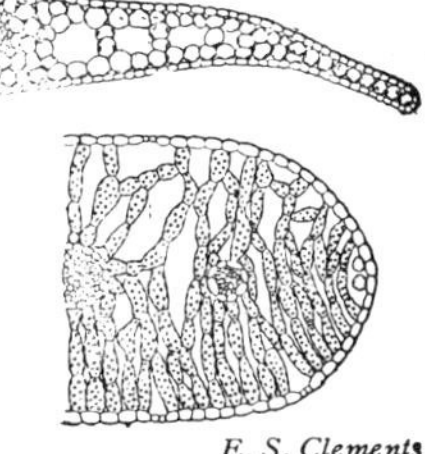

E. S. Clements

FIG. 44. Vertical sections of leaves of *Hippuris*, a water plant. The upper figure shows an aërial leaf, the lower figure a submerged leaf. The aërial leaf is much thinner and the tissues more compact. In the submerged type the guard cells form but the stomata do not open.

Intense light and drought decrease the size and number of leaves and increase the compactness of the mesophyll, the amount of cutin, mucilages, hairiness, and mechanical tissue. Incidentally, these changes retard transpiration. Hence these external factors, by producing changes in the physical and chemical processes within the plant, indirectly modify transpiration.

The higher the temperature is, the greater the rate of transpiration, not only because the water in the mesophyll cells changes to vapor more rapidly, but because the vapor particles move out of the leaf faster.

Submerged and floating leaves. An examination of a submerged leaf on any pondweed shows that it has no stomatal openings. Sometimes the guard cells are formed but do not separate (Fig. 44). The floating leaves of water lilies and other pond plants have stomata only on the upper surfaces. Being completely surrounded by water, submerged plants have no transpiration. It is also certain that they get their carbon dioxide directly from the water through the epidermis, for carbon dioxide is found dissolved in pond waters, often in larger proportion than in the air.

In water-lily leaves the upper surface is covered by a cuticle that is not readily made wet, and it has stomata that do not open until the leaf is above water. If the leaves are raised entirely above the surface of the water, as sometimes happens when the plants are crowded, both surfaces develop stomata.

Desert plants and water storage. In the desert, where the air is very dry and the scanty rainfall is confined to one or two periods in the year, plants have great difficulty in securing water. The perennial plants have various ways of conserving water from one

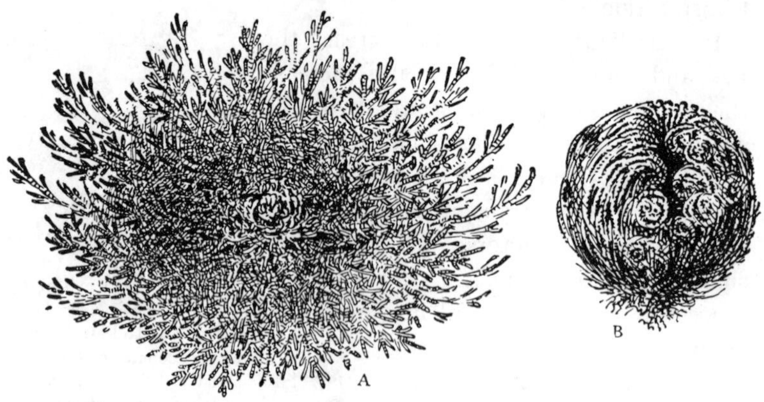

FIG. 45. Resurrection plant (*Selaginella*) of Texas and New Mexico. During the rainy season the plant spreads out and grows as a rosette. When drought comes, it dries out and curls up into a ball, as shown at the right.

rainy period to the next. The barrel cactus has no leaves at all, and the stem is a thick cylinder composed largely of water-storage tissue; it may live without additional water for two years or longer. Some of the desert shrubs have leaves during the rainy periods only, and these fall as soon as the drought comes. Still others, like the agaves, have very thick, leathery leaves with much internal water-storage tissue and a very low transpiration rate (Fig. 48).

Adjustment to desert conditions by ability to withstand drying. Another group of plants is adjusted to desert conditions by being able to withstand complete drying. The resurrection plant of Texas is an example of this group (Fig. 45). During the rainy season it is green and has its many scale-leafed branches spread out, making possible food manufacture and growth. When drought comes, the plant dries out and its branches curl upward until it is in the form of a ball. In this condition it may be blown about by the wind and remain dormant for weeks or months.

all its physiological processes having been reduced to a minimum. When the plant becomes wet it unfolds, and its processes become active again. In the eastern United States we find plants of this same type in the lichens, mosses, and a few small ferns that grow on the bark of trees and on bare, dry rocks.

CHAPTER ELEVEN

PHYSICAL PROCESSES INVOLVED IN THE MOVEMENT OF MATERIALS IN PLANTS

SINCE the earliest times students of plants have been trying to find out just how gases, water, and minerals get into plant cells and how they pass from one cell to another; also how the soluble foods move from one organ to another. Not all of these questions have been satisfactorily answered, but there are certain physical processes that at least help to explain all of them. We must not be misled into thinking that the plant does all these things because, being alive, it can take in substances, move them where they are needed, and throw off those that are not needed. All investigations indicate that these processes take place, not because plants exert some peculiar vital force, but because plant cells possess those particular physical and chemical properties and structural arrangements which, even in a non-living piece of physical apparatus, induce these same processes. Although we cannot now imitate perfectly all the processes concerned in the movement of materials within the plant, it is fair to predict that we shall be able to do so in the not distant future.

Solution. All substances, whether gaseous, liquid, or solid, must be dissolved in water before they can pass into, or out of, a plant cell. By solution is meant the dividing of a substance into invisible particles that distribute themselves throughout a liquid. Carbon dioxide, for example, occurs in the air as a gas. Its particles strike the water surface of the cell and go into solution. Mineral substances coming in contact with the water in the soil do the same thing, and it is only after they are dissolved that they enter the root. Sugar, likewise, can move out of or into a cell only when it is dissolved in the water of the cell. Solution is the first of four physical processes which are important in the movement of substances in plants.

Diffusion. If a small dish of ether is exposed in a room, in a few minutes the odor of the ether may be noticed in all parts of the room. Even if there were no air currents, the ether would evaporate; that is, particles of ether would rise from the surface of the liquid, pass out of the dish, and move through the room in every direction. This is an example of the diffusion of a vapor. The vapor is concentrated in the dish and the particles move outward into the room where there is none; that is, the particles move from the place where their concentration is greater to where it is less. After the ether has evaporated, the vapor tends to become evenly distributed throughout the room. Solids like camphor and naphthalene might be used in place of the ether.

Similarly, if a few crystals of copper sulfate are placed in the bottom of a vessel of water, particles of the copper sulfate diffuse through the water. The crystals are blue in color, and as diffusion proceeds, the water in the vessel gradually becomes blue. The direction of the movement is again from the place where the diffusing substance is most concentrated to where it is less concentrated. The particles pass from the place where they are most abundant to where there are fewer of them, and this process is continued until they are evenly distributed throughout the water.

Diffusion of a gas or vapor is very rapid. Diffusion of a dissolved substance in a liquid is slow, but the distances that substances must travel in plant cells are very small. Oxygen and carbon dioxide, when once dissolved in the water of cells, move about partly by diffusion. The soluble foods in plants move from one part to another by diffusion. Soil salts enter the root by diffusion and are not carried into it by water. Diffusion may occur under special conditions, and it is then conveniently spoken of as *imbibition* and *osmosis*.

Imbibition. The process of imbibition may be illustrated by placing a sheet of gelatin in water. Dry gelatin is a hard, brittle, partly transparent solid. After it has been in water for a few minutes, it will be found to have increased in weight and in length, breadth, and thickness. The gelatin, instead of being

W. S. Cooper

FIG. 46. The barrel cactus (*Echinocactus cylindraceus*) of the Colorado desert, California. These plants are highly resistant to water loss because of the presence of mucilaginous carbohydrates which imbibe and hold water.

brittle, is now soft and pliable; it is also more transparent than it was.

The increase in size and weight is explained by the fact that particles of water have diffused into the gelatin and have forced the particles apart. Since the gelatin particles have been forced farther apart, the gelatin is more pliable and the particles cling to one another less firmly. The cell walls of a plant take up water in the same way. Hence when a piece of dry wood is put in water, it imbibes water and swells. When dry seeds are placed in water, they imbibe water and increase in size. Indeed, most organic substances have the property of imbibing water and swelling. *Imbibition is a form of diffusion that results in swelling.* Compare the size of a sponge when dry with its size after it has been soaked in water and squeezed as dry as possible.

When a piece of wood becomes saturated, it stops taking up water. If, however, the water were being removed from the inside, more would continue to pass into the wood. This is exactly what happens in the root of a living plant. The external cells of the root are in contact with the water of the soil. Inside the root the water is being used and removed by being drawn up through the stem to the leaves. More water then passes into the cell walls and protoplasm to take the place of that which is drawn away, and this tends to keep the amount of water in the plant nearly constant. Imbibition becomes very powerful in plants that have mucilage, gums, and pectic compounds in their tissues, both in absorbing water and in holding it against evaporation.

Osmosis. A third form of diffusion that aids in the absorption of water is osmosis. If an animal membrane, as a piece of bladder, is tied over the broad end of a thistle tube and the bulb of the tube is immersed in water, the water will gradually pass through the membrane. The membrane is permeable to water; that is, it allows water to pass through it. The water continues to move through until its level is the same inside and outside (Fig. 47).

When the water level is the same inside and outside the tube, one might think that the water particles were at rest. This is not the case. Water particles are still passing both into the thistle tube and out of it through the membrane. The rate is the same in both directions, however, and so the water level within the tube remains unchanged.

If we put a little sugar into the thistle tube, something different happens, as is shown by the fact that the liquid in the tube begins to rise. Evidently, more water is passing through the membrane into the tube than is passing out, and this change has been brought about by the presence of the sugar. Perhaps we can get a mental picture of what causes this difference from the diagram in Figure 48. The membrane (*C*) allows water molecules to pass through it freely, but it permits scarcely any of the sugar molecules to pass. The outer side of the membrane is completely covered with water molecules, tending to diffuse through the membrane. The inner side is only partly covered with water molecules, since part of the area is occupied by sugar molecules. Consequently there are fewer water particles on the inside tending to diffuse outward. Sugar dilutes the water in the

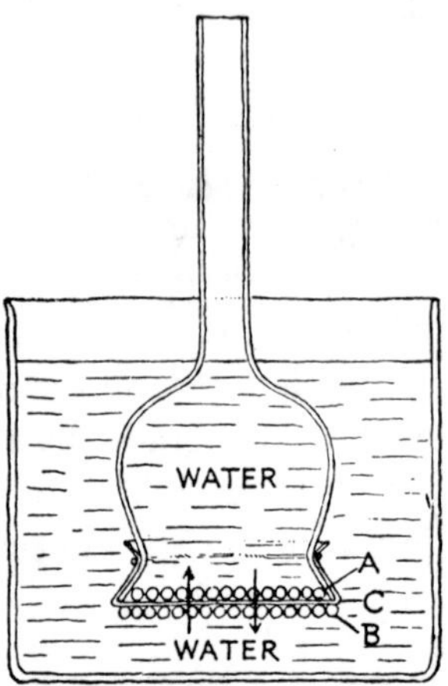

FIG. 47. Diagram to illustrate the passage of water through a membrane: *A* represents a molecule of inside water, *B* a molecule of outside water, and *C* the membrane. Equal numbers of water molecules are in contact with the inside and outside of the membrane, and the rate of movement through the membrane in both directions is the same. Hence the level of the water in the tube remains unchanged.

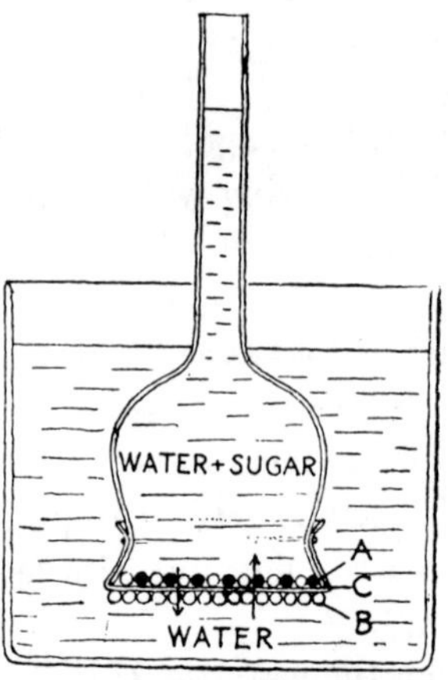

FIG. 48. Diagram to illustrate osmosis: *A* represents a sugar molecule, *B* a water molecule, and *C* a differentially permeable membrane. The sugar in solution dilutes the water so that fewer water molecules are in contact with the inside than with the outside of the membrane. Hence water passes in more rapidly than it passes out, and the level of the water in the tube rises.

tube. Consequently the water is more concentrated outside the tube than inside, and in keeping with a general law of diffusion the water passes from the place of greater concentration to the place of less concentration.

Moreover, sugar is a highly soluble substance; that is, it has a great affinity for water, and the sugar particles tend to hold the water particles in contact with them inside the thistle tube. The sugar, like the water, tends to pass from the place of greatest concentration but is restrained by the membrane from moving outward.

Differentially permeable membranes necessary for osmosis. When a membrane permits water or other substances to pass through, it is said to be *permeable* to that substance. For example, animal membranes are permeable to water and to various dyes. A membrane that allows one substance to pass through it, but retards the passage of another substance, is said to be *differentially permeable.* The membrane on the thistle tube is differentially permeable, because it allows the passage of water but restrains the sugar that is dissolved in the water. The diffusion of water through a differentially permeable membrane to the side where it is less concentrated is called *osmosis.*

Osmotic pressure. If we close the upper end of the thistle tube, the water will continue to rise and compress the inclosed air. If a large amount of sugar is put inside the tube, the water will rise rapidly and exert great pressure. If only a small amount of sugar is present inside the tube, the water will rise slowly and exert but little pressure. The pressure which is developed by diffusion under these conditions is called *osmotic pressure.*[1]

The plant cell as an osmotic apparatus. In living plant cells the cytoplasm lining the cell walls is the differentially permeable membrane. The cell walls of some tissues are also differentially permeable.

The cells contain sugars, salts, acids, and other substances dissolved in the water of the vacuole, just as the sugar is contained within the thistle tube in the experiment described above. The

[1] Experiments to show osmosis and osmotic pressure are best performed with paper, or collodion, diffusion shells, or with specially prepared porcelain cups. The thistle tube and animal membrane are used in this discussion because of the simplicity of the apparatus.

There is no agreement among scientists as to the complete explanation of osmotic pressure. The explanation given above leaves out of account some of the factors, principally electrical, involved in the process. This simple explanation is introduced merely to help the student to form a mental picture of the mechanics of osmosis as it occurs in plant cells.

cells of the plant are in contact either with water or with other cells containing water. Cells may, therefore, take up water either from adjacent cells or from their surroundings.

The conditions for osmosis as it occurs in plants, then, include a differentially permeable membrane between two masses of water, one of which is capable of developing a higher osmotic pressure than the other. The water passes from the mass of water or adjoining cell in which *it* is most concentrated to the cell in which dissolved substances are most concentrated and the water is least concentrated.

Turgor. When cells contain dissolved substances, or substances that swell greatly in water, osmosis and imbibition lead to the taking up of water and the stretching of the cell walls. A cell that is thus distended is said to be *turgid*. Cells with an inadequate water supply may be only partly filled, and the cell walls are consequently not stretched. Cells in this condition are *flaccid*. A condition of turgor is present in actively growing tissues. Cells are flaccid when a plant shows the familiar signs of *wilting*.

The movement of the guard cells of stomata (page 74) is brought about by turgor pressure, being open when the pressure is high and closed when the pressure is low. The changes in the position of leaves possessing a pulvinus are also due to changes in the turgidity of a part of the cells of the pulvinus.

Materials move by various combinations of physical processes. The physiological processes involved in the movement of materials in plants are various combinations of the four physical processes briefly outlined. Oxygen and carbon dioxide move into a cell by solution and diffusion. They pass out of a cell by diffusion when the cell contains a greater concentration than the surrounding air or the adjacent cells. Water passes from one cell to another by diffusion, particularly the forms of diffusion called imbibition and osmosis. As a result of imbibition and osmosis, pressure develops in the cell, stretching the walls and resulting in turgidity. Water is also held in cells against evaporation by the conditions that give rise to osmosis and imbibition; that is, water loss from cells is retarded by the same internal conditions that facilitate water absorption. Salts and sugar pass from cell to cell by simple diffusion. We can account for the rate of movement of soluble food substances only in part, but diffusion is certainly one of the physical processes involved.

PROBLEMS

1. Apply the principles of diffusion discussed in this chapter to:
   a. The opening and closing of stomata;
   b. The movements of leaves having pulvini;
   c. The wilting and recovery of plant tissues;
   d. The interchange of gases between a land plant and the surrounding air:
   e. The interchange of gases between a water plant and the surrounding water.
2. Why do red beets retain their red color when placed in cold water, but lose it when placed in boiling water?

CHAPTER TWELVE

THE WATER BALANCE IN PLANTS

Of all the factors that influence the growth of plants and modify the form, size, and structure of leaves, the water content of the cells is the most obvious. Abundant water permits a plant to grow to its greatest height, and permits the leaves to attain their largest size and number. Long-continued internal drought during the growing period may cause the plant to be dwarfed and the leaves to be small and few in number. In the river bottom the bur oak may develop into a magnificent tree 100 feet in height, while on the dry uplands it may attain only a stunted growth of less than 15 feet. An average leaf on a large tree will have twice the area of a leaf on a stunted one, and the number of leaves on the larger tree will be many times the number on the smaller.

The balance between transpiration and absorption. The amount of water in the cells of the plant as a whole is determined largely by two processes: (1) the rate of absorption — taking of water from the soil; and (2) the rate of transpiration. The relation between these two rates determines the water balance of the plant. If the transpiration is rapid and absorption is slow, internal drought results and the plant wilts. If the transpiration is slow and the water intake is rapid, the cells will be filled to their utmost capacity.

In summer, when the soil is dry and the air is hot, transpiration may cause the leaves to lose water so rapidly that they droop, and we say that the plant is wilted. Water has passed out of the cells of the leaf faster than the water-conducting tissue has brought in water to replace it, and the cells are no longer distended and firm. They are like a football that is only partly inflated. After a heavy shower the plants quickly recover, because the water available in the soil has been increased and more water is taken into the plant. The shower has also covered the leaves with a film of water and made the air moist around them, and this reduces the water loss. Under these conditions, the cells of the plant quickly become turgid and the leaves recover their firmness. The leaves of many plants like lettuce, pumpkin, and ragweed, that have little or no woody tissue in them, depend for their firmness almost entirely upon the turgidity of the leaf cells. *The balance between the rate of water supply and the rate of water loss is the most important water relation of the plant.*

The water balance illustrated. The presence of a water balance may be clearly seen by cutting the stem of a potted plant in two and then connecting the two pieces by a T-tube filled with water. To the side arm of the T-tube connect a U-shaped tube containing mercury and water. Then allow the U-tube to dip into a beaker of water. If transpiration is more rapid than absorption, the mercury will be drawn toward the plant. If absorption is more rapid than water loss, the mercury will move up the outer end of the tube. Figure 50 shows a sketch of the experiment. Adding water to the soil, placing it in full sunlight or in darkness, or moving it from a high to a low temperature will soon change the position of the mercury in the tube.

The water balance can be further illustrated by using the apparatus shown in Figure 49. This consists of a porous cup at the top, connected by a glass tube, with two bulbs, to a porous cup filled with a solution of sugar. The middle of the tube is further connected with a U-tube containing mercury. The whole interior is filled with water. When the lower cup is placed in a beaker of water, absorption begins and evaporation takes place from the upper cup. By placing the apparatus in different conditions, the changes in internal water balance will be shown by movement of the mercury in the tube.

Transplanting and the water balance. When the skillful gardener transplants a tree, he cuts off a number of branches to reduce the number of leaves, in order that the plant may not dry out before new water-absorbing roots are developed. Before lettuce, tomato, and cabbages are lifted for transplanting, the plants should be watered and allowed to become turgid; water should be poured into the holes in which they are placed before the soil is closed in around the plants. It is customary also to cover the plants for a day or two with boards or paper covers so as to reduce the transpiration. Maintaining the water balance in transplanted plants may prevent the loss of many of them and may save weeks of delay in the maturing of the crop. When herbaceous plants are propagated by cuttings, pieces of the stem a few inches in length, with one or two of the uppermost leaves, are taken and the lower half of the cutting is put in wet sand. In a few days, or weeks, depending on the kind of plants, roots are developed and a new plant is established. The leaves are left on the cutting so that some photosynthesis may go on. Most of the leaves are removed, so that the transpiration will not be sufficient to dry out the stem. The cuttings are kept moist, so that the absorption will be sufficient to keep all the tissues turgid; that is, to maintain the water balance. Cuttings of woody plants that root with difficulty are successfully started by painting the parts exposed to the air with melted paraffine in order to keep them from drying out.

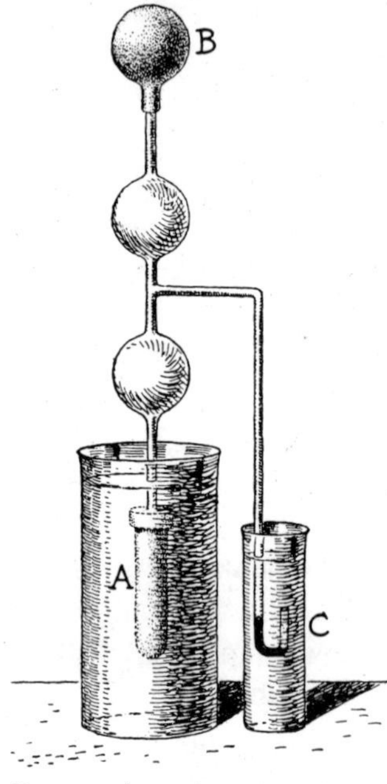

FIG. 49. Apparatus set up to perform an experiment to show the principles involved in the maintenance of the water balance in a plant. The entire apparatus is filled with water, and *A* and *C* are immersed in water. The water is evaporated from the porous cup *B*, and other water to take its place is absorbed by osmosis into the porous cup *A*. If the rate of evaporation is faster than the rate of absorption, the mercury falls in the outer end of the tube *C*, as is shown in the illustration.

The water balance and plant habitats. The place where a plant grows naturally is called its *habitat*. The willow grows beside a stream and the cactus grows in the desert, each in its natural habitat. If we put the willow in the desert and the cactus on a wet stream bank, both die. This means that the conditions that make up each habitat are favorable to one kind of plant and not to another. The conditions include not only the kind of soil and the amount of soil water, but also the evaporative power of the air. In selecting plants that may live in a particular habitat, the great importance of the dryness or the moistness of the air is to be kept in mind. Plants whose leaves are soft and transpire water rapidly can succeed only in moist air, while those that have a low transpiration rate maintain a suitable water balance even in a dry atmosphere. This is one of the reasons why on a southern slope we find a set of plants that are different from those on the northern slopes.

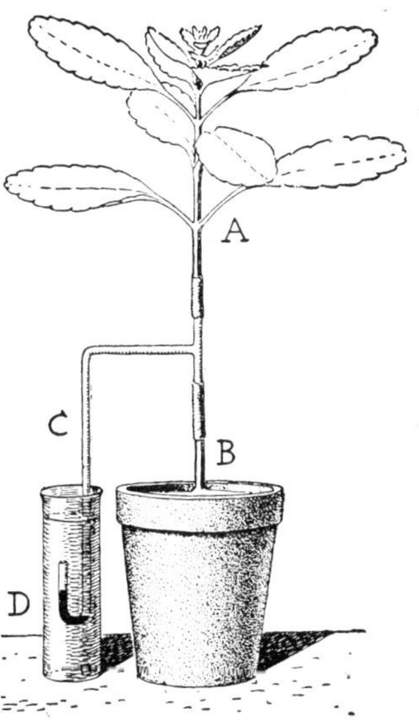

FIG. 50. A plant with its stem cut in two and connected again with a tube similar to that shown in Figure 49. If the roots absorb water more rapidly than the leaves transpire it, the mercury at *D* is pushed away from the plant, as in the illustration. If the plant is set in bright sunshine, the transpiration will be increased and the mercury will then almost immediately be drawn toward the plant.

Recent studies have shown that the leaves of plants growing near the bottom of a ravine transpire water ten to fifty times as fast as do those of plants growing higher up on an adjoining southern slope. Doubtless, each year seeds of plants that grow in the low ground germinate on the upper part of the slope; but each year the plants that spring from those seeds are eliminated through their inability to get the water needed by their higher rate of transpiration. There are plants like the dandelion that become adjusted to both these conditions. Most plants, however, are not so readily modified, and those with a high transpiration rate die off on a dry hillside, while those with a low transpiration rate survive. This indicates only one of the factors which must be taken into account in attempting to explain the distribution of plants in nature and in the selection of plants for particular habitats. Other habitat factors will be considered in later chapters.

Water balance and crop yields. In view of their large water requirements, it is easy to understand why droughts are so disastrous to crops. When the rainfall is slight, not only is the amount of water that can be secured by the plant from the soil reduced, but the sunshine is brighter and the air is usually drier, so that transpiration from the plant is increased. It is in part because of the water requirement of crop plants that bottom lands — lands along streams in the bottoms of valleys — are more valuable for growing crops than are uplands. There the underground water is nearer the surface and keeps the supply for plants more nearly constant.

Irrigation is a method of artificially maintaining a constant supply of soil water for crop plants in dry regions. It prevents the slowing down of the plant processes during the growing season, thus enabling the plant to work at its highest efficiency and produce its greatest yield. For example, at the Utah Experiment Station an acre of corn without irrigation produced 26 bushels; with 15 inches of irrigation water added, 52 bushels; with 38 inches, 82 bushels. An acre of wheat produced 4½ bushels without irrigation and 26 bushels when 30 inches of irrigation water was added. One of the highest recorded yields of corn on a small plot (225 bushels an acre) was obtained in Colorado with irrigation.

Plants classified according to their water relations. In preceding chapters we have pointed out the importance of water in the physiological processes of plants, and how plants are modified in size and structure by growing under different conditions of water supply.

Most plants cannot be grown to maturity in all kinds of wet and dry conditions. Each kind of plant has a rather definite water requirement for its best development. Hence in nature plants live only in those situations in which they are able to maintain a suitable water balance. Three great classes of plants are distinguished on this basis:

(1) The plants that naturally live where the evaporative power of the air is intense and the available water is limited are called *xerophytes* (Greek: *xeros*, dry, and *phyton*, plant). These are the plants that are adjusted to a nearly continuous dearth of water; the cacti, agaves, yuccas, and sagebrush of our Western plains and deserts are striking representatives. In the eastern United States there are less marked examples of xerophytes in the plants that live on dry cliffs and sand beaches, and in the mosses and lichens that grow on trees and rocks.

(2) The plants that live partly or wholly submerged in the water are known as *hydrophytes* (Greek: *hudor*, water, and *phyton*, plant). These plants have an excessive water supply, and transpiration is reduced or entirely wanting. In this class are included the water lilies, pondweeds, cat-tails, bulrushes, and many sedges. They are the common plants of fresh-water ponds, swamps, and marshes throughout the world.

(3) Between these extremes are the *mesophytes* (Greek: *meso*, middle, and *phyton*, plant), by far the largest class of seed plants. They have a medium rate of transpiration and grow best with a moderate water supply. In this group are included the plants that yield most of our garden, field, and meadow crops; also most of the forms that are found in the maple, beech, and elm forests of the Eastern states and in the fir and spruce forests of the canyons and bottom lands of the Western states.

Xerophytes, hydrophytes, and mesophytes are readily distinguished as groups because of their great difference of habitat and appearance. But it is not always easy to decide whether a particular plant is a xerophyte, hydrophyte, or mesophyte, because we find all gradations of form among plants of the three classes. Nevertheless, these terms are useful in describing the water relations of most plants.

PROBLEMS

1. Why do plants that are wilted in the late afternoon of a hot summer day recover their firmness during the night, even though there is no rain?
2. Where, near your home, do mesophytes, xerophytes, and hydrophytes occur?
3. In what regions of the United States are mesophytes most common?
4. In what parts of the United States are xerophytes abundant?
5. In what parts of the United States are hydrophytes common? What states have very few hydrophytes?
6. What xerophytes furnish useful products to man and animals?
7. Are there any economic plants that are hydrophytes?

CHAPTER THIRTEEN

THE GROWTH AND FALL OF LEAVES

WE are all familiar with the fact that when a live seed is planted in the soil it germinates, and that from it there develops a seedling which continues to enlarge for a longer or shorter time, depending on the plant and the conditions for growth. The period of growth may be a month, as in the radish in midsummer, or it may be hundreds of years, as in some trees. In the process of growth vast quantities of food are consumed. During the early stages of a plant's development most of the food it manufactures is used in this way. In order to grow, a plant must make new protoplasm, develop new cell walls, and thicken and strengthen old cell walls. Growth requires not only food but energy as well. Indeed, a part of the energy derived from respiration is used in certain chemical changes involved in growth. We might expect assimilation, respiration, and food consumption to be most active in young growing parts, and that this is the case has many times been verified by experiment. Growth takes place through the enlargement of cells already present in the plants, through cell division, and through modification of cells without enlargement.

Conditions for growth. The conditions most favorable for growth are abundant water and oxygen supplies and warm temperatures, such as normally occur in summer. For the growth of the plant as a whole, moderately strong light is favorable because it increases the supply of food. For the growth of leaves in particular, medium light is generally most favorable. In darkness the blades of many plants do not expand, and in very intense light they do not expand fully because of the retarding effect of the light itself and the excessive water loss.

The growing regions of leaves. By watching the development of leaves on any common herb, or on the trees in spring, we can see that growth takes place rapidly; also, that growth ceases when the leaves have developed to a certain rather definite size. The leaf starts as a small protuberance on the side of the apex of the stem. The mass of cells that make up this protuberance are all similar. As growth proceeds, cell division, cell enlargement, and cell differentiation take place and the five tissues of the leaf are formed. At first, then, all parts of the leaf are growing. After the leaf is mature, further enlargement will not take place, no matter how favorable to growth the external conditions may be. The question arises, do all parts of the leaf mature at the same time, or does growth continue in some parts longer than in others? There is one characteristic of growing tissue that will help us in answering this inquiry: young tissue is very tender and easily broken, while old tissue is stronger and firmer (Figs. 51 and 72).

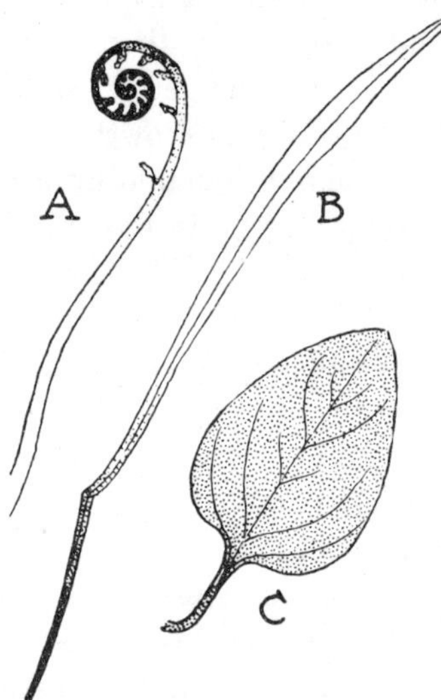

FIG. 51. Growing regions (shaded portions) of leaf of fern (*A*), grass (*B*), and sunflower (*C*). The fern leaf grows at the apex; the leaf of the grass, as is common in parallel-veined leaves, grows at the base; the sunflower leaf, like other net-veined leaves, grows in all its parts.

Fern leaves grow at the apex. The fern leaf is one that may be studied in this connection, for the growing portion is not only tender but coiled up, and its unfolding may be noted from day to day by marking with India ink the successive positions of the coil. In the Boston fern, which is so commonly cultivated as a window plant, the leaf may continue to unfold for weeks, if the water supply is adequate and other conditions are favorable. Evidently in the ferns the last growing region is at the apex and the older part of the leaf is the base. If the tip of a fern leaf is injured, further growth is stopped.

Growth of leaves of seed plants. The flowering plants have either parallel-veined leaves or net-veined leaves, and the place of growth in these two types of leaves is different. In parallel-veined leaves, like those of the members of the grass family, the growth continues longest at the base. If you have pulled a growing leaf from any of the taller grasses like wheat or timothy, you will recall that it broke near the base, and if you put the broken end in your mouth that it had a sweet taste.

The breaking near the base, and the sugar there, indicate that the final growing region of the grass leaf is at the base. A more exact determination of the growing region may be made by marking a young grass leaf into equal spaces with India ink. This will show that as the leaf develops, it is continually pushed upward and outward from the node where it is attached. This mode of growth is characteristic not only in grasses but in many other plants having *parallel-veined leaves* (monocots, page 130). It is a great advantage in maintaining pastures, but on the other hand we are obliged to mow our lawns more frequently because of it.

In plants with *net-veined leaves* (dicots, page 131) development is different from that in either ferns or grasses. The growth of a young leaf of this type — for example, a geranium or nasturtium leaf — may be studied by marking it off into equal squares by means of two series of parallel lines at right angles to each other. After several days it will be seen that the only change has been an increase in size of the squares. The lines in each direction are still roughly parallel. This indicates that all parts of the blade are growing equally. Note that all parts of the blade seem equally firm, which indicates that they are all of the same age. All parts of needle leaves of pine develop at the same time.

These facts regarding the growth of leaves may be summarized in a somewhat different way. In the ferns the last part of the leaf to mature is the apex. In parallel-veined leaves a region near the base is still in growing condition after the other parts are mature. In net-veined leaves all parts of the blade mature at the same time.

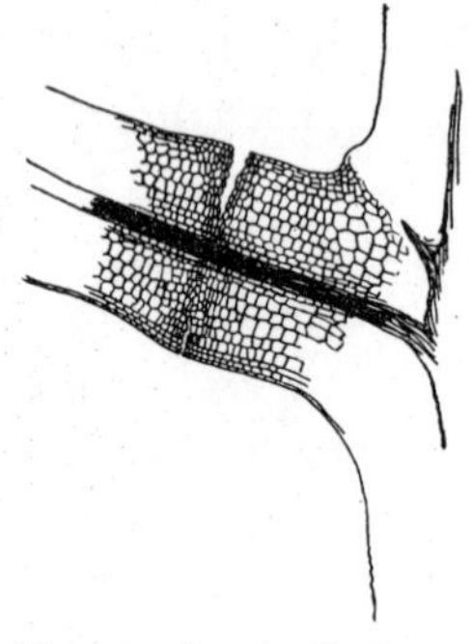

FIG. 52. Longitudinal section of base of a petiole, showing abscission layer. The dropping of the leaf is due to the softening of the cell walls in this layer.

Leaf fall, or abscission. In the life of a leaf the final stage is abscission, or the fall of the leaf. On many temperate plants the leaves remain only 5 to 8 months. On evergreen shrubs and trees the leaves are attached from 3 to 8 years. The first part of the process of leaf fall is a phase of growth, and we shall see that the tissue which makes abscission possible is constructed long before the leaf falls (Fig. 52).

The causes of leaf fall. There are two distinct stages in the process by which plants drop their leaves: (1) the formation at the base of the petiole of two or more plates of thin-walled cells, known as the *abscission layer:* this takes place during the development of the leaves and may require weeks or months for completion; and (2) the actual separation of the cells of the abscission layer, which is brought about by the softening or dissolving of pectic compounds in the middle layer of the walls of the abscission cells: this stage of the process may take place within a few hours, or at most within a few days.

The plant is protected from disease and water loss at the scars left by the falling leaves through the addition of woody and corky materials to the cell walls beneath the abscission layer. This corky layer is formed in some plants before the leaf drops, in other plants after the leaf has fallen.

Conditions promoting leaf fall. *After* an abscission layer has developed, there are many climatic and soil conditions that may accelerate the falling of the leaves. Among these are low temperature, reduced light intensity, and any disturbance of the water relations of the plant which results in internal drought. Disease and insect injuries to the blade frequently bring about abscission.

Leaves contain food materials when they fall. The materials used in building the cell walls in a leaf are lost to the tree when the leaf falls, and the fallen leaves still retain considerable amounts of starch, sugar, protein, and other nutrient material which leach back to the soil. In the autumn, however, photosynthesis declines, and the amount of food lost by a deciduous tree through leaf fall is small in comparison with the quantity that has accumulated in other parts of the plant. Even the part that falls to the ground is not entirely lost to the plant. It is used by other plants and animals, which in turn produce substances that are of great importance to the original plants. Forest trees are in this way benefited by the leaves that fall to the ground. In agriculture the leaves, stems, and roots of one crop are frequently plowed under to improve the soil for succeeding crops.

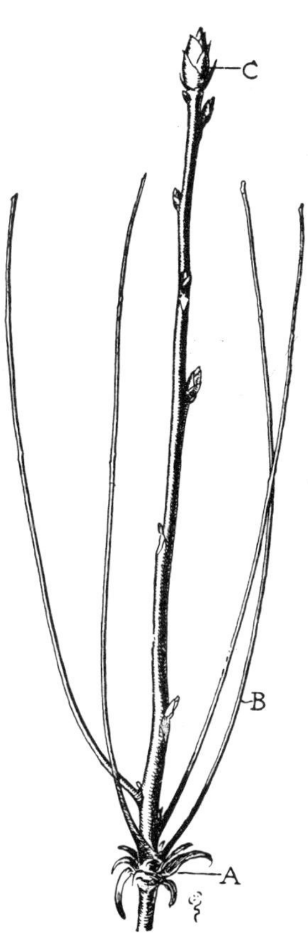

FIG. 53. Shagbark hickory twig. *A* is the bud scales of the terminal bud of the previous year; *B*, several petioles remaining attached after leaf fall; and *C*, the terminal bud that will develop the following spring. Drawn from a specimen collected in December.

Abscission in compound leaves. In many compound leaves, like the horsechestnut, ash, and hickory, abscission first takes place at the base of each leaflet. Later the petiole is cut off from the stem in the same way. Consequently the leaflets fall first and the petioles later. In the king-nut hickory and occasionally in the shagbark also, the petiole remains attached to the tree through the following year (Fig. 53).

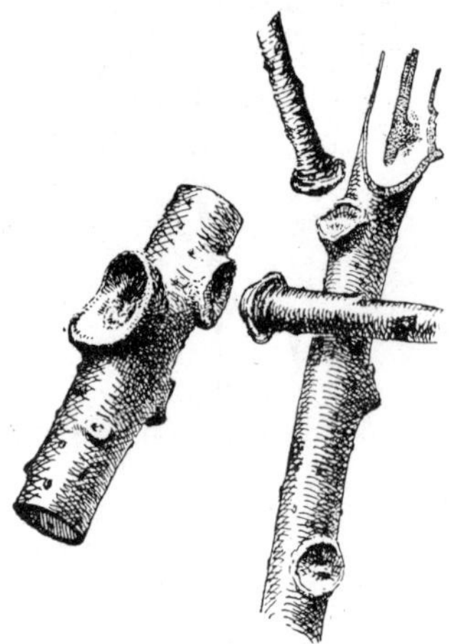

FIG. 54. Abscission of branches of the cottonwood. Twigs and small branches, as well as leaves and fruits, are cut off by the formation of abscission layers.

Self-pruning. A large number of our common trees, like the cottonwood, maple, and elm, develop abscission layers which cut off twigs and sometimes branches an inch in thickness. In these trees we have twig fall as well as leaf fall. The falling of flowers, and of fruits like apples and nuts, is also due to the softening or dissolving of abscission layers formed across the stem at the point of attachment. Sometimes abscission is of advantage to the plant, sometimes it is quite disadvantageous.

Evergreen and deciduous trees. In the Northern states many persons have come to think that the evergreen habit is associated only with needle leaves, because in the North the evergreens are mostly of the needle-leafed type. But in the Southern states there are many broad-leafed trees, like the magnolia, rhododendron, and holly, that are also evergreen. Moreover, the tamaracks of the North and the bald cypress of the South furnish examples of needle-leafed trees that are deciduous. If we include the shrubs, there are many broad-leafed plants, both in the North and in the South, that have the evergreen habit. In the tropics most of the trees are evergreen, and almost all have broad leaves. It must be noted that even in the case of evergreens individual leaves remain on the trees for only a limited number of years. The leaves of the evergreens of temperate regions are quite different structurally and physiologically from the leaves of deciduous trees. The evergreens of temperate regions must be able to withstand freezing and thawing, and also the dry winds of winter, which cause water loss even when the ground is frozen. Their usual transpiration rate is very low in comparison with that of deciduous trees, and in the autumn they undergo changes which reduce the water loss nearly to that of deciduous trees that have dropped their leaves.

Evergreen versus deciduous habit. In temperate regions, where there are great changes in temperature and moisture, the deciduous and the evergreen habit each has certain advantages. The advantages of the evergreen habit are: (1) that the leaves can manufacture food even when the temperature is low; (2) that with their low water requirement, evergreens can withstand drier conditions throughout the year; (3) that the tree does not use so much material each year in the construction of a complete set of new leaves. The disadvantages of the evergreen habit are: (1) that the heavy cuticle and compact tissues which aid in conserving water interfere with rapid photosynthesis; (2) that the lower rate of food manufacture prevents rapid growth; (3) that the leaves lose in efficiency by their longer service on the trees.

The advantages of the deciduous habit are: (1) that the leaves, being renewed each year, are more efficient organs of food manufacture; (2) that the leaves, with less cuticle and with tissues less compact, are better fitted for rapid food manufacture; (3) that the total leaf area may be much larger than in the case of the evergreens; (4) that the trees are better fitted to withstand the winter drought, because at that season the entire tree is covered with cork. The disadvantages of the deciduous habit are: (1) that the food-manufacturing season is only from 5 to 8 months, as compared with from 8 to 10 months in the evergreens; (2) that each year a large amount of food material is needed to make an entirely new set of leaves.

Finally, we must observe that there are in trees all gradations between the deciduous and evergreen habits. In the rainy tropics there are many delicate-leafed evergreens. In the dry tropics the evergreens have thick, fleshy leaves, or they may be quite leafless. Some plants, like the holly and the Virginia creeper, may have the deciduous habit in the North and the evergreen habit in the South. Some deciduous trees, like the cherry, when planted within the tropics become evergreen ; while the magnolia, which is evergreen in the Southern states, becomes deciduous when grown in a colder climate. Evidently leaf habits are in part responses to climatic conditions, especially to conditions of temperature and moisture.

PROBLEMS

1. What are the commonest evergreen trees and shrubs of your locality?
2. What trees of your vicinity develop leaves earliest in the spring? What trees develop their leaves last?
3. What trees develop flowers before the leaves?
4. What trees drop their leaves first in autumn? What trees drop their leaves last in autumn?
5. What trees shed part of their leaves during the summer?
6. During how many months does each of these trees carry on food manufacture?
7. During how many months do the evergreen trees of your locality manufacture food?
8. For how many months do the three leading crop plants of your locality carry on photosynthesis?
9. Do all specimens of the same species of trees put out new leaves at the same time? How much variation is there?
10. Do those which develop leaves first drop them earliest in the autumn? This information is important in selecting trees for street planting and general landscape effects.

CHAPTER FOURTEEN

THE STEMS OF PLANTS

THE stem forms the axis of the plant and bears the leaves, flowers, and fruits. Plants showing all degrees of stem branching are found, from the unbranched palm and corn to the finely divided asparagus and elm. In most plants the stems are upright, aërial structures; but in some plants they lie on the surface of the soil, in other plants the main stem is underground and only the branches rise above the surface, and in still other plants the entire stem is underground. The upright stem is the common type and has many advantages over a horizontal stem.

Upright stems. The photographer uses light to effect chemical changes in photographic papers and plates. Light also brings about chemical changes in the green tissues of the plant. The photographer who uses sunlight for his work usually locates his studio at the top of a tall building, because there he avoids the shadows of near-by buildings and secures a more constant exposure to light. The same advantages come to the plant that has its leaves raised well above surrounding plants; the leaves are in less danger of being shaded, and each day they are exposed to the sunshine during a longer period (Fig. 55).

The tall plant has an additional advantage in being able to expose to the light a greater leaf area over a given space of ground, because it can display several or many layers of leaves one above the other. The rosette of leaves formed by the burdock illustrates the possibilities of leaf display near the soil. A large sunflower plant covers no greater soil area than a burdock, but it is able to expose to the sunlight several times as great a leaf area because the sunflower leaves are placed at several different levels. Trees have the greatest stem development and the greatest leaf display. *Rosette plants*, like the dandelions and plantains, represent the opposite extreme of slight stem development, small leaf area, and a relatively poor leaf display. One advantage in a tall, upright stem is that it holds the leaves up to the light and thereby makes possible a greater leaf display. Under certain conditions upright stems may facilitate pollination and seed dispersal (Fig. 55).

The advantages of the upright stem are all dependent on its capacity to support other organs. The stem must be strong enough to support leaves, flowers, and fruits. The city skyscraper needs first of all a strong framework about which the building is constructed. The stems of tall, erect plants must be correspondingly strengthened by a mechanical structure. The base of a tree is much smaller in proportion to its height than that of the tallest and narrowest building, and it is possible for trees to reach great heights only because their stems are composed in large part of supporting tissues of great strength and pliability.

The largest upright stems. Stems attain their best development under medium conditions of moisture, light, and temperature. Such conditions are found in the eastern United States, in the canyons of the western mountains, and along the Northern Pacific slope, and there the plants are characterized by large leaf area. In the East the vegetation culminates in the forests of the rich, well-watered soils of the river valleys. Here may be found oaks, walnuts, elms, maples, sycamores, and magnolias which reach heights up to 100 or 180 feet and have trunks attaining diameters of from 4 to 14 feet.

FIG. 55. Sunflower and burdock, showing advantage of the tall, upright stem. The sunflower covers no greater soil area but it displays more leaves to the light.

In the moist canyons of the Sierras of California the giant

sequoia reaches heights of from 250 to 320 feet above the ground, with extreme trunk diameters of 35 feet. A cypress tree near Oaxaca, Mexico, has a trunk 50 feet in diameter. These trees are the largest and probably the oldest of all living things. The redwood, a near relative of the sequoia, grows in the fog-abounding ravines of t e Coast Ranges north of San Francisco. Its trunk diameter may be not over 28 feet, but it surpasses the giant sequoia in height. You can better appreciate the size of these trees if you will pace off from an ordinary tree a distance equal to the diameter of a sequoia trunk and will calculate how many times the height of the tallest tree in your locality a giant sequoia is, and then try to imagine how a sequoia would look growing beside it.

FIG. 56. Group of Big Trees (*Sequoia gigantea*) on western slope of the Sierra Mountains, California. An idea of their size may be gained by comparing them with the man at the left. Formerly sequoias were widely distributed over the northern hemisphere, but now they are practically restricted to a few localities in California.

Climbing stems. Among mesophytes are many vines with exceedingly long, slender stems. The Virginia creeper, wild cucumber, poison ivy, and grape have stems 50 to 300 feet long, and usually only a fraction of an inch, or at most a few inches, in diameter. These long, slender stems grow rapidly and enable the plants to spread their leaves quickly over the tops of large trees. With a proper support these stems have all the advantages of upright stems, without having to use so mucn material in building woody supporting tissue.

FIG. 57. Tendrils of wild cucumber. Note the coiling of the tendril by which the plant is drawn nearer the support and the reversal of the spiral in different parts of the tendril. Is there always such a reversal?

Some climbers — for example, the morning-glory — gain support by twining. Others, like the grape, have tendrils, specialized organs developed in place of branches or leaves. A tendril responds to contact with a support by coiling tightly about it (Fig. 57). When attached, mechanical tissues develop within, greatly increasing their strength. Then, due to unequal growth on the two sides, the tendril twists spirally and draws the plant close to the support.

In some vines, like the Boston ivy (Fig. 59) the tendrils have at their tips sensitive disks which, when rubbed against a support, secrete a sticky substance and become cemented to it. This

FIG. 58. Climbing stems on a tree trunk in a tropical forest. This type of stem does not require the use of so much material in building woody supporting tissue, and if it comes in contact with a support, it has all the advantages of an upright stem. Its disadvantage is that unless a support is found, the leaves are poorly displayed.

type of tendril is especially effective in taking hold of the bark of trees, rock cliffs, and walls. Still other climbers, like the trumpet creeper and poison ivy, have aërial roots that become fastened to a support by growing into cracks and crevices; or, by the cementing action of the outer pectic layer of the epidermal cell walls, they may become attached to quite smooth surfaces. In the moist tropics, climbing stems may attain a length of more than 1000 feet. Thus the water transpired by terminal leaves has been carried about a fifth of a mile within the plant. A cross-section of such a stem will show not only that most of the stem is occupied by water-conducting tissue, but that the individual tracheæ are very large and long when compared with those of upright stems.

Horizontal stems. Horizontal stems have little woody tissue, and they display leaves to the light advantageously only when they grow in the open. There are advantages in stems of this type, however, because by growing horizontally on the soil or beneath the surface of the soil they spread the plant; because they are in contact with the soil and may take root at frequent intervals; and because they are better protected than upright stems during the winter and other unfavorable seasons.

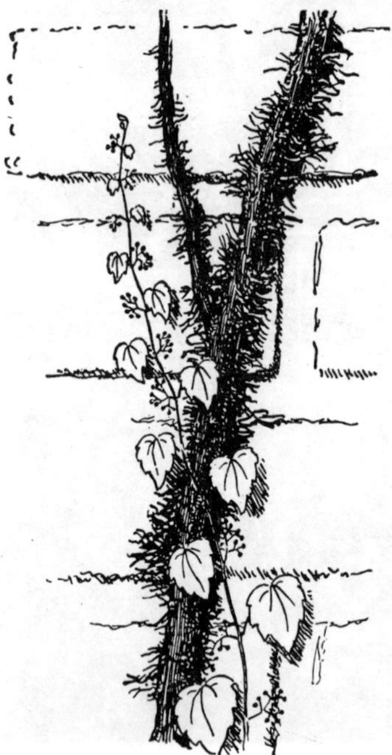

FIG. 59. Young and old stems of Boston ivy. The young stems are held by means of tendrils, the older stems by means of adventitious aërial roots.

Underground stems. Many plants, both herbaceous and woody, possess underground stems. They are particularly useful as places of food accumulation and in vegetatively spreading and multiplying the plant. Their position renders

W. S. Cooper

FIG. 60. Sand-reed grass (*Ammophila*) planted on dunes near Casmalia, California, to prevent further movement of the sand. The bare areas are dunes formed since the planting. The underground stems of the grass bind the sand and aid in preventing its movement by the wind.

them nearly free from transpiration, from injury by fires (an important matter on the prairies), and from the destructive effects of winds. Yet when plants having underground stems *only* come into competition with those having erect stems, they are quite likely to be overshaded; at any rate they cannot compare with erect stems in leaf display. In competition with annuals, however, they are highly successful by occupying all of the space and thus preventing the young seedling from getting a start. One need only look at the plants in old meadows, pastures, prairies, and swamps to see the result of such competition.

The commonest type of underground stem is the *rootstock* or *rhizome*. Rootstocks are horizontally growing stems, from which the aërial stems arise. They may be slender, or thick and fleshy. Usually they have small scale leaves and buds at the nodes, as well as roots that arise from the nodes or from the entire under surface. The presence of nodes is the external feature of underground stems that distinguishes them from roots (Fig. 62).

In many of the grasses and grass-like plants, rootstocks develop rapidly in all directions, sending up erect branches at short intervals. The rootstocks and their accompanying roots soon become mixed with those of adjoining plants, finally forming a closely interwoven mat which is the "turf" of lawns and meadows. Turf-forming grasses are often of great value for holding in place the soil of embankments, dikes, and levees. In these plants the rootstocks are mainly useful in spreading or extending the plant. Bermuda grass and Johnson grass are troublesome weeds in the Southern states because of their extensive rootstock systems. On the other hand, this same feature makes some plants of great use to man. The sand-reed grass (*Ammophila*) has been planted extensively in Europe and in America to hold drifting sand in place and to prevent the sand from invading towns and cultivated fields. This grass may also be used as a soil binder in starting forests in sandy places.

In plants like the May apple, Solomon's seal, and yellow water lily, the rootstock not only causes the plant to spread, but it also accumulates a part of the food manufactured each season and thus serves as a storage organ. The underground stem of asparagus is a storage organ, and as it

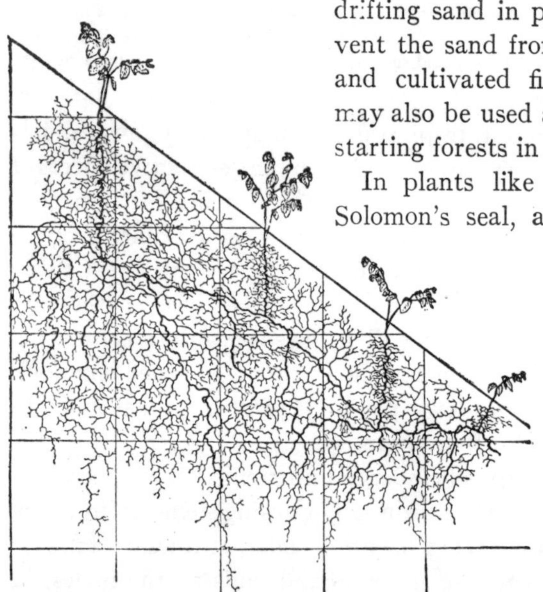

FIG. 61. Vertical section of a gravel slide, showing dogbane (*Apocynum*) with underground stems (rootstocks) connecting the several shoots, and the much-branched root system. Section divided into one-foot squares.

increases in thickness each year, the upright branches used as food become larger and more succulent. It is this store of food

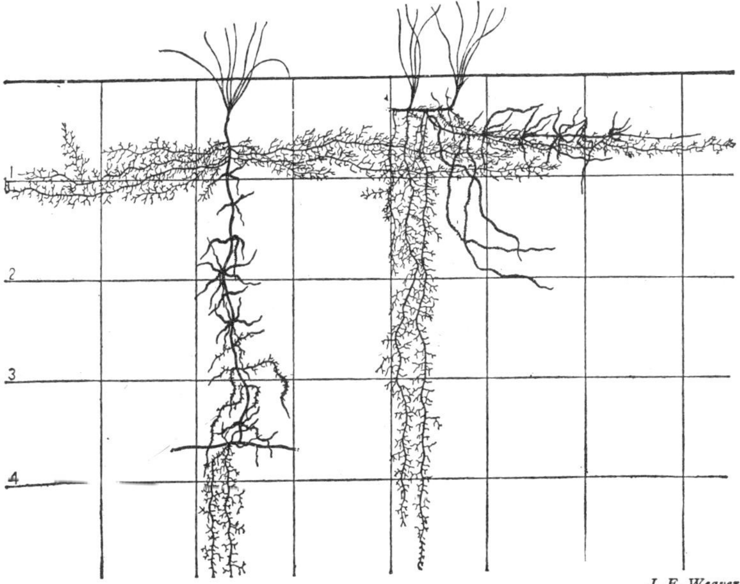

FIG. 62. A common grass (*Redfieldia*) in the sand hills of Nebraska, showing parts of the extensive system of rootstocks and roots. Sections divided into one-foot squares. The underground parts of the plant were carefully dug out, and their horizontal extent and depth in the soil were found to be as shown in the illustration.

and the readiness with which the rootstock sends up shoots, that make the bindweed and perennial morning-glories so difficult to eradicate from cultivated fields.

A short, upright, fleshy rootstock, like that of the jack-in-the-pulpit, caladium (elephant's ear), or gladiolus, is called a *corm*. Corms contain large amounts of food, and by the development of their lateral buds may serve to reproduce the plant as well as to carry it over the winter. The dasheen, a tropical plant which resembles the caladium, and which has recently been introduced into the United States, has an edible corm that is an important source of food (Fig. 63).

A *bulb* is a fleshy underground bud, made up of a short stem covered with several layers of thick scales in which food is stored. Tulips, hyacinths, and onions are commonly propagated by means of bulbs. Some kinds of onions also produce small bulbs ("sets") in place of flowers, and some lilies develop them in the axils of their leaves.

By planting bulbs of the tulip in autumn, we can have flowers early in the following spring, whereas if we planted the seeds, we should have to wait several years for flowers. It requires two or three seasons of photosynthesis to accumulate sufficient food for flower production. Furthermore, tulips do not grow well except in a very moist climate, and the development of large, vigorous bulbs is impossible in most parts of the United States. For this reason nearly all our tulip bulbs are brought from Holland. The importation of bulbs from countries where they grow particularly well is an important industry and enables us to have many flowers which cannot be as successfully propagated in our climate.

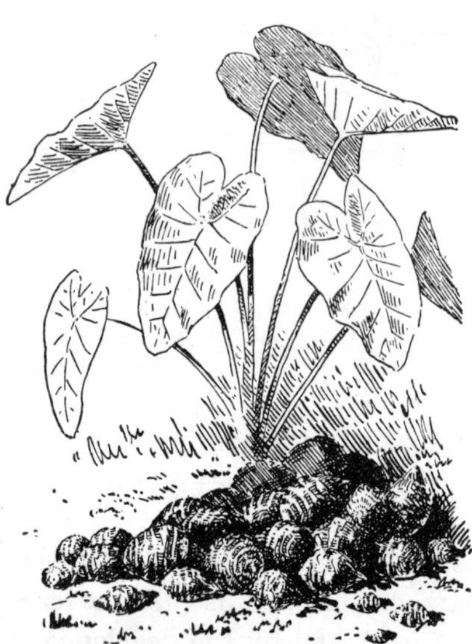

FIG. 63. Dasheen and edible corms produced by it. The dasheen is related to the common "elephant's ear" or Caladium, and is extensively grown in the tropics for food. In the states along the Gulf Coast it is being introduced as a food plant.

Tubers are the enormously thickened portions of short underground stems. The potato and the Jerusalem artichoke are the most familiar plants forming tubers. The scale leaves of the ordinary rootstock are in tubers reduced to ridges, and the buds themselves to mere points. The scales and buds together form the eyes of tubers. Tubers, like other fleshy underground stems, accumulate surplus food and multiply the plant. The potato tuber has become one of the most important sources of food for man.

Summary. Stems vary greatly in structure, size, and position. Each type of stem has certain advantages and gives the plant characteristic habits of growth. Each of these stem types also fits into certain environments better than into others; consequently there are great differences to be observed in the kinds of plant stems in different habitats and different regions.

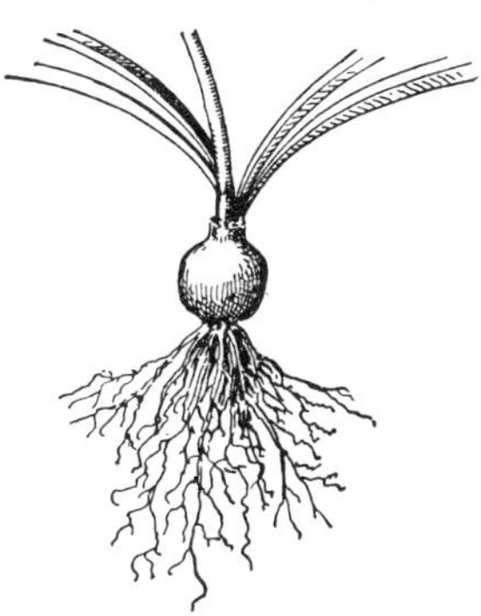

FIG. 64. Amaryllis bulb. A bulb is a fleshy underground bud made up of a short stem covered with several layers of thick scales in which food is stored.

PROBLEMS

1. What advantage in resisting wind have tall, columnar tree trunks over equally tall smoke stacks or monuments? What disadvantage?
2. What are the best trees for street planting in your locality? What trees now planted there are objectionable? Why?
3. Compare a tree growing in an open field with one of the same species growing in the woods. Account for the differences in arrangement of branches and leaves.
4. Which will furnish the better lumber, a tree grown in the open, or one grown in the forest? Why?
5. What commercial products are derived from each of the several types of stems described in this chapter?

CHAPTER FIFTEEN

THE EXTERNAL FEATURES OF STEMS

On a woody stem, nodes, leaf scars, buds, and lenticels may be seen. The *nodes* are the places where the leaves arise, and they are usually the most prominent external feature of stems. The arrangement of leaves at the nodes has already been discussed (page 62). In addition to the leaf, the node gives rise to one or more buds, just above the place of leaf attachment, in the so-called axil (Latin: *axilla*, armpit) of the leaf. The part of a stem between two nodes is called an *internode*. The *leaf scars* are markings on the stem where leaves have fallen. At intervals along the stem ring-like markings (*bud scars*) may be found. These show where a terminal bud was formed at some previous time. The *lenticels* are small, dot-like elevations scattered over the surfaces of the internodes.

Buds. Stems and branches produce leaves only once. We are accustomed to speak of deciduous trees clothing themselves with a new set of leaves each spring, as though the branches of the previous year put forth a new set of leaves to replace those lost the preceding autumn. As a matter of fact, when we look at a deciduous tree in winter, we see branches and twigs, all of which have borne leaves and none of which will ever bear leaves again. The possibility of producing new foliage lies in the development of new branches and twigs. This is the function of the buds; from them the new growth of each year arises (Fig. 65).

The buds of many tropical plants are like those we see at the tops of the stems of garden vegetables. A bud of this kind consists of the stem's growing point and the undeveloped leaves, with no special covering of any kind. These *naked buds* occur also on the underground stems of some of our herbaceous plants. A simple sort of bud covering, which is common in the tropics, is made by the folding together of the stipules. This type of bud covering may be seen in the tulip tree and the magnolias of temperate climates. The buds of most temperate perennials are covered with specialized scale leaves. Frequently the outer or the exposed parts of scales die with the approach of winter. Not infrequently the scales are further covered with matted hairs and secretions of wax and resin. These all tend to make the bud coverings impervious to water. By these coverings the tender growing parts are protected from excessive loss of water during the winter and during the still more critical stage in early spring when the buds are opening. Bud scales do not protect the growing point of the stem from low temperatures. During zero weather all the tissues of buds and twigs are frozen solid.

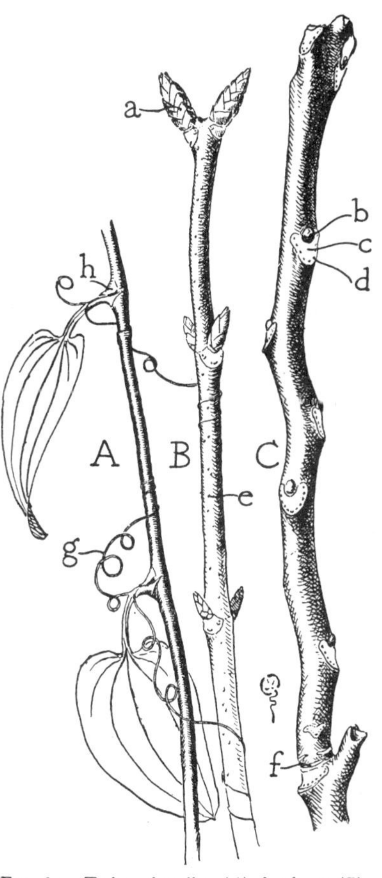

FIG. 65. Twigs of smilax (*A*), buckeye (*B*), and tree-of-heaven (*Ailanthus*) (*C*). The bud scales are designated by *a*; *b* and *h* are buds; *c* is a leaf scar, *d* a bundle scar, *e*, a lenticel, *f* a terminal bud scar, and *g* a tendril.

We are likely to think of buds as being formed at about the time when the leaves fall from the trees. A good observer, however, will have noted that the buds begin to develop when the leaves unfold in spring, and that they grow all summer long. Because of the prominence of the leaves, the buds are obscured somewhat during the summer months and become conspicuous only after the leaves have fallen from the trees.

FIG. 66. Date palms in fruit, on an oasis in the Algerian desert. The strong terminal bud, and the failure of the lateral buds to develop, leads to an unbranched stem. An unbranched stem is more common among monocots than among dicots. (*From photo U. S. Dept. of Agriculture.*)

The opening of buds. When the warm weather of springtime comes, the innermost bud scales begin to grow and expand. Sometimes the outer scales are pushed off; sometimes they elongate and grow like the inner ones. But the scales quickly reach their full growth, and soon they are cut off by the formation of an abscission layer at the base of each. In the buds of a few plants all the scales are dead and are pushed off by the growth of the stem and leaves inside. The expansion of bud scales and leaves takes place partly through cell multiplication and partly through the enlargement of cells already formed the preceding year. Within the bud the minute leaf cells absorb water and develop large vacuoles. The expansion of these cells results in the enlargement and spreading of the leaves. Material for the study of the different habits of bud expansion may be secured in winter by bringing branches of different kinds of trees into a warm room and placing them in water until the leaves expand.

Contents of buds. Every bud contains the growing point of a stem. In addition, most buds contain the beginnings of foliage leaves; that is, the leaves have already begun to develop on the sides of the young stem within the bud. These are called *branch buds*, because when they grow they produce a new leaf-bearing branch. Some buds, as for example many of those on the maples and elms, contain the beginnings of flowers (*flower buds* or *fruit buds*). Other buds, like some of those of the catalpa and the horse-chestnut, contain both leaves and a flower cluster (*mixed buds*). Bulbs are really a special underground form of bud, and they are similar in structure to other buds.

Bud development and plant form. Buds which occur at the ends of stems are called *terminal buds;* those which occur at the nodes are called *lateral buds*. This classification is useful because only a part of the buds on a stem ever develop and because the form of a plant depends on which set of buds develops more freely and grows more rapidly. In most plants the terminal bud simply extends a stem or branch; the lateral buds produce new branches. Plants with very strong terminal buds tend to become columnar in form, like the large, unbranched sunflowers of the garden or like the spruce and palm (Fig. 87) among trees. Plants with strong lateral buds usually branch continually and become bushy in form, like the lilac and hydrangea. There are all gradations between these extremes, in the development of the terminal and lateral buds and in the resulting plant forms.

In many roses the shoots from the base of the stem develop only through their terminal buds the first year. The shoot is thus extended to great length by the season's growth. The following year the lateral buds develop, and the long shoot becomes highly branched. As these lateral branches bear the flowers and produce them abundantly only once, we can promote flowering in these roses by trimming away each year all but the long, unbranched shoots. In many other shrubs, as spirea, barberry, and privet, a few strong lateral buds at the surface of the soil

W. S. Cooper

Fig. 67. Fir and spruce forest on slope opposite Mt. Aberdeen, Alberta. The excurrent stems and spire-like form of the trees result from the continued growth of the terminal buds and the slow development of lateral branches.

FIG. 68. A hackberry (*Celtis occidentalis*), showing deliquescent stem. This tree is growing on the open prairie in Illinois. Through the development of the lateral buds, the central stem has been lost in the branches.

develop each year. This accounts for the basal branching of these plants.

Excurrent and deliquescent stems. When trees have strong terminal buds, the main stem extends to the top and is called *excurrent* (Latin: *excurrens*, running out). The spruce has a strong terminal bud, and just beneath it a whorl of several smaller lateral buds (Fig. 71). The terminal bud grows upward, and the lateral buds grow outward, forming a whorl of branches at the base of the season's growth. This is repeated each year, the terminal shoot lengthening the stem and the lateral buds adding a new whorl of branches. Consequently each year's growth is marked by a whorl of branches, and the age of a tree may readily be estimated by counting the number of whorls on the stem. Since the oldest branches are nearest the ground, thev are the longest, and the tree becomes cone-shaped as it grows.

The terminal buds of the elm tree seldom survive the winter. The lateral buds develop, and the main stem divides and sub-

FIG. 69. An American elm (*Ulmus americana*). The terminal buds of the elm seldom survive the winter, and the development of the lateral buds causes the main stem to divide and subdivide until it dissolves into the branchlets that form the crown. This tree is growing in the Berkshire Hills, Massachusetts. It probably developed in a forest which was afterward cut down.

W. S. Cooper

FIG. 70. A deliquescent monocot (the tree yucca, *Yucca arborescens*), photographed at Cajon Pass, California. The deliquescent type of stem is unusual among the monocots.

divides until it is lost in the crown of the tree. The gradual dissolving of the trunk into a spray of terminal branchlets suggested the name *deliquescent* (Latin: *deliquescens*, dissolving) for this type of stem (Fig. 69).

We see, therefore, that the excurrent type of stem depends on the continual development of terminal buds, while the deliquescent type depends on the growth of lateral buds. The form of plants in cultivation may be modified by trimming them, and so forcing the growth of certain buds. Lawn trees and shrubs are grown either for shade or for ornamental effects. We secure shade by trimming off the terminal buds and so causing many of the lateral buds to develop into branches and thus form a denser crown. Ornamental effects are secured by trimming plants so that they will be in artistic harmony with their surroundings.

Fruit trees and grapes have been found to produce more fruit, and fruit of a better quality, when the number of branches is limited. A smaller number of branches on a tree secures an open crown and permits the sunlight to penetrate to every leaf, and the removal of some of the branches forces the development of flower buds which might remain dormant if the terminal and branch buds were allowed to grow uninterruptedly. In grape culture, only four or five branches are allowed to remain on a vine each year, and these branches are shortened. This insures full development for a few of the lateral flowering branches and the production of the best quality of fruit.

Black raspberry bushes produce fruit only at the ends of branches. Hence the object in pruning is to develop the maximum number of short branches. Each year new shoots develop from the base of the stem. If the tips are cut off when they are 18 inches high, five or six lateral buds immediately start growth and by the end of the season have formed branches. If these lateral branches are also trimmed back to a length of eight inches the following spring, each will develop several side branches which will be terminated later by flower clusters and fruits. After fruiting, the old much-branched "canes" should be removed and new shoots should be pruned for the next year's fruit production.

Leaf scars and bud scars. The leaf scars on some plants are round; on others they are narrow lines; on most plants they are crescent-shaped. Usually they are smooth, except for small, dot-like markings. These markings are *bundle scars;* they show where the bundles of conductive and mechanical tissue extended outward from the stem into the petiole and thus into the veins of the blade. The shape of the leaf scar and the arrangement of the bundle scars are so characteristic for many kinds of trees that they may serve to identify the tree in winter.

The bud scales also leave scars when they drop. These scars are usually numerous and so closely crowded that they form a roughened ring about the stem. The terminal-bud scars occur at intervals, surrounding the stem or branch. The lateral-bud scars are found only at the bases of the branches and the twigs.

FIG. 71. Plantation of Norway spruce, showing whorls of branches at base of each year's growth.

Determining annual growth of shoots from terminal-bud scars. Since the terminal bud marks the end of each year's growth, the terminal-bud scars mark off a perennial stem into segments, each of which represents the growth of a single year (Fig. 65). Often an interesting life history is suggested by the varying length of the intervals between the bud scars on a particular stem. By a study of these intervals we can determine the seasons that were favorable and those that were unfavorable because of drought, excessive rain, attacks of insects, or some other cause.

In the pines and spruces the year's growth is marked off not only by the bud scars, but also by whorls of branches. Differences in the color of the bark and in its texture will also help to distinguish successive annual stem segment in most trees.

On many varieties of apple and pear trees, flower buds are usually borne on the ends of " spurs " or short, slow-growing twigs. When the spur ends in a flower, the further growth of the shoot depends upon the development of a lateral bud. Commonly the spurs produce flowers in alternate years. When once a branch produces flowers, it continues to do so and its growth rate is usually much slower than that of the vegetative branches. Because of the successive development of lateral buds, spur branches are crooked and the intervals between terminal bud scars are short.

Lenticels. All living cells require energy. This is mostly obtained from respiration. Therefore, in addition to a constant food supply, the cells of the stem must have access to oxygen. As in the leaves the oxygen is supplied through the intercellular spaces, so in stems there must be sufficient intercellular spaces to permit oxygen to diffuse inward and carbon dioxide to diffuse outward. There must also be openings through the epidermis or bark to connect these intercellular spaces with the outside atmosphere.

The young green stems of all plants have stomata. Perennial stems, however, soon develop a corky layer beneath the epidermis, which cuts the cells in the interior of the stem off from the stomata. While this layer is developing, masses of round, loose cells form beneath some of the stomata, pushing out and tearing the epidermis above them. These open places are the lenticels. They permit gas exchanges, and in older stems take the place of the stomata. The lenticels of most twigs of trees and shrubs are closed in the late autumn by the growth of a thin layer of cork beneath them. The following spring loose cells are again formed at the same point, the cork is burst open, and the lenticels again permit gas exchanges. Apparently water influences the development of open lenticels. If a willow twig is placed in water, the submerged lenticels enlarge greatly. Perhaps a similar condition effects the opening of lenticels in the spring.

In the cherry and birch the lenticels persist for many years and become elongated transversely, forming rough granular rings part way around the stem. In the trunks of thick-barked trees the lenticels occur in the furrows of the bark.

PROBLEMS

1. Find out how your local gardeners trim their grapevines, berry bushes, and fruit trees. Secure definite information for five of these plants, and determine the reasons underlying the practices.
2. Remove the scales from various buds in December and determine how long the unprotected buds live. What weather conditions are most unfavorable to uncovered buds?

CHAPTER SIXTEEN

THE STRUCTURE OF STEMS

If we study the development of a stem from a bud, we find that the growing point is made up of very minute cells, all of which are practically alike. These cells divide, making other cells like themselves, and the lower ones begin to enlarge. In this way the growing point is pushed forward and the diameter of the stem increased. Then certain groups of cells begin to take on special forms. The cells that are to form the bundles elongate and the cross-walls between some of the cells disappear, forming the *water-conducting tissue*. Some of the cells develop thick woody walls (*wood tissue*). Others elongate but remain thin-walled, and these form the *food-conducting tissue*. Just outside the food-conducting tissue very slender elongated cells with thick walls develop (the *bast*). The other tissues of the stem are composed of cells which have enlarged and have become rounded or variously angled through mutual pressure and which have their walls more or less thickened. These cells form the pith or soft inner part of the stem, and the cortex or outer portion. In this way the various tissues of stems composed of a variety of cells arise from the small uniform cells of the growing point.

This tissue composing the growing points of leaves, stems, and other organs is called meristematic tissue, or simply *meristem*. The tissues that are formed from the meristem are composed of a variety of cells, which have been classified into several types to facilitate the description of plant organs.

Parenchyma and prosenchyma. The cells that make up the epidermis and mesophyll of leaves, the softer parts of herbaceous stems, and the fleshy fruits are either spherical or spheres that have been compressed. Their length and breadth are not very different and they are arranged in rows and layers, with their walls touching each other. Cells of this type make up *parenchyma* **tissue.** In the veins of leaves and the bundles of stems there are

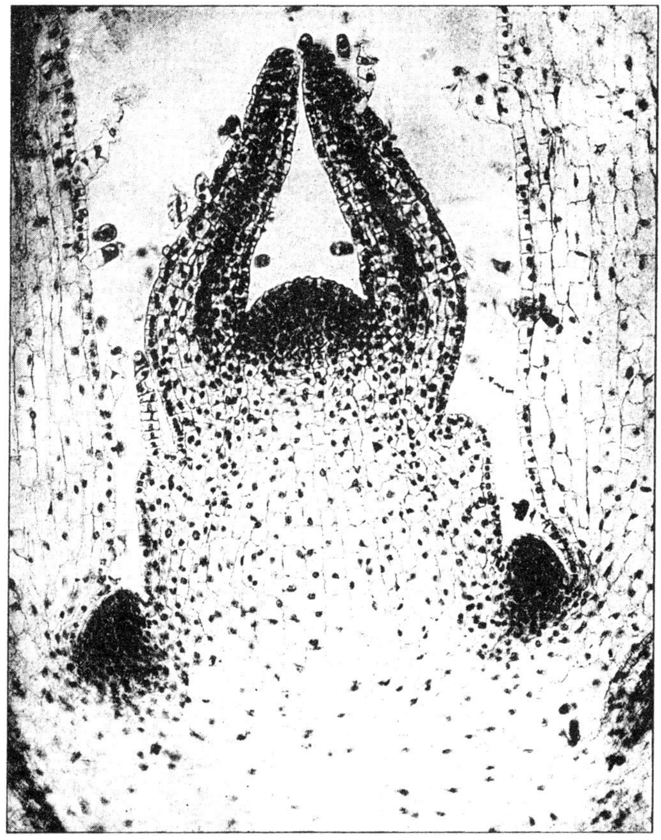

FIG. 72. Photograph of a stem tip of coleus, an opposite-leafed herb. The growing point is in the center above, surrounded by two young leaves. Just below are two shoulders representing the second node, the leaves of which are at right angles to the first pair and do not show. The leaves of the third node (partly shown in photograph) are quite large and have growing points in their axils. Note change in form and size of cells as growth proceeds.

cells which are greatly elongated — the length is many times the breadth. These cells touch the adjoining cells along their sides, but their ends are pointed or wedge-shaped and fit in between the cells above and below them. A tissue composed of this type of cells is called *prosenchyma*.

Sometimes the cell walls of both parenchyma and prosenchyma become thickened by the deposition of additional layers of cellulose, and the cellulose may be hardened by the addition of lignin. Cells with thick, hard walls are said to be *sclerotic* (Greek : *scleros*, hard). Sclerotic parenchyma and prosenchyma are often grouped together under the term *sclerenchyma*. Thus the stone cells found in pear fruits, and in the shells of nuts, may be called sclerotic parenchyma. Bast fibers and wood cells are sclerotic prosenchyma. If we wish merely to call attention to them as strong, hard tissues, we may call them sclerenchyma. When sclerenchyma cells are mature, they are usually devoid of protoplasm and are filled with either air or water.

Collenchyma. Another kind of tissue widely distributed in plants and closely related to parenchyma is known as *collenchyma*. This tissue differs from parenchyma in having the corners, or edges, thickened where three or more cells come together. These thickened edges give rigidity to the tissue, and for this reason collenchyma is often placed among the mechanical tissues.

Stem structures and plant groups. There are three groups of seed plants that we wish to distinguish at this time, because the stems of the plants that belong to these groups differ fundamentally. These groups are: (1) the *conifers*, or cone-bearing trees, like pines, spruces, firs, and cedars, that have scale or needle leaves and are for the most part evergreen; (2) the monocotyledonous plants (*monocots*), or plants with parallel-veined leaves, like the grasses, lilies, cannas, orchids, and

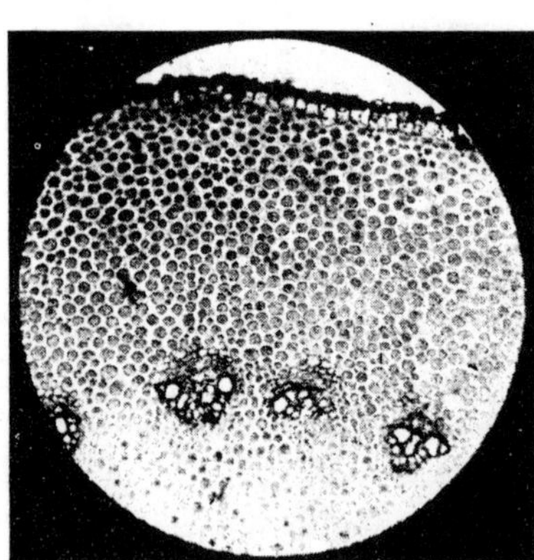

FIG. 73. Photograph of a cross-section of the outer part of calamus rootstock. The tissue forming the background is collenchyma, in which starch accumulates.

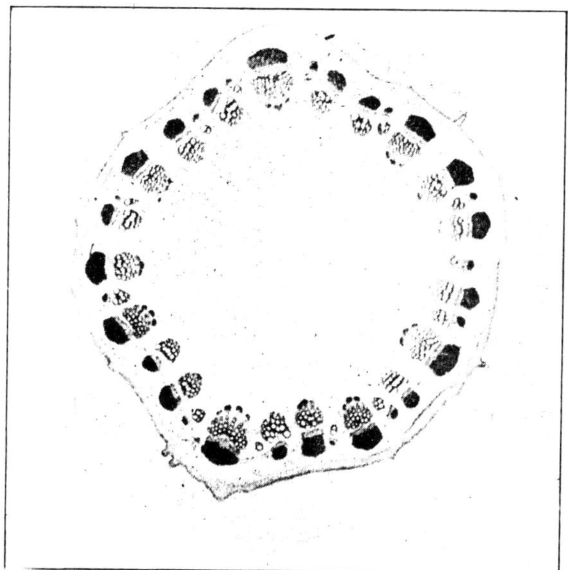

FIG. 74. Photograph of a cross-section of a young sunflower stem, showing arrangement of the bundles. Locate the several tissues.

palms; and (3) dicotyledonous plants (*dicots*), or plants with net-veined leaves, like oaks, maples, sunflowers, asters, and clovers.

The stems of the plants belonging to these three groups differ in (1) the kinds of tissues and cells making up the bundles, and (2) the arrangement of the bundles in the stem. We shall first study the bundles and the arrangement in a dicot stem, and then we shall learn how the stems of monocots and conifers differ from those of dicots.

The structure of a dicot stem. When a young dicot stem is cut across, the *bundles* are seen to be arranged in a ring. The core of tissue lying inside the *bundle cylinder* is the pith; outside the bundles is the *cortex;* and covering the cortex is an *epidermis* very similar to that of leaves. In older and harder stems the epidermis disappears and the outer cortical cells may be replaced

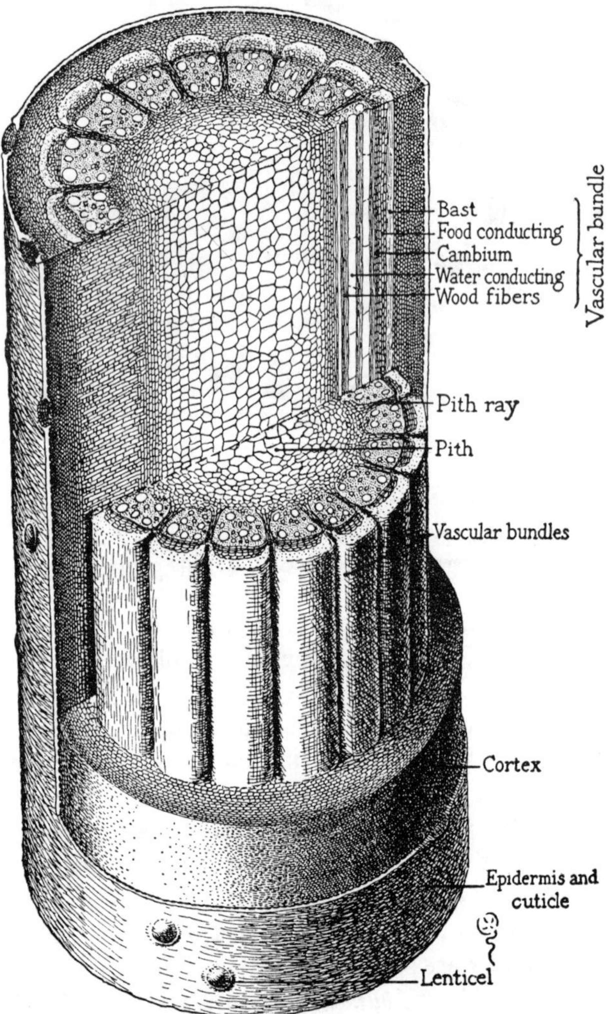

FIG. 75. Stem of moonseed vine, showing tissues and their arrangement. This stem is typical of a herbaceous dicot.

by soft layers of cork cells or by layers of sclerenchyma. The pith and the inner part of the cortex are made up of parenchyma. In annuals and young perennials the cortical parenchyma contains chlorophyll and resembles the mesophyll of the leaf in appearance and function. It is this tissue that forms the inner "green bark" of twigs and gives the green color to the stems and branches of herbaceous plants.

There are, then, four distinct layers in dicot stems: (1) on the outside is the epidermis; (2) from the epidermis to the bundles is the cortex; (3) inside the cortex is the bundle-cylinder; (4) the pith forms the axis of the stem, filling the space inside the cylinder of bundles (Fig. 75).

Between the bundles of the dicot stem there are strands of parenchyma cells that connect the pith parenchyma with the cortical parenchyma. These are the *pith rays*. They convey food across the stem, and with the other parenchyma cells form a complex tissue system in which foods accumulate and from which they later move to other parts of the plant.

General structure of the dicot bundle. The bundles in a plant stem are continued above in the veins of the leaves, and below in the bundles of the roots. In the dicot stem these bundles contain four tissues; (1) the water-conducting tissue, (2) the food-conducting tissue, (3) the cambium, and (4) the mechanical tissue. The *cambium* is a layer of thin-walled cells that lies lengthwise in the bundles and separates the water-conducting tissue from the food-conducting tissue.

The *water-conducting tissue* contains long, tube-like vessels made up of cylindrical cells joined end to end, often for considerable distances without end-walls between them. These tubes (tracheæ) usually have heavy walls marked by spiral and lattice-form thickenings. When mature they are empty of protoplasm. In other words, they are the coverings of dead cells joined together, forming tubes usually several inches, more rarely several feet, in length. Mixed with them are smaller and shorter tubes and cylindrical living cells. All together these tissues form the passageway for the movement of water and mineral salts to all parts of the plant. The general direction of the water movement in this tissue is upward, because the lifting of the water is brought about principally by transpiration from the leaves.

The simplest land plants are very small and grow flat on the soil in wet places. They are constantly in contact with the moist soil, and their cells can be supplied almost directly with water and mineral salts. In such plants a conductive system is not necessary; but if the leaves of a plant are to be raised into the air, water lost by transpiration must not only be supplied to them continuously, but at times it must be supplied in great quantity. Because of this fact, a plant that raises its leaves even a few inches above the soil must possess conductive tissues, and when large numbers of leaves are raised 200 or 300 feet into the air, a very extensive water-conducting system is necessary.

The *food-conducting tissue* differs from the water-conducting tissue in being composed of smaller, thin-walled cells, all of which retain their living protoplasm. The largest of these cells are set end to end, and the end-walls have holes in them like the top of a salt shaker. These rows of cells, therefore, form tubes with sieve-like cross-walls in them, and on this account they are called *sieve tubes*. Through the openings in the sieve plate the protoplasm is continuous from cell to cell, and through these tubes the foods pass from one part of the plant to another. Surrounding the sieve tubes are smaller living cells called *companion cells*. Because the cells of the stem and root are supplied with food manufactured in the leaves, it is often said that the movement of foods is downward in a plant. In reality, the direction of the food current is not so fixed as is that of the water current. Food moves toward any part of the plant where it is being used or being accumulated.

The roots and stems require a continuous supply of food for nourishing old cells and for building new ones. Since the foods

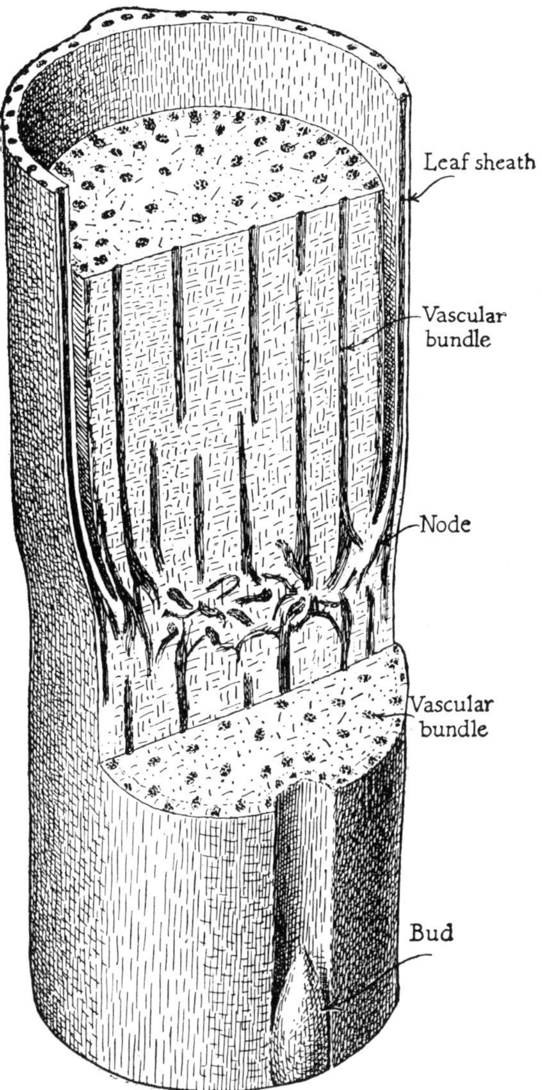

FIG. 76. A solid grass stem (*Panicum*), showing arrangement of the tissues in a typical monocot stem.

are manufactured primarily in the leaves, there must be food-conducting tissues that are adequate to carry them to all parts of the stem and roots. The food-conducting tissues also transfer food from the leaves to the seeds and growing parts, and when food has accumulated in the stem or roots it may pass up through the conductive tissues of the stem to other parts of the plant.

The *mechanical tissue* is made up of cylindrical or spindle-shaped cells with very heavy walls. Indeed, the walls at maturity may be so thick as to render the cells almost solid. Ordinary cellulose is not very hard, but the walls of the mechanical tissue are hardened and thickened by a deposit of *lignin*, a substance composed of cellulose and certain aromatic compounds. The difference between hard and soft woods is for the most part due to the thickening of the walls of the mechanical cells; secondarily it is due to chemical changes in the walls themselves (lignification).

Mechanical tissue is found on both the water-conducting and food-conducting sides of the bundles. On the food-conducting side it lies outside the food-conducting tissue, and is made up of long, exceedingly slender, nearly solid, spindle-shaped cells. These cells are called *bast fibers*, and the tissue that is made up of them is called the *bast*. Bast may be seen in the stringy fibers on a grapevine or in the bark of trees. It is the bast fibers from flax, hemp, jute, and other dicotyledonous plants that are used in the manufacture of thread and cordage.

The cells of the mechanical tissue on the water-conducting side of the bundle are somewhat shorter and thicker than the bast fibers. They are known as *wood fibers*, and make up what is properly called the *wood*. In most dicots the wood fibers (Fig. 91) are mixed with the water-conducting vessels and living thin-walled cells called *wood parenchyma*, and the whole inner part of the bundles is known as *wood*. In woody dicots this mechanical tissue is present in abundance and forms the bulk of the stem. The lumber that is obtained from dicotyledonous trees is derived from the inner parts of the bundles and is made up of wood fibers and water-conducting tissues. Examine a smooth piece of oak and you can readily see the small wood fibers and the larger water tubes. You can also see the pith rays that extend radially in thin layers at right angles to the wood fibers.

FIG. 77. Photograph of a cross-section of corn stem, showing arrangement of fibro-vascular bundles.

The *cambium* is a layer of soft tissue between the two sides of the bundles. It is the growing tissue which results in the increase in thickness of the dicot stem. Growth takes place by the longitudinal division of the cells. New cells formed on the inner side by the division of the cambium layer change into water-conducting cells or wood fibers; on the outer side they change into food-conducting cells or bast fibers. In this way the bundles of perennial dicots enlarge from year to year, and this causes the *stem* to increase in thickness. In a tree, cambium cells form a continuous layer between the wood and the bark, and the diameter is increased by the addition of successive layers of tissues built by these cells. These layers are the *annual rings* that one sees at the ends of logs. At the apex of the stem the cambium terminates in the growing region. At the lower end of the stem it connects with a similar tissue in the root.

Every one who has made willow or hickory whistles has become acquainted with the cambium. In early spring the cambium cells are dividing actively, and the cambium layer can be broken by tapping on the bark. The whole bark can then be readily stripped from the wood.

As the trunks and branches of trees age, secondary cambiums arise in the cortex that develop secondary layers of cork, or hard cells, or even bast fibers. These cambiums are usually irregular in their position in the bark, and in their extent. Secondary cambiums are in part responsible for the plate-like peeling of the sycamore and shagbark hickory, and the knobs and ridges that occur on cork oak, cork elm, hackberry, and sweet gum.

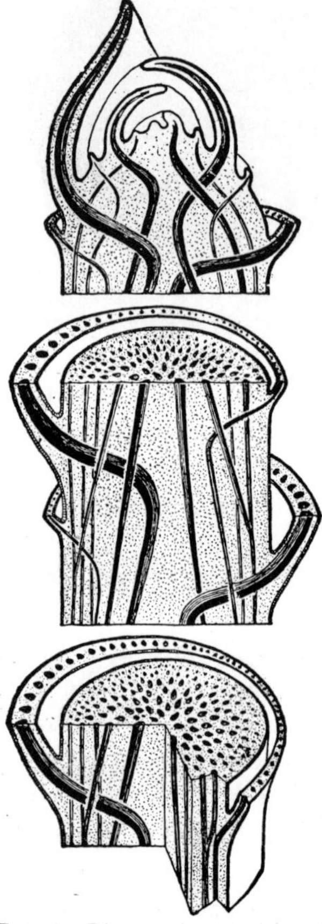

FIG. 78. Diagram showing the path of the fibro-vascular bundles in stems of the palm type. The bundles of each leaf arise by the growth outward of the innermost and largest bundles of the stem at that point. Lower down, these bundles connect with the outer bundles of the stem. (*From "Plants and Their Uses," by Frederick Leroy Sargent; Henry Holt & Co.*)

The monocot stem. The monocot stem, like dicot and conifer stems, is bounded externally by an epidermis which closely resembles that of the leaf. The groundwork of the stem is made up of parenchyma, which is commonly called the *pith*. The parenchyma is usually composed of thin-walled cells, and is the principal tissue in which the temporary accumulation of foods occurs; from it the sugar solution is obtained when the stems of sorghum and sugar cane are crushed. In a monocot stem the bundles are scattered, instead of being arranged in a cylinder as they are in a dicot stem. In stems of grasses that are hollow (Fig. 79) they are scattered through the cylinder of parenchyma tissue; in a cornstalk, a shoot of asparagus, or the trunk of a palm they are distributed through the whole stem. As in the dicot and conifer bundles, the water-conducting tissue is on the side next the center of the stem, and the food-conducting tissue is on the side toward the epi

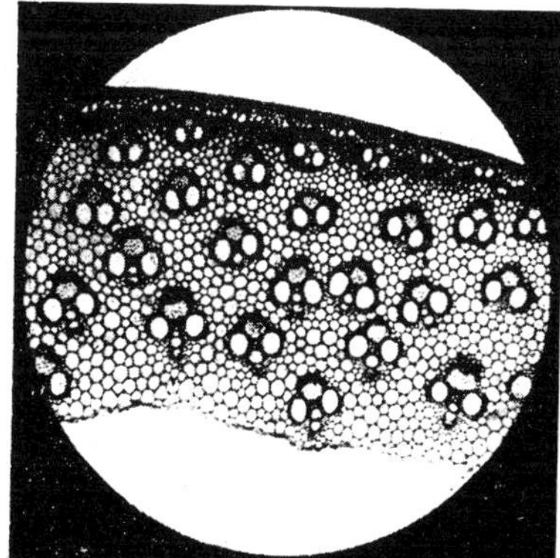

FIG. 79. Photograph of part of a cross-section of a bamboo stem, showing thick-walled mechanical tissues massed in the outer layers of the stem. The large openings in each bundle are water-conducting tubes.

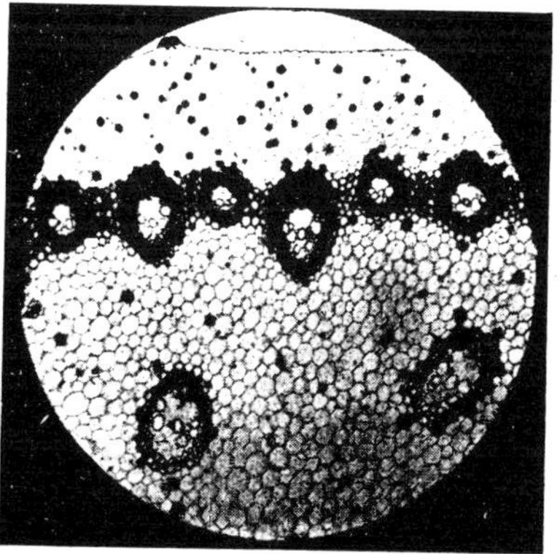

FIG. 80. Photograph of part of a cross-section of a rattan stem, showing bundles. The dark ring surrounding each bundle is the mechanical tissue. The scattered dark cells contain crystals of calcium oxalate.

dermis. The scattered arrangement of the bundles in the pith may easily be seen in a stalk of corn.

The monocot bundle. The monocot bundle differs from the dicot bundle in that it lacks a cambium layer. It is frequently called a *closed* bundle because, in the absence of cambium tissue, the bundle cannot increase in size and there can be no growth in diameter of the monocot stem through the multiplication of cambium cells. The dicot bundle, on the other hand, is spoken of as *open*, because there is a cambium layer between its water-conducting and food-conducting tissues and the bundles can increase in thickness. The monocot bundle differs further from the dicot bundle in that its mechanical tissues form a complete sheath about the food- and water-conducting parts. It is as though the bast of the outer part of the dicot bundle and the wood of the inner part were joined at the sides of the bundle, forming a sheath about the conducting tissues.

The fibers like sisal and Manila hemp (Figs. 81, 235) that are derived from the monocots are usually coarser than the fibers derived from dicots, because the monocot fibers are entire bundles, while the dicot fibers are made up of only the strands of bast cells from the food-conducting side. The bundle sheaths are usually thicker in the bundles near the outer part of the monocot stem. In fact, in some monocots, like rattan and bamboo, the sheaths of adjacent outer bundles may join each other and

FIG. 81. Plantation of abacá, a species of banana, from the petioles of which Manila fiber is obtained. Abacá flourishes only in the Philippines. The fibers are used chiefly in the manufacture of ropes.

Bureau of Agriculture, P. I

FIG. 82. Stripping abacá for fiber. The long petioles are pulled under toothed knives which scrape the soft tissues from the bundles. Abacá is a monocot, and the fiber is composed of an entire bundle.

thus form a hard layer beneath the epidermis (Fig. 79). Some monocot stems, like the palms and dragon tree, increase in thickness a number of years during their early life. This is accomplished by the development of secondary cambiums in the pith between the bundles. The cells of the secondary cambiums divide, *forming new bundles* between the older ones. In this way the stems increase in diameter, without forming annual rings.

Dicot stems are enlarged by the development of *new layers* of cells between the wood and the food-conducting tissue. It follows that there will be annual rings in such stems.

The structure of conifer stems. The conifers, like the dicots, have their bundles arranged in a cylinder. In structure these bundles are somewhat similar to those of dicots, except that the wood and water-conducting tissues are not distinct. The wood cells form the water-conducting tissue as well as the mechanical tissue. In keeping with their double function, these cells (tracheids) are thick-walled and spindle-shaped, with numerous thin places, or pits, in two of the walls. Because of this structure, the stem retains its rigidity and still permits the ready passage of water and mineral salts.

The stems of some conifers, such as pine, spruce, and fir, have resin ducts distributed more or less irregularly in the wood. In cedar, hemlock, sequoia, cypress, and arbor vitæ, resin ducts are absent. Resin ducts are not tubes like the tracheæ of dicot stems, but are intercellular spaces in which resin accumulates.

CHAPTER SEVENTEEN

LONGEVITY OF HERBACEOUS AND WOODY STEMS

Every one who has occasion to grow plants needs to know something about the length of life of the plants he is concerned with, and he must know also whether they have herbaceous or woody stems. For example, suppose a farmer wishes to determine whether it will be more profitable to grow sweet clover or alfalfa in a certain field. Before planting either of these crops, he should know that one of them is a biennial and the other perennial, because all his plans for handling the crop will depend on this information. Or suppose that another man wishes to have a permanent border of flowering plants about his lawn to obstruct the view of some unattractive fields or buildings. He can choose wisely from among the hundreds of plants listed in nursery catalogues only when he has definite information about the longevity of the plants and as to whether they are herbs, shrubs, or trees. A clear understanding of the classification of plants on the basis of their length of life, their woodiness, and their habit of forming single large trunks or a number of smaller stems is helpful; also in any study of the structure and processes of stems. Plants differ greatly in their length of life. To indicate the length of the natural life periods, the terms *annual*, *biennial*, and *perennial* are commonly applied to plants.

Annuals. Most of our common garden vegetables and field crops are started from seeds in early spring. The seeds germinate; roots and shoots develop; and by midsummer or autumn, flowers and fruits are produced and new seeds, which contain the beginning of another generation of plants, are formed. Then the plants die. The period from seed germination to seed production and death is called the *life period*. If it is completed within a single growing season, the plant is called an annual (Latin: *annus*, year). Corn, lettuce, radishes, beans, pumpkins, morning-glories, and ragweeds are familiar annual plants.

Biennials. During the first season some plants develop only leaves and roots and a very short stem. The root is usually large and accumulates a considerable amount of food. In the second season growth is renewed, and there is developed an elongated stem with leaves, flowers, fruits, and seeds. These plants which pass a winter season during their vegetative development, and whose life period includes two different growing seasons, are called *biennials* (Latin: *biennium*, space of two years). The seeds of some common weeds, like the shepherd's purse, evening primrose, and wild lettuce, germinate in August or September, and a little rosette of leaves is formed close to the ground. Food accumulates in the root until winter comes. The following spring the plants make rapid growth, and by midsummer they have blossomed, produced seed, and died. In spite of the fact that their whole life is passed within a twelve-month period, these plants are called biennials, because their life period covers parts of two growing seasons.

FIG. 83. Wild carrot (*Daucus carota*), showing the plant as a seedling, at the end of the first growing season, and as a mature plant during the second growing year. The life history shown above is typical of biennials.

The term *annual* or *biennial* as applied to plants, therefore, does not imply any definite length of life in months. Wheat may be grown either as an annual or as a biennial, depending upon whether it is planted in the spring or in the fall. Shepherd's purse and wild lettuce not infrequently live as annuals in nature. The commonest biennials of the garden are beets, carrots, parsnips, turnips, and cabbage. In the first four, large amounts of food are accumulated in the roots; in the cabbage the food is stored in the enormous terminal bud, the "head." These stores of food are used in the production of seeds the following year. Usually biennials and annuals are herbs. Biennials, like annuals, are comparatively small in size, and die after flowers and seeds have been produced.

Perennials. Perennials (Latin: *perennis*, lasting through the year) are plants that live for a number of years. Some of them, as for example certain grasses, produce seed during the first and succeeding years. Other perennials, like alfalfa, form seed at the end of the second and succeeding seasons. Trees and shrubs usually require several seasons' growth before seeds are produced. The century plant of our Southwestern deserts develops vegetatively for 25 or 30 years before it produces a flowering stem and seeds. Then it behaves like an annual or a biennial, for as soon as the seeds are mature the whole plant dies. This calls our attention to the interesting fact that in annuals, biennials, and a few perennials there is no well-marked period of senility or old age. They die suddenly at maturity, immediately after their period of greatest vigor. Trees and shrubs, on the contrary, have a distinct period of old age in which the physiological processes are slowed

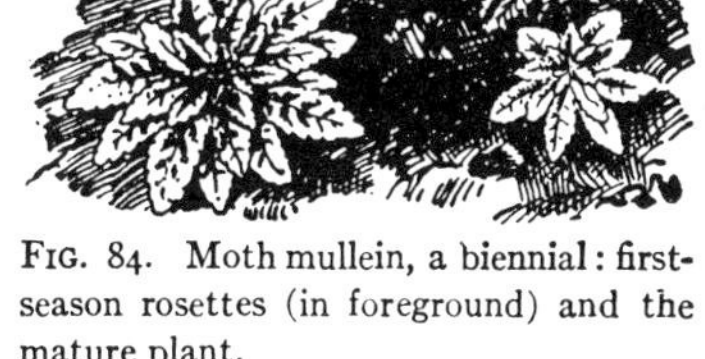

FIG. 84. Moth mullein, a biennial: first-season rosettes (in foreground) and the mature plant.

down gradually until the plants succumb to diseases and unfavorable conditions which they could have withstood in youth.

Perennials classified according to the persistent parts. All perennials add new leaves, new stems, and new roots each year; but they may be classified roughly according to the parts that persist from one season to the next.

Evergreen trees and shrubs are perennial in all parts of the plant body. Deciduous trees and shrubs are perennial in their stems and roots. Many herbaceous perennials, like the cattails, grasses, mints, peonies, trilliums, and bananas, have annual above-ground stems but perennial underground stems and roots. Dahlias and sweet potatoes have perennial roots. Potatoes and the Jerusalem artichoke (a kind of sunflower) have perennial thickened underground stems (tubers). Tulips and hyacinths have perennial underground stems and buds (bulbs). These examples show that perennial plants have many different ways of living over unfavorable seasons like periods of cold or drought.

There seems to be no limit to the length of life of some perennial herbs, like ferns, the May apple, Solomon's seal, and certain grasses and mints. The older parts die each year, and new parts form at the growing ends of the underground stems. The plants change their locations slightly each year, one end of the stem growing forward and the other end dying away. There is no apparent reason why such plants should not live indefinitely, perhaps longer than the oldest tree; but no one part of the plant lives for a long time.

Herbs, shrubs, and trees. Shrubs and trees have woody stems. The stems of herbs have comparatively little woody tissue. Our garden and field crops are all herbaceous plants. Their stems contain little woody tissue, and in temperate climates the above-ground parts live only during a single growing season.

The principal difference between shrubs and trees lies in the fact that shrubs develop numerous slender above-ground stems from a single base, while trees develop a single stem or trunk.

This distinction may be expressed in another way by saying that shrubs branch underground, while trees branch only above ground. Most shrubs are less than 10 feet in height, but some, like the staghorn sumac, may reach a height of 20 feet, or, like the bamboo, 40 feet. Most trees are between 25 and 200 feet in height.

However, the distinction between herbs, shrubs, and trees is not one of size. Herbaceous plants, like the sunflower, may reach a height of 20 feet, and in the tropics corn and bananas a height of 30 feet, while some shrubs are only a few inches in height and some of the dwarf trees of Japan that are a century old are less than 5 feet in height.

FIG. 85. Japanese dwarf pine. Some of these small potted trees are a century old.

Some of the oldest trees known were seedlings 3000 years ago. Many trees now standing are over a thousand years old. The average age of the older trees in our Eastern forests, however, is much less than this, ranging from one to three hundred years; in some of the Western forests, three to five hundred years.

Plant characteristics and the plant-producing arts. The differences in the habits of growth, longevity, and materials stored by plants has led to specialization among those who grow plants. For many evident reasons the most important art of growing plants is *agriculture.* The farmer deals entirely with herbs and largely with annuals, though biennials and perennials may be grown for forage crops. He is for the most part concerned with plants that accumulate foods in a highly concentrated form in seeds. He transforms some of this food into meat and dairy products by feeding it to animals. But the basis of all animal industry is the growing of plants.

Horticulture embraces a wider range of plants, but in actual practice a horticulturist usually specializes on plants having somewhat similar habits. The growing of food-producing shrubs and trees represents one division of horticulture. The object sought is the production of fruits containing pleasantly flavored substances stored in cells with the thinnest possible cell walls. The truck gardener specializes on annuals and biennial herbs that accumulate both food and flavors, and to a less extent on perennials, like asparagus, strawberries, rhubarb, and berries of various kinds. *Floriculture* deals with all classes of plants and has for its object the production of attractive flowers and foliage. It reaches its highest development in *landscape architecture*, in which masses of vegetation are arranged to beautify a landscape with effective arrangements of foliage, and with varied texture and color effects at different seasons of the year.

Silviculture is the art of growing trees to create forests. The silviculturist specializes on growing trees of many different types for a great variety of uses, such as lumber, pulp wood, bark, rubber, cork, and fuel.

CHAPTER EIGHTEEN

THE GROWTH OF STEMS

THE limit of growth of stems is not so definite as that of leaves. The length and the diameter of a stem depend largely upon the conditions under which the plant lives, the available water supply, amount of light, the length of daylight, the temperature, and quality of the soil. Along a dry roadside a ragweed may complete its development with a stem less than 6 inches long, while in a rich bottom-land field the same plant might have reached a height of 15 feet.

Growth in length. The growth in length takes place at the apex of a stem, the *growing point* being located in the terminal bud. The *growing region* extends back from the tip, sometimes for only a fraction of an inch, more rarely, as in rapidly growing vines, a foot or two. If we mark the upper portion of a growing stem into equal spaces, we may observe on the following day that the uppermost spaces have elongated the most. The adjoining spaces below are less and less elongated. This indicates that the greater part of the cell division takes place near the tip (the growing point), but that some cell division and most of the enlargement of cells occurs in the adjoining part of the stem (the elongating region). During enlargement the minute cells of the growing point absorb water and increase their volume from one hundred to two thousand times. In the growing point the nuclei and the surrounding cytoplasm completely fill the cell walls. In the elongating region the cytoplasm forms merely a thin layer lining the inside of the cell wall, most of the internal space being occupied by cell sap.

The above description holds for most stems. In grasses and some other monocots, however, the process is slightly modified. In these plants the tissue at or just above each node continues to grow for some time after the tissue of the upper part of the internode is mature. In these plants the growing point (primary meristem) develops nodes, and short internodes that continue growing independently. Instead of a continuous growing region extending back from the growing point, there is a series of shorter and shorter growing regions at the base of each internode. It is this fact which explains why growing corn breaks easily just above the nodes, and why it grows into an upright position again when blown over during its period of development.

Diameter growth of annuals. Annual stems increase in thickness until the plant matures. This increase in size is brought about by the enlargement of cells and by the formation of additional cells by the cambium. In many annuals, like mustard, zinnia, onion, squash, and corn, the stem thickens by increase of the size of cells. In the sunflower the cambium continues to form woody tissue and bast for a considerable part of the growing season, so that very large plants have stems 1 to 2 inches in diameter near the base.

Growth in diameter of trees and shrubs. Shrubs and trees increase in thickness each growing season. This is often called *secondary growth;* as we have seen, it is brought about by the continued growth of the cambium. This layer of cells produces new water-conducting tissue and wood fibers on its inner side, and it produces food-conducting tissue and bast fibers on its outer side. As growth proceeds from year to year, *annual rings* mark the successive additions to the wood. The bark also develops annual layers, but in most woody plants these are much thinner and less conspicuous than the annual layers of the wood. Further, since growth takes place inside the cortex, the cortex is continually being split and broken. The outer layers may die and after a few years will be gradually weathered off. *Secondary cambiums* in the cortex may develop additional layers of cork. The ridges and grooves of the bark show how much too small the outer bark is to cover the more recently formed wood. Smooth, thin-barked trees lose their bark very rapidly. Trees

Bureau of Agriculture, P. I.

FIG. 86. Rafting large bamboo stems to market in the Philippines.

with bark that is thick and has large ridges are the ones that hold their bark more tenaciously. But in all large trees the bark contains only a part of the layers formed by the outer side of the cambium; much material has scaled off and fallen away. It should be noted that as a tree gains in diameter, the annual rings of wood in the stem are each year farther removed from the corresponding annual layers of the bark. That is, the wood rings nearest the pith are nearest in age to the outermost rings of the bark.

Diameter growth of perennial monocot stems. Most perennial monocots, like the bamboo and asparagus, have horizontal underground stems to which new and thicker stem segments with a larger number of bundles are added each year. The aërial, erect branches never increase in size after they are once mature; but the erect branches from old underground stems are from the beginning much thicker than those from young plants. Consequently, no little bamboo rod could ever grow into a bamboo beam. No large bamboo beam was ever a slender rod. These aërial branches come out of the ground nearly as thick as they will be when mature. Asparagus plants are several years old before the underground stems become large enough to send up thick, upright branches suitable for marketing.

The increase in thickness of stems of the palm type has already been described. Usually the growth in thickness ceases after a time, and the further growth of the stem takes place only in the terminal bud. Such stems taper for a short distance at the base, but above they are quite cylindrical.

Annual rings. The wood derived from conifers and dicots have certain characteristics which are due to the variation in growth during spring and summer. It is the difference in size of cells and thickness of walls laid down at different times of the year that make the annual rings visible. In such woods as the oak, ash, and yellow pine the annual rings are conspicuous because of the difference in texture of the spring and summer wood. Beech, birch, and redwood have quite inconspicuous annual rings because growth is quite uniform throughout the growing season. In all trees, however, there is a perceptible slowing down toward the end of summer and the wood cells are smaller near the outer edge of a ring.

The width of rings is primarily dependent upon the amount of carbohydrate furnished by the leaves, and secondarily on the water supply. In wet seasons the rings are wider, in dry seasons narrower. Indeed, the width of annual rings shows such perfect correlation with wet and dry years, that the rings of our oldest trees are being measured and studied to determine periods of excessive rainfall and drought during prehistoric times, and to estimate changes in climate.

Classification of woods. The structure of the wood of any species of tree is so characteristic that any piece of wood may be identified by a careful examination. Woods are primarily classified as *ring porous*, *diffuse porous*, and *non-porous*. The " pores " refer to the openings among the wood cells made by the tracheæ or water tubes. Since these are not present in conifers, conifer woods all belong to the third class.

FIG. 87. The "bottle" palm along a river in Cuba. An unusual type of palm stem, in which secondary thickening occurs above the base but not throughout the entire length of the stem.

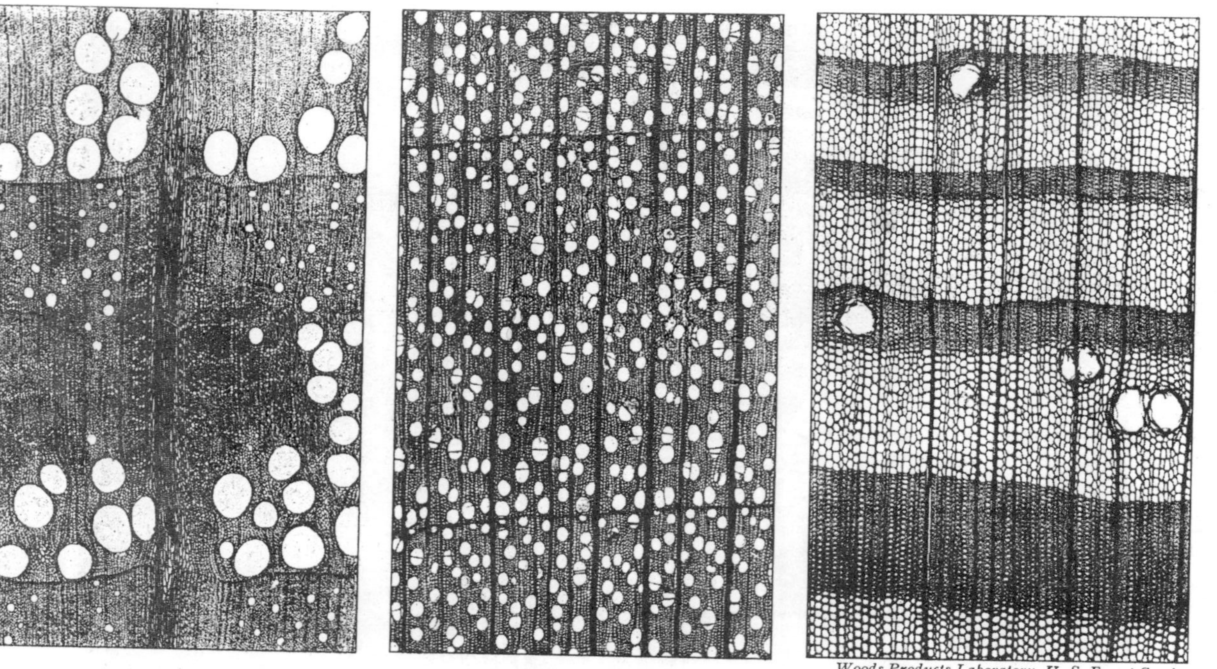

Woods Products Laboratory, U. S. Forest Service

FIGS. 88, 89, and 90. Thin cross-sections of various woods, highly magnified, showing the three principal characters by which woods are classified. At the left is red oak (*Quercus rubra*), a ring-porous wood; in the center is broadleaf maple (*Acer macrophyllum*), a diffuse-porous wood; and at the right is yellow pine (*Pinus palustris*), a non-porous wood with conspicuous resin ducts.

(1) Ring-porous woods have larger and more numerous tracheæ in the spring wood, and denser summer wood. To this class belong ash, catalpa, locust, elm, chestnut, oak, and hickory.

(2) Diffuse-porous woods have the pores about equally distributed through both spring and summer wood, and annual rings inconspicuous. Here belong walnut, cherry, cottonwood, beech, maple, holly, birch, gum, and basswood.

(3) Non-porous woods may have conspicuous rings when the wood cells, or tracheids, are large and thin walled in the spring wood, and smaller and heavier walled in the summer wood. Yellow pine and hemlock exemplify this type. Red cedar, spruce, and arbor vitæ have inconspicuous rings, marked only by the slight decrease in size of cells near the outer edge of the ring. Some of the non-porous woods have prominent resin ducts, in others ducts are wanting.[1]

Structure determines usefulness. For manufacturing purposes the differences in wood structure just outlined are of the greatest importance, because the quality of the products is largely a matter of wood structure. Spruce, because of its soft texture and freedom from resin, is used for paper pulp. Its uniform grain makes it desirable for sounding boards of musical instruments and in the manufacture of airplanes. Walnut, because of its color and the ease with which it may be polished, is prized for gunstocks and furniture. Hickory is the best wood for tool handles and the spokes and rims of wheels. Oak is valuable for flooring, interior finish, and furniture. Cypress and redwood are especially noted for their durability in the soil or under conditions where other woods decay. Ash is notably strong, elastic, and

[1] Further information on the identification of woods may be obtained from *Guidebook for the Identification of Woods Used for Ties and Timbers* (United States Forest Service).

not very heavy, and consequently is extensively used for implement handles, for wagons, automobile bodies, and railway-car frames. The best baseball bats are made of second-growth ash, because the layers of dense summer wood are thicker in more rapidly growing shoots. These are but a few of many interesting relations that exist between the structure and use of wood.

Attention must also be called to the effects of the environment on the woody tissue. The quality of any kind of wood is modified by the conditions in which the tree lived. Oak lumber from rich, well-drained, moist uplands is very different from oak that grew in relatively sterile, poorly drained lowlands. Consequently, lumber from certain localities is far more valuable than the same kind of lumber from others.

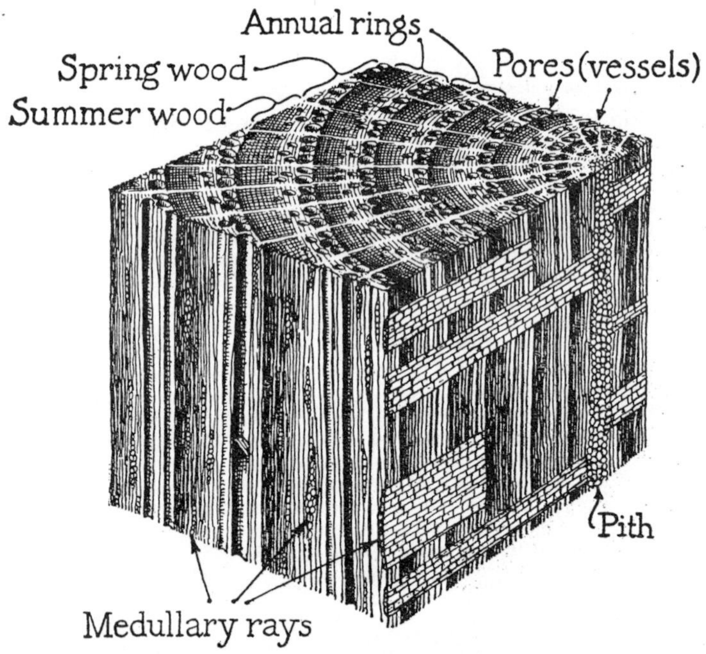

FIG. 91. Diagram of a block of oak wood, magnified to show the arrangement of the various tissues that produce the patterns on polished wood surfaces.

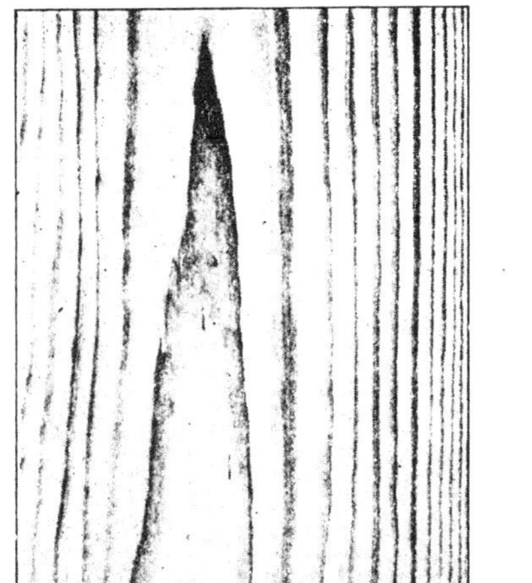

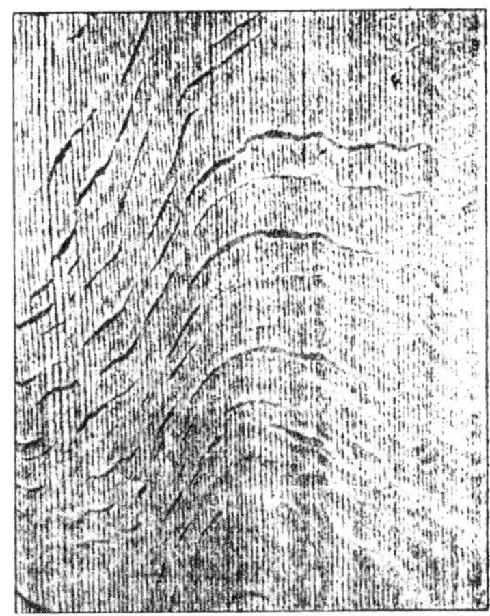

Southern Lumberman

FIGS. 92 and 93. Patterns formed by annual rings and medullary rays. The board at the left is longleaf yellow pine, and the one at the right "quarter-sawed" white oak. In the pine the markings are due to the rings. In the oak the edges of the rings appear as longitudinal lines and the pith rays as irregular cross markings. Can you explain exactly how each board was cut to show these markings?

Heartwood and sapwood. As the trunks of trees increase in thickness, all the cells toward the center of the stem gradually die. The wood usually changes in color after the death of these cells. In a peach tree only the outer three or four annual rings may be alive. In a walnut trunk 2 feet in diameter, all but the outer 2 inches may be dead. The dead wood still helps to support the enormous weight of the tree top, but it has nothing to do with the conduction of water and substances in solution. This inner dead wood is called the *heartwood;* the outer living wood is called the *sapwood.* The heartwood in many species of trees is much more valuable than the sapwood for lumber, because of its color and greater durability.

Grafting and budding. In the propagation of many varieties of fruit trees it has been found that seeds are not satisfactory. Most of our cultivated fruit trees are so highly variable that their seedlings are not like the parent plants in quality of fruit. Horticulturists long ago learned to overcome this difficulty by grafting a twig from the desired variety of tree on a seedling of a similar

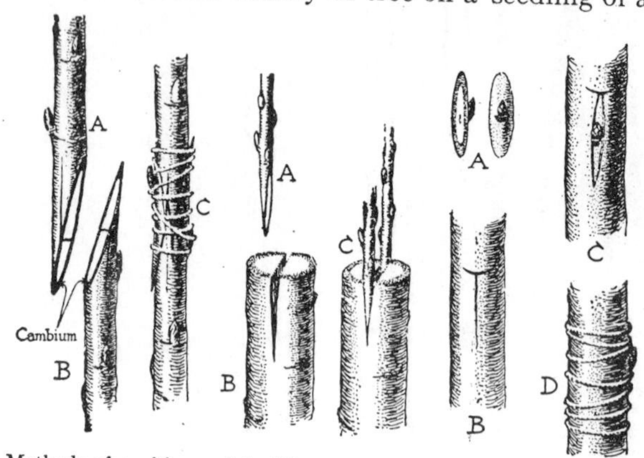

FIG. 94. Methods of grafting and budding. At the left, whip grafting; in the middle, cleft grafting; at the right, budding. *A* is the cion, and *B* the stock. *C* shows the cion and stock joined. In both grafting and budding, success depends on bringing the cambium of the cion into contact with the cambium of the stock.

tree. The grafted-in branch then becomes the top of the tree, and the fruit it bears is like that of the tree from which it came.

In grafting, the plant that furnishes the root is called the *stock.* The twig that is attached to it is called the *cion.* In cleft grafting, the top of the stock is cut off. The stock is then split and two cions with chisel-shaped ends are placed in the cleft, one on either side, *so that the cambium of the cion is in close contact with the cambium of the stock.* The wound is covered with wax to prevent the drying out of the tissues. If the cambium tissues are in perfect contact, they will soon unite. New tissue will grow under the wax and finally cover the wound. If both cions grow, the weaker one is removed. In grafting nut trees the cions, after being set, are painted with melted paraffine to protect them from drying while the union between stock and cion is taking place. This practice will increase the number of " takes " in fruit-tree grafting also.

Whip grafting is the common method of uniting cions to small seedlings. Usually this is done at or below the surface of the soil. Both cion and stock are cut obliquely, and each is split. The upper half of the oblique end of the cion is pushed into the cleft of the stock and is bound firmly in place with raffia or twine. Again, *the success of the graft depends upon the contact between the cambium of the cion and the cambium of the stock.*

In budding, a T-shaped cut is made on the side of the stock, through the cortex, down to the cambium. A bud from a tree of the desired variety, with a small oval piece of wood and bark attached, is slipped down inside the cortex of the stock and tied firmly in place. This places the two cambium layers in contact and the two pieces unite. The stock is then trimmed and the bud develops into a branch. When the branch is well started, the original stem is cut off just above the base of the branch.[1]

Grafting is commonly done in the spring; budding, in the early fall. The fruit produced on grafted or budded trees is usually like that of the cion, regardless of the variety of stock. However, there are cases in which the cion is modified by the stock. Discussions of these cases may be found in books on horticulture. Grafting is usually possible only between closely related species of plants. Sometimes, however, plants that are more remotely related may be grafted on each other, as for example tomato, potato, and nightshade, or the pear, apple, and quince.

The essential features of budding and grafting are relatively simple, but in practice there are details and refinements which are of the greatest importance. The selection of stock and cion and the best method of operation vary not only with the species but with climate and soil. The best results can be obtained only through profiting by the experience of others and the results attained by scientific experiment.

[1] In budding some hardwoods like hickory and walnut, better results are obtained by cutting about a small patch of the bark and allowing the formation of wound callus about the cut edges. As soon as the callus forms, the patch is removed, and a patch of the same size bearing a bud is fitted accurately into its place. The method is called "patch budding."

CHAPTER NINETEEN

THE MOVEMENT AND ACCUMULATION OF MATERIALS IN STEMS

ASIDE from growth the most important processes going on in stems are those connected with the transfer of water, foods, and other materials. The living cells of the stem secure energy for chemical processes through respiration. They also assimilate foods, and may temporarily, or permanently, accumulate food and other substances.

The lifting of water in stems. Nothing concerning the physiology of plants has interested more people than the transport of water from the soil to the topmost leaves of trees. Yet in spite of much observation and experiment, the process is still only partially explained.

There can be no doubt that one of the principal factors in the rise of sap is the evaporation from the cells of the mesophyll in transpiration. The water thus lost is replaced by more water passing into these cells from the adjoining water-conducting tissue of the veins by osmosis. This is brought about by the sugar and other substances in solution in these cells, as we learned in Chapter XI.

Water inclosed in tubes has a high cohesive power; that is, it holds together like a solid. If a pull is exerted on the upper end of a column of water in the vessels of a tree, the column holds together like a cord or wire, and the whole column is pulled upward. As the water at the upper end of the water-conducting tissue moves into the mesophyll cells, additional water is pulled upward into the vessels of the blades, petioles, and stems.

Transpiration is greatest and the largest amounts of water are being lifted in trees during the summer. If at this season a hole is bored into the trunk of a tree and an air-tight connection made between this hole and a tube that has its lower end in a vessel of water, the water is drawn into the stem, not forced out. This indicates that there is more pull on the water from above than there is pressure from below. It is known also that there may be currents moving downward in one layer of the wood and upward in another, although the general direction of water transport is upward to the leaves. The movement of water in the tracheæ and tracheids is a mass movement, similar to the flow of water in a pipe, in spite of the fact that it frequently must pass through the cross-walls which divide the vessels at intervals. It is certain that the roots alone do not force water up into the tops of trees.

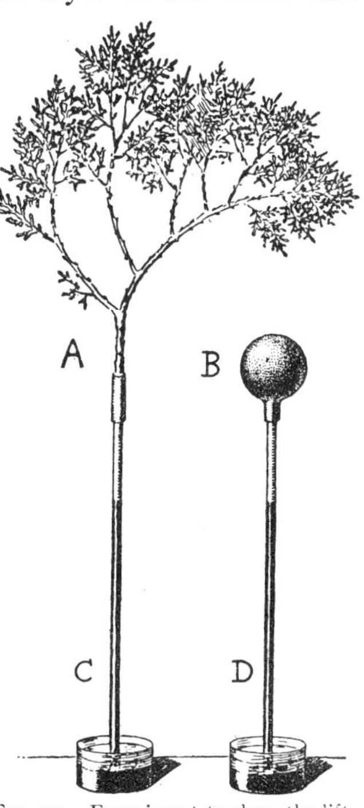

FIG. 95. Experiment to show the lifting power of transpiration and evaporation. Both tubes were filled with boiled water and placed in a dish of mercury. In *C* the mercury has been drawn up by transpiration from a branch of arbor-vitæ (*A*); in *D*, by evaporation from a porous cup (*B*).

The primary factor, then, in the rise of sap is transpiration; the second factor is the movement of water from the water-conducting tissue to the mesophyll cells, replacing that lost through transpiration; the third factor is the cohesion of water columns in the long strands of water-conducting tissue, which makes it transmit the pull from the mesophyll cells all the way down to the roots. In Chapter XXI we shall learn how the water passes from the soil into the roots, and to what extent the roots aid in the lifting of water. (See Fig. 95.)

The pulling up of water by transpiration is exemplified when cut flowers are placed in a vase containing water. That water is drawn into the flowers may be shown by placing the stems of white flowers in water colored with red ink. Try this experiment with a white carnation or a chrysanthemum. Repeat, using dilute ammonia instead of red ink.

Accumulation of water. The stems of many plants are succulent; that is, they accumulate water. Numerous examples might be cited among desert plants of the cactus type, but they also occur among mesophytes; for example, purslane, begonia, and certain orchids. Water retention depends in some instances upon the presence in the cells of substances like pectic compounds and mucilage, in other cases upon high osmotic pressures; but for still other cases no explanation can be given at the present time.

The flow of sap. The water in stems may contain a small amount of sugar in addition to mineral salts. In the maple, during early spring when the days are warm but at night freezing still occurs, quantities of sugar pass into the water-conducting tissue. This sugar comes from the medullary rays and other tissues where it accumulated in the form of starch during the preceding growing season. With the coming of warmer weather the starch is changed to sugar and diffuses into the water-conducting vessels.

The earlier sap is the richer and apparently comes largely from the upper parts of the trunk. The last sap is more dilute and probably comes from the roots. The positive pressure that produces the flow occurs usually during the day but may occur during warm nights. When the temperature falls to the freezing point at night, the pressure becomes negative and the sap flow ceases. The causes of the pressure are only partly known. A portion of it is due to the expansion of gas bubbles within the tree, but this gas expansion accounts for only a small part of the pressure.

Whether the flow shall continue for weeks or stop after a few days is determined by weather conditions; but just how the several weather factors (like changes in temperature or rainfall) bring about the increase and decrease of pressure, is unknown. The flow continues longest when the night temperatures are below the freezing point, and the day temperatures above. Even under the most favorable conditions it is not possible to draw out of a tree more than 5 per cent of the sugar that it contains.

A flow of sap somewhat similar to that in the maples occurs in the spring in some other species of trees, as in the birch, butternut, and hornbeam. In the birch the flow is more regular and continues until May; but the rate of flow and sugar content are less than in the maple.

The movement of sugar in the water-conducting tissues of stems is rather exceptional; its usual path lies in the food-conducting tissues.

Movement of foods. In previous chapters we learned that the vascular bundles of stems contain food-conducting tissue. This tissue is composed of thin-walled elongated cells and extends from the veins of the leaves, through the stem, into the extremities of the roots. The larger of the vessels are the sieve tubes. We also learned that dissolved substances alone can move from one cell to another, and that the movement is by diffusion from regions of greater concentration to regions of less concentration.

Applying this information to the movement of food in stems, it is evident that in annuals the general direction is from the leaves to the stems and roots during the vegetative period. When flowering and the development of seeds begin, a considerable part of the excess food moves into the reproductive structures.

In trees food moves into the branches and trunk during the summer and autumn, and accumulates in the food-conducting tissues, in the pith rays, and in young stems in the pith also. In the spring there is a great increase in the amount of soluble foods, and these move both into the roots and into the twigs. At this time new roots are developing, and buds are growing into new leafy shoots. When growth has stopped and photosynthesis is active, the movement is again from the twigs toward the trunk.

Digestion. Before insoluble foods may be moved from one part of a plant to another, they must be changed to soluble substances. This process may be illustrated by the changes that take place in starch within the plant. Starch is insoluble in water. It does not dissolve in the cell-sap, and the starch within the cells is not divided into particles small enough to pass through the cell walls. The process of changing starch into a soluble substance has been carefully studied; and we know that it is first converted into the sugar, maltose, and that the maltose is further split into the simple sugar, glucose. Both of these sugars are readily soluble in water and consequently can pass from cell to cell and thus to any living part of the plant. *The changing of insoluble substances* like starch *into simpler soluble substances* like glucose *is called digestion.* Unlike animals, plants have no special organs of digestion. All their living cells are capable of digesting the insoluble substances that are required for their nutrition.

Digestion brought about by enzymes. Digestion is a chemical process and is brought about in cells by the catalyzers called *enzymes.* These are produced by the living protoplasm of the cells. A large number of different kinds of enzymes have been recognized in plants; each enzyme usually digests only one particular kind of food, and there must be a different enzyme to digest each kind of food within the cell. The enzyme which changes starch to maltose is called *amylase,* and the enzyme which changes maltose to glucose is called *maltase.* A mixture of these two enzymes is called *diastase.*

The digestion of fats and proteins. The enzyme which digests fats is called *lipase.* Many seeds are rich in fats. More than one third of the weight of a peanut seed, for example, is fat. When the seed germinates, this fat is digested and changed to fatty acids and glycerin, which are soluble in the cell sap and which may move to other cells. Furthermore, the fats are chemically stable substances, while the fatty acids are active substances that may enter into a great variety of chemical processes and are, therefore, more readily used in the life processes that go on within the cells. There are other enzymes, *proteases,* which act upon the insoluble forms of proteins and render them soluble. Proteins are mostly inactive storage materials and bear much the same relation to the simpler and more active nitrogen compounds (that is, *amides* and *amino acids*, produced by protein digestion) that starch bears to the simple sugars, or fats to the fatty acids.

It seems probable that enzymes are concerned in the principal activities of all living cells. Without them there could be none of the rapid chemical changes in foods that are necessary for the transfer of foods within the plant and for carrying on the other processes. Both the building up of the complex food molecules from simpler ones and the splitting of these large molecules again is brought about by enzymes that are within the cells.

Properties of enzymes. It is interesting to know that if an enzyme is put into a test tube with the appropriate food substance, under favorable conditions it will bring about digestion the same as if it were in the living cell. This proves that digestion is not directly carried on by the living protoplasm, and that to be digested, foods do not need to be in contact with living matter. It requires but a very minute quantity of enzyme to digest a large amount of the particular food upon which it acts; for example, a preparation of an enzyme extracted from the pancreas of an animal was found to digest 2,000,000 times its weight of starch. The amount of diastase in a mesophyll cell necessary to transform the starch in that particular cell to sugar is so small that it cannot be measured. Its presence is inferred from the observed changes in the starch.

In most forms of digestion water chemically is added to the original substances. This seems to weaken the bonds between the different parts of the molecules and to bring about a splitting into simpler compounds. For example:

$$\begin{array}{cccccc} 2(C_6H_{10}O_5)_n & + & n(H_2O) & \longrightarrow & n(C_{12}H_{22}O_{11}) & \\ \text{starch} & + & \text{water} & \longrightarrow & \text{maltose} & \end{array}$$

and

$$\begin{array}{cccccccc} C_{12}H_{22}O_{11} & + & H_2O & \longrightarrow & C_6H_{12}O_6 & + & C_6H_{12}O_6 \\ \text{maltose} & + & \text{water} & \longrightarrow & \text{glucose} & + & \text{glucose} \end{array}$$

Accumulation of food. A healthy plant usually manufactures more food than it uses immediately, and this food may accumulate in various parts of plants. In the potato, surplus food passes to underground stems, the tubers, where it accumulates. Turnips and beets are examples of plants that accumulate excess food in their roots. In the maple, the food accumulates in the branches, trunk, and roots. In cabbage, food is stored in the cluster of leaves at the top of the stem. In corn and cereals, most of the food finally accumulates in the grain. In the century plant, a considerable part of the excess food is stored in the thick, fleshy leaves; the process of accumulation may go on from 20 to 30 years, and the total quantity of food stored may amount to many pounds. In nature such accumulated foods may be used during the next season's growth of the plant or in starting the growth of the offspring.

When soluble substances pass into and accumulate in the cells of the stem, they are largely transformed again into an insoluble form. This makes possible the continued entrance of the soluble material. For example, starch formed in potato leaves is transferred through the plant to the underground tubers in the form of glucose and maltose, and there it accumulates in the cells in the form of starch. It is believed that the same enzyme which changes the starch to maltose, under suitable conditions changes the maltose back again to starch, and that in general the enzymes that digest foods are the agents that under slightly different conditions build them up again into the more complex insoluble forms. Enzyme activities may be reversed by changes of temperature, acidity, and water content of the cell.

Kinds of food accumulated. In any given plant in which food is accumulated, protein, carbohydrate, and fat are all present. Depending on the plant, however, the amount of any one of these may be very great or it may be so small as to be practically negligible. In the sugar cane and sugar beet the excess food occurs mainly in the form of cane sugar (sucrose). In the potato it is almost wholly starch. The grains of wheat, oats, and rice contain mostly starch, but also some protein. In sweet corn there are both sugar [1] and starch; in field corn the excess food is mostly starch. In both sweet and field corn there are considerable quantities of protein and oil. In the soy bean and peanut there are large quantities of both protein and oil. In the seeds of the coconut, flax, and cotton there is a large proportion of oil.

Summary. The upward movement of water from roots to leaves through the water-conducting tissue is due mainly to pull of transpiration. Water moves from the water-conducting tissues to other tissues of the stem by imbibition and osmosis.

The movement of foods takes place mainly through the sieve tubes and companion cells. Diffusion is known to be important in this process, but it is inadequate to account for the rapid transfer which occurs in many plants.

Before insoluble foods move out of or into a cell, they must be digested. This is done by enzymes, and there are probably as many kinds of enzymes as there are classes of food substances. The products of digestion are soluble substances.

Both soluble and insoluble foods accumulate in stems and other organs of plants. Insoluble foods are built up out of the soluble foods that enter the cell by the same enzymes concerned in their digestion. The reversal of the process is accounted for by certain changes of conditions in the cell.

PROBLEMS

1. Why can a hollow tree continue to live for many years?
2. Why do sprouts not develop from stumps of trees that were girdled a year before being cut down?

[1] Unless sweet corn is cooked almost immediately after its removal from the plant, it rapidly loses its sweetness. This is because the enzymes in the grains constantly convert the sugar into starch. Peas and some other vegetables lose their sweetness after being gathered, for the same reason. The enzymes work more slowly at a low temperature, and the vegetables will lose their sweetness less rapidly if kept in a refrigerator.

CHAPTER TWENTY

ECOLOGICAL TYPES OF STEMS

ATTENTION has already been called to the variety of stems, and the advantages of upright, horizontal, climbing, and underground stems. In a sense these are ecological types of stems, since each of them bears a slightly different relation to the environment. All the kinds of stems discussed, however, occur among mesophytes, and the descriptions of stem structures that have been given were also based on the stems of plants that live under medium moisture conditions. Since the great bulk of plants live as mesophytes, we may look upon these structural arrangements as typical of the plants living under the most favorable circumstances.

In this chapter the peculiar features of the stems of water plants (hydrophytes) and of desert plants (xerophytes) are described. Only a comparatively small number of flowering plants are included in these groups.

How drought modifies mesophytes. When mesophytes are grown under very dry conditions, the stems are reduced in size, the stem tissues are more compact, and the cell walls are heavier. Some plants also develop thorns, spines, and hairy coverings under these conditions. Plants that have a tendency toward succulence become thicker and more succulent when subjected to drought. The water-holding mucilages which they contain are increased, and their water-holding capacity is enlarged. The leaf area of mesophytes is greatly reduced by drought; consequently, photosynthesis and transpiration are also reduced. Under the same conditions the mechanical tissues of woody plants are increased, their stems become more rigid, and the bark becomes thicker and more impermeable to water. If we keep all these facts in mind in studying the stems of xerophytes, we may be able to understand better the causes of the peculiar features of this ecological group of plants.

FIG. 96. A group of xerophytes, including species of *Cereus*, *Opuntia*, *Yucca*, *Aloe*, *Euphorbia*, and *Agave*. Such plants are characterized by compact form, little or no leaf area, thick cuticle, and water absorbing and retaining substances in the cells.

Xerophytes. The xerophytes are the characteristic plants of deserts and dry plains, but they are by no means confined to these regions. They occupy sand dunes and sand plains along the Atlantic coast and on the shores of the Great Lakes. They may even be found locally on rock cliffs and on dry, exposed hilltops. In fact, they may occur in any situation in which a reduced water supply in the soil is accompanied by atmospheric conditions that promote rapid transpiration, or in which the plants are periodically, or continuously, subjected to drought. The stems of plants that thrive in these habitats are reduced both in size and in the amount of branching. The leaf area is reduced, or temporary, or leaves may be entirely wanting.

The cactus type. The cactuses represent the extreme type of drought-resistant, succulent plants. Leaves are wanting except during the early stages of growth, and then they occur only as small scales at the nodes. The stems are columnar, often ridged and fluted, and always thick and fleshy. The photosynthetic work in cactuses is done by the chlorenchyma of the cortex. As the green surface is small compared with the green surface in mesophytic plants, food manufacture is slower and growth is correspondingly less. Yet some of the cactuses of Mexico attain heights of 50 feet (Fig. 97). The cactus form points clearly to one of the most characteristic features of desert plants; namely, water storage. A single large plant may contain from 1 to 25 tons of water. As the plant loses moisture so slowly, it may continue to live for several years without an additional supply of water. At every node there are usually clusters of spines and spicules. A heavy cuticle and one or several layers of epidermal cells prevent rapid transpiration.

The shrub type. The mesquite, greasewood, and sagebrush represent a second type of extreme xerophyte with much-branched, hard woody stems. These plants are characteristic of semi-deserts and of those parts of deserts in which soil moisture is more constant. As a rule these plants are deeper rooted than

C. J. Chamberlain

FIG. 97. Giant cereus of south-central Mexico in bloom. This specimen is 35 feet in height; occasional plants attain a height of 50 feet. In the pulpy interior of a cactus plant of this size 15 to 25 tons of water may accumulate.

those of the cactus type. The stems are covered with cork, heavy cuticle, and sometimes wax and hairs. In the "palo verde" leaves are absent, and the cortex contains chlorophyll. Many of these shrubs have spines and thorns.

Short-stemmed type. These plants are often called stemless because the leaves occur seemingly at the top of the root. In reality the stem is a disk, or a rounded cone of nodes topping a fleshy root. Internodes fail to develop, and the result is a rosette of leaves, either flattened against the ground as in the evening primrose, or raised and forming a hemispherical group of radiating fleshy or bayonet-like leaves, as in the yucca and agave. Some yuccas develop very rapidly, and in spite of the short internodes the stem rises a few feet above the ground. In one species the stem branches and forms a tree with rosettes at the tips of the branches (Fig. 70).

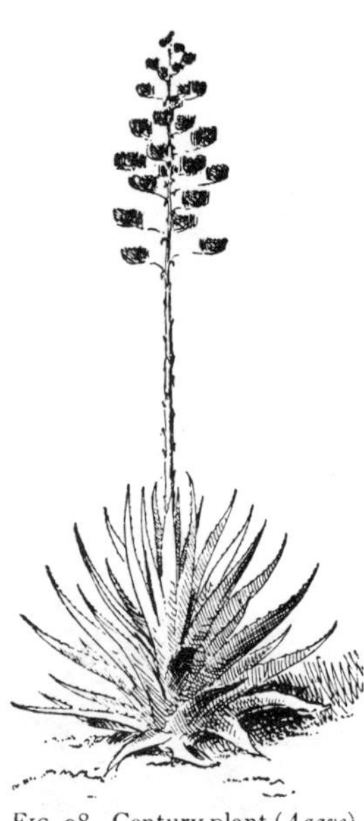

FIG. 98. Century plant (*Agave*), showing the rosette of fleshy leaves and the flowering stalk. It is a perennial, but, like an annual or a biennial, it dies when it flowers and fruits. The century plant is an example of the short-stemmed type of xerophyte.

Annual-stemmed type. These include many herbaceous perennials with underground rootstocks, tubers, corms, bulbs, and fleshy roots. The aërial stems are developed during the moist periods, and after supporting leaves and flowers die down to the ground. To this group belong the conspicuous grasses and perennial flowering herbs. Many of the aërial stems show the characteristic structures which reduce transpiration.

Summary of xerophytes. Stems of xerophytes, then, are either (1) succulent with a high water-holding capacity, or (2) hard, water-proofed, and woody, or (3) short underground supports for evergreen or temporary rosettes of leaves, or (4) temporary stems arising from underground structures. The stems of the leafless forms possess chlorenchyma. Many of these stems have coverings of hairs, spines, and thorns.

A word of caution concerning the origin of xerophytes may not be out of place in this summary. It is not to be assumed that xerophytes may be formed from mesophytes by the direct action of the environment, though some plants are temporarily modified in this way when grown under conditions of drought. The characteristics of the more pronounced xerophytes are hereditary qualities which may have arisen as variations entirely aside from the influence of a desert environment. Likewise, the peculiar features of the hydrophytes, about to be described, must not be considered as *necessarily* due to the direct influence of the environmental factors to which they are exposed.

Effect of submergence on mesophytes. When mesophytes are grown under very moist conditions, one of the first changes observed is the great increase in air spaces among the cells. Furthermore, when stems of mesophytes are submerged they may develop large air cavities by the breaking down and separation of certain groups of cells in the cortex. While the stems consequently increase in diameter, there is a decrease in the amount of wood tissue. Certain terrestrial plants with heavy cuticle and hairy coats are smooth and without cuticle when grown under water (Fig. 5).

Stems of hydrophytes. The most distinctive feature of submerged stems of hydrophytes is the presence of large air chambers extending throughout their length. When the stems are broken open, the tissues are seen to occupy much less space than the air cavities. We may properly speak of " intercellular spaces " in mesophytic stems; but in describing hydrophytes, the term " air cavities " is more appropriate. These air cavities buoy up the plant and provide an internal atmosphere in which gas exchanges between the leaves and roots take place. Frequently the cells are so distinctive in form and arrangement that we have as a result a special tissue which is called *aërenchyma*. Living cells of plants are slightly heavier than water, and the ability of any plant to float is due to the air contained in the intercellular spaces.

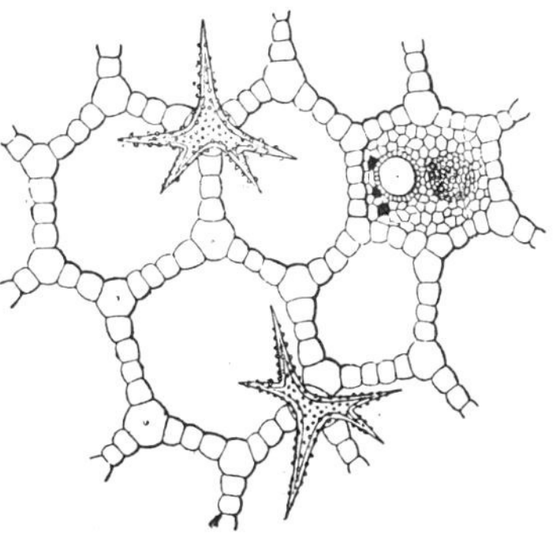

FIG. 99. Air cavities and giant cells in petiole of yellow water lily (*Nymphæa*). The most characteristic feature of the stems of hydrophytes is the presence of large air spaces in the tissues. (*After Frank.*)

Floating type. Many hydrophytes are free-floating. Among

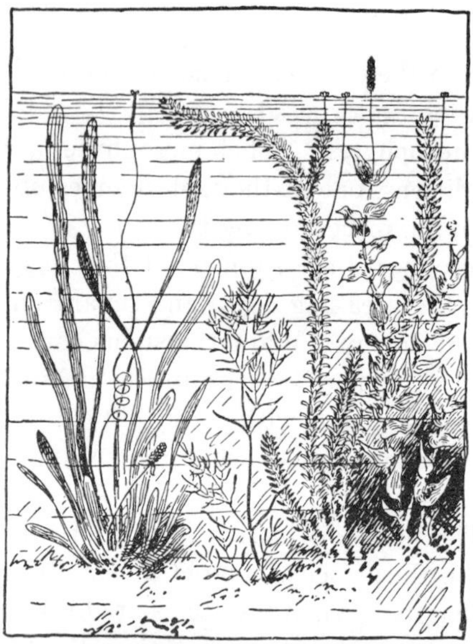

FIG. 100. Submerged-rooted plants. From left to right: eelgrass (*Vallisneria*), naiad (*Najas*), water weed (*Elodea*), and pondweed (*Potamogeton*). In such plants the mechanical tissue and the conductive system are poorly developed.

these free-floating forms are the duckweeds, the Salvinia (a fern), and the water hyacinth. These plants float because of the internal air cavities. In the duckweeds stem and leaf are not differentiated — the plant consisting of a flattened globular green mass of cells with a pendant root. In other floating plants the stem extends the plant by branching and by forming new plants at intervals. The duckweed plants that form in cold water in late autumn are constructed with more compact tissues than those formed earlier in the season. These later plants, being heavier than water, drop to the bottom of the pond and remain there during the winter. Only the late fall plants survive; the earlier ones that are lighter remain afloat and are frozen.

Submerged-rooted type. Other hydrophytes, like the pondweeds and water lilies, are rooted in the soil, and their stems bear submerged or floating leaves. The stems have little or no mechanical tissue. As compared with land plants, the conductive system is much reduced. Most hydrophytes develop horizontal underground rootstocks and tubers. For this reason the plants commonly grow in masses. Contrary to the usual opinion, even wholly submerged seed plants obtain their water and mineral salts from the soil, and not from the water surrounding the upper part of the plants. Thus plants growing on sandy bottoms make poor growth when compared with the same plants growing on rich humus-covered bottoms in the same lake. In submerged plants there is a definite movement of water through the water-conducting tissue just as in land plants. At the upper ends of the leaves the water is given off through water pores.

Emersed hydrophytes. A third group of hydrophytes, like the cat-tails, rushes, bulrushes, and sedges, may have their roots, stem bases, and underground stems under water, while the upper parts are exposed to the air. These plants have both the conductive and mechanical tissues well developed and are therefore able to grow erect without being supported by the water. The stems and leaves are exposed to the action of wind and wave and to the conditions that bring about normal transpiration.

The importance of the aërial portions of these plants and the

FIG. 101. Water lilies. In the foreground, species of *Castalia;* near the middle, the giant water lily of the Amazon (*Victoria regia*). Stomata are found on the upper surfaces of the leaves, and, as in land plants, there is a definite upward movement of water through the plant.

internal air cavities is shown by the fact that the plants may be exterminated by cutting off all the rootstocks at the water's edge and cutting the erect stems below the water level. Even water plants may be drowned.

Summary of hydrophytes. Among hydrophytes, then, there are plants with upright stems, some with floating or buoyant stems, and others with underground stems. Hydrophytes resemble the mesophytes in most respects but differ in having aërenchyma, or air cavities, as well as in having a somewhat reduced water vascular system and less mechanical tissue. Hydrophytes spread rapidly by growth and branching of horizontal stems. Many of them survive the winter, through the development during the lower-temperature period of autumn of short, heavier-than-water shoots and buds.

CHAPTER TWENTY-ONE

THE FORMS AND STRUCTURES OF ROOTS

In preceding chapters we have learned that leaves manufacture food in the presence of light; that their exposure to air facilitates the entrance and exit of carbon dioxide and oxygen; and that their efficiency is increased by being raised and displayed on erect stems.

Tall, erect stems with sufficient strength for the maximum display of leaves are made possible by the development of mechanical tissue in the stems. The display of leaves high above the water supply of the soil requires a conductive system capable of raising water and mineral salts from the roots and of permitting the movement of food away from the leaves.

Stems, in turn, must be firmly anchored, and they must be supplied with water and mineral substances from the soil. Anchorage and absorption are the particular functions of roots, though they carry on other processes also, such as conduction of water, transfer of food materials, accumulation of food, respiration, assimilation, and growth.

FIG. 102. Stages in the development of a corn seedling. *P* is the primary root, *S* a secondary root, and *A* an adventitious root from the base of the stem.

Primary and secondary roots. The root system of a well-developed bean seedling will show the essential features of roots. The *primary root* extends downward from the base of the stem. On its sides are numerous *secondary roots* which extend at right angles, or grow obliquely downward (Fig. 102). Unlike stems, roots possess no definite nodes from which branches arise. A secondary root may originate at any point on the primary root. Some common plants, like parsnip and carrot, have primary roots that thicken above and taper gradually, extending deeply in the soil. These are called *tap roots* (Fig. 104).

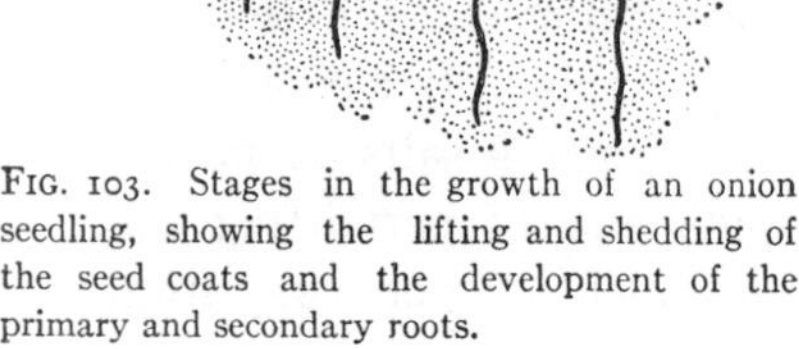

FIG. 103. Stages in the growth of an onion seedling, showing the lifting and shedding of the seed coats and the development of the primary and secondary roots.

Adventitious roots. In many seedlings there are also roots that develop from the first node of the stem. All roots arising from stems and leaves are called *adventitious roots*. The "prop roots" that develop from the lower nodes of corn stems and the roots that grow from "cuttings" are familiar examples. The adventitious roots of corn, sorghum, wheat, and many other grasses are far more important than the primary and secondary roots. In many instances only the adventitious roots remain at maturity (Fig. 108).

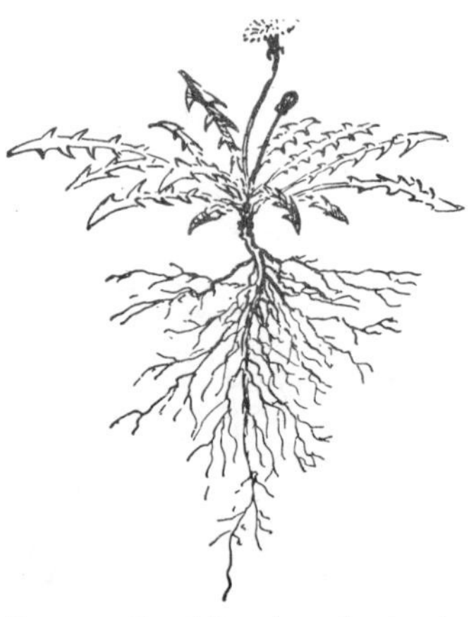

FIG. 104. Dandelion plant, showing the primary tap root and its branches, the secondary roots.

Adventitious roots develop also from the stems of many plants like

FIG. 105. Prop roots of a rubber tree in Florida. As the trees age, these roots thicken and become secondary trunks.

Bureau of Science, P. I.

FIG. 106. A mangrove swamp along the seashore of an island in the Philippines. The prop roots support the plants in the soft mud.

the poison ivy and trumpet creeper (Fig. 111) and act as *holdfasts* in supporting these climbers on trees and walls. Adventitious roots may arise also at any point on a primary or secondary root, following injuries. For example, when we plant pieces of horse-radish or dandelion roots, adventitious roots develop. During the dry season in deserts the younger parts of the root system of some plants like the cactuses are dried out and killed, and when the next wet season comes, many adventitious roots develop from the parts of the root system still alive. In desert plants these new adventitious roots do the absorbing work during the moist period.

In plants that propagate by " runners " and by underground stems, all the roots that have developed from the horizontal branches are adventitious. All plants that are commonly started by bulbs, like the tulip, hyacinth, crocus, onion, and lily, have only adventitious roots. Probably all the plants of the potato, sweet potato, yam, sugar cane, banana, dahlia, and peony that you have ever seen had only adventitious roots, because seedlings of these plants are either unknown or are rarely grown except by plant breeders.

Deep and shallow root systems. The root systems of plants are distributed in the soil in a variety of ways. In a soil that is deep, readily penetrated, and sufficiently moist to permit growth, some plants have roots widely distributed just beneath the surface, while others have roots penetrating to great depths. There are also plants that combine these two habits; that is, their roots are spread in a surface group and a deep group (Fig. 107).

Each of these habits has its advantages. The shallow roots are in contact with water after every rain, and in dry regions where the rainfall of summer showers penetrates but a few inches these plants may be better supplied than others. Deep roots have a distinct advantage in anchoring the stem firmly. Their absorbing surfaces are in contact with the ground water that accumulates from all the rains of the region. This supply is usually a more permanent one in moist regions, but in dry seasons it may fail for longer periods than the supply near the

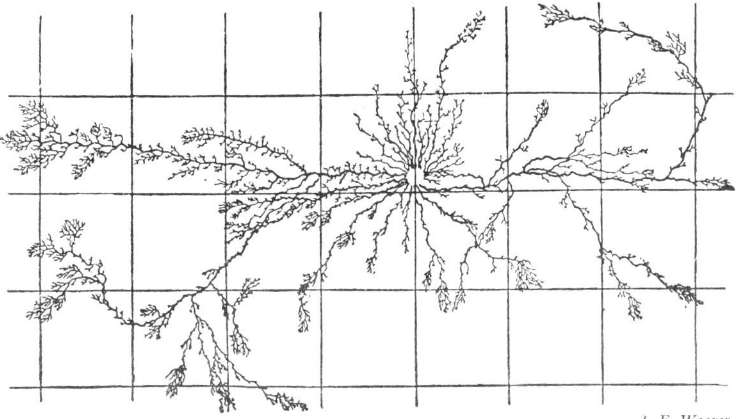

FIG. 107. The principal absorbing root system of a prickly-pear cactus, *Opuntia camanchica*, as seen from above. Beneath the stem is a group of vertical anchorage roots not shown in the figure. The diagram is divided into one-foot squares.

surface. Plants whose root systems are a combination of the two obviously have the best possible arrangement with reference to soil water.

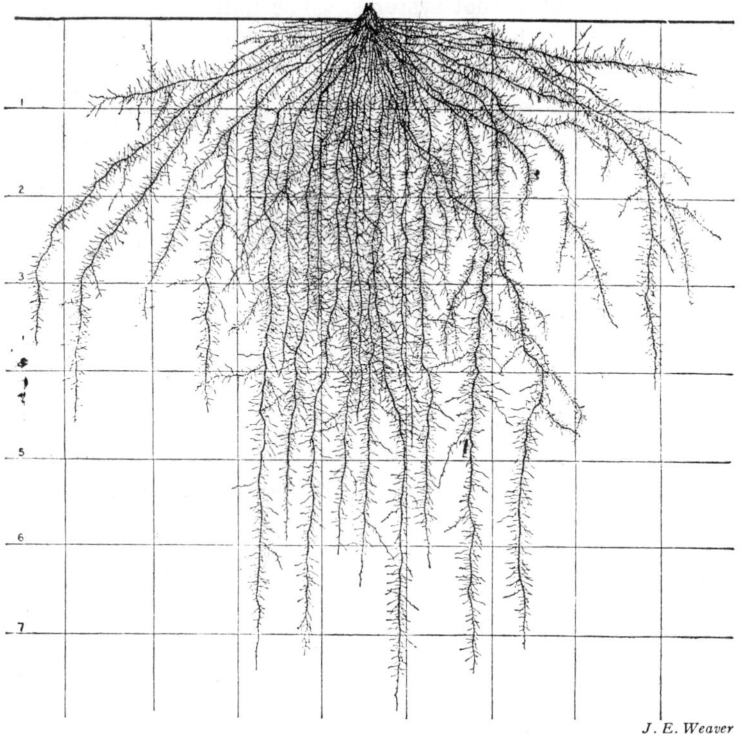

FIG. 108. Vertical section showing the extent of the root system of mature corn grown at Peru, Nebraska. The section is divided into one-foot squares.

Corn, cabbage, mock orange, blue-grass, and buffalo grass are examples of shallow-rooted plants. Cottonwood, oak, hickory, alfalfa, sugar beet, and California poppy have long tap roots which may penetrate to the water table. Many of our tall grasses, particularly the " bunch grasses," have great masses of long, fibrous roots — the underground part of the plant being far greater than that above ground. However, the form of the root system of any plant may be greatly modified by soil conditions.

Structure of roots. In the root, the xylem, made up of water-conducting and wood tissues, forms the central core, or axis. In seedlings and in a few exceptional plants the root may possess a central pith like the stem. The water-conducting tissue of the root, like that of the stem, is composed of tracheæ, and tracheids which die at maturity. Surrounding the xylem is the phloem, which is composed of the food-conducting and bast tissue. In perennial roots a cambium tissue lies between the central axis and the food-conducting tissue. Outside the phloem there are usually many layers of parenchyma cells, making up the cortex. The innermost layer of the cortex is often composed of smaller or thick-walled cells and is called the *endodermis*. The outermost layer or layers may likewise be modified in various ways, forming cork or collenchyma. All young roots are bounded outwardly by an epidermis. In roots like the radish that thicken rapidly this is soon broken. The strips of epidermis remain attached to the growing root for some time after it is broken and may be readily seen on the radish.

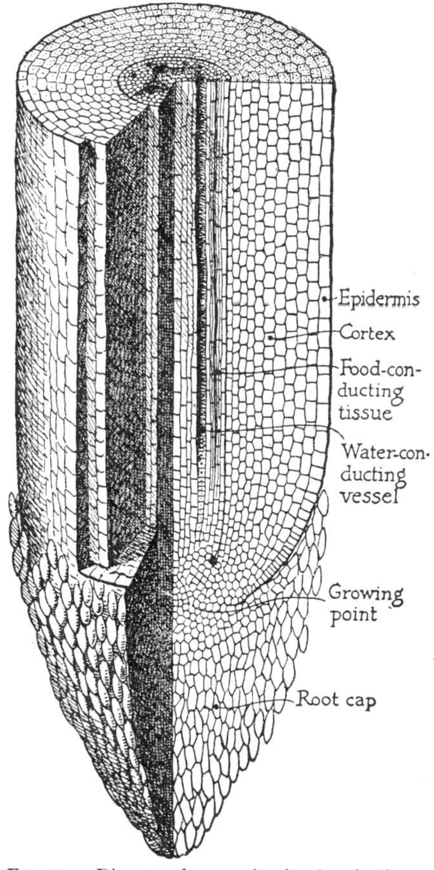

FIG. 109. Diagram of a root tip, showing the tissues and their arrangement.

In fleshy roots like the beet, which form in a single growing season, the thickening takes place through the continued formation of new cells by the cambium. As the cambium produces conductive tissue, alternating rhythmically with parenchyma tissue, the mature root appears to be composed of concentric layers.

Perennial roots. The perennial roots of shrubs and trees increase in length from year to year, and the older roots increase in thickness by the formation of annual rings. These roots soon lose their epidermis, and later the cortex also dies, and a dead bark similar to that of tree trunks is formed. In some instances new cambiums arise in the cortex, which produce a layer of cork enveloping the roots.

Growth of roots. Roots develop from growing points near the tip. The growing point, however, does not lie at the surface, as in stems, but some distance below, being covered by a *root cap* (Fig. 109). As it moves, or forces its way between the particles of the soil, it is in this way protected from abrasion. The cells of the growing point are alike, but at a very short distance back from the growing point they become differentiated into the vascular axis, with elongated cells, cortex, and epidermis. The growing region of soil roots is usually very short. *Lateral roots* arise from growing points that develop on the vascular axis and push outward through the cortex, breaking it as they elongate.

Root hairs. The young roots of land plants generally bear root hairs. These are delicate elongations of the epidermal cells of the root. They are especially concerned with absorption of water and mineral salts, and their presence increases the absorbing surface of the root from two to one hundred times. Since the rate of absorption depends in part upon the surface area in contact with the soil water, the advantage of root hairs is evident (Fig. 110).

Root hairs are usually short-lived structures, their duration being best measured in days. Their walls are not only thin, but they are composed of pectic material which causes them to adhere to the soil particles and brings them in intimate contact with the water films on the soil particles. They begin to develop at a short distance from the tip of the root. Farther back they have attained full length, and beyond this they are in a dying or dead condition. Thus from day to day the zone of root hairs moves forward with the growth in length of the root, by the continual production of new root hairs just above the elongating region of the root. This brings new root hairs continually into contact with new supplies of water and minerals in the soil. As a plant enlarges, its root system becomes more complete through repeated branching and the elongation of the branches. Most of the absorption occurs in the root-hair zone, and this is continually moved farther and farther from the base of the stem. In large trees this zone may be many feet from the base of the trunk. The fact that the outside of root hairs is composed of pectic substances is of great interest, because these substances have so strong an attraction for water. It may be of great importance, on this account, in absorbing the last particles of available water.

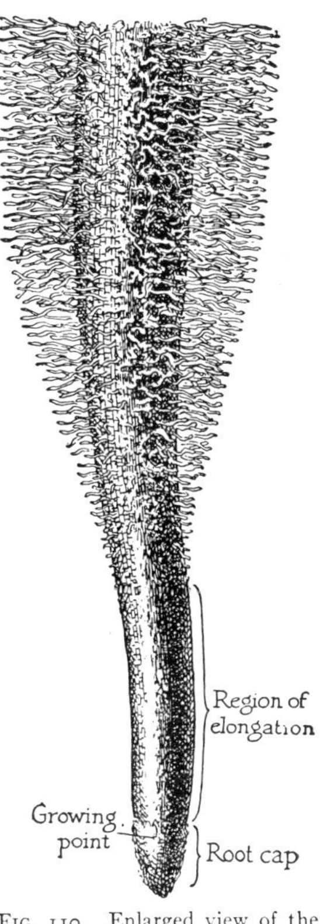

FIG. 110. Enlarged view of the end of a root, showing root cap, growing region, and root hairs.

Root contraction. As roots mature, they may contract in length and so draw the base of the stem a slight distance into the soil. In this way crevice plants on cliffs are continually held firmly in place, in spite of the wearing away of the cliff face by erosion. In the same way the crowns of clover and plantain roots that have been lifted up by frosts may be drawn into the soil, and small bulbs and tubers, many of which are formed at higher levels than the parent bulbs, may be pulled deeper into the soil by root contraction.

Root duration. Roots are annual, biennial, or perennial. Perennial plants may have either annual or perennial roots, just as they may have either annual or perennial aërial stems. Plants with bulbs, tubers, or corms grow a new set of roots each year. Plants with rootstocks, like the May apple and Solomon's seal, generally have roots that last for several years. Shrubs and trees also have perennial roots. We must be sure to understand, however, that even in perennial roots the work of absorption is for the most part done by the youngest portions of the new roots which are added each year. Most biennials, like the common evening primrose and wild carrot, have fleshy roots in which food accumulated during the first year. This food is used in the rapid development of the plant during the second season.

Roots of hydrophytes. Most of the root characteristics thus far described are those of the roots of mesophytes. In hydrophytes the roots are notably smaller and less branched than in mesophytes. They absorb water and mineral substances from the soil, even when the plants are totally submerged. The roots of hydrophytes, like the leaves and stems, are remarkable for the presence of internal air cavities.

When the roots of land plants (mesophytes) extend into well-aërated water, they develop innumerable branches, differing in this respect very markedly from the roots of hydrophytes. On account of this fact, roots of trees, especially those of willow and cottonwood, that enter drainpipes and tiles often develop masses of fine branches that obstruct the flow of the water even when the entering root is not thicker than the lead in a pencil. The banks of streams are often protected from erosion by the mat of roots developed along the water's edge. This is why willows are planted on levees. When roots of mesophytes grow in water, they also develop air cavities in the cortex.

Holdfast roots. Climbing plants, like the Virginia creeper, poison ivy, Boston ivy, and trumpet creeper, develop holdfast roots which help to support the vines on trees, walls, and rocks. By forcing their way into minute pores and crevices, they hold the plant firmly in place. Usually the roots die at the end of the first season, but in the trumpet creeper they are perennial. In the tropics some of the large climbing plants have holdfast roots by which they attach themselves, and long, cord-like roots that extend downward through the air until they strike the soil and become absorbent roots.

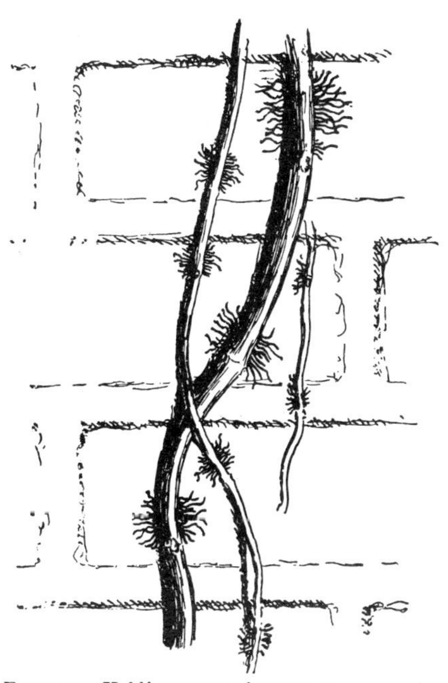

FIG. 111. Holdfast roots of trumpet creeper, developed from the nodes. These roots are perennial and may lengthen and branch for several years.

Epiphytes. A plant that lives perched on another plant is an *epiphyte* (Greek: *epi*, upon, and *phyton*, plant). Mosses and lichens are the most common epiphytes in temperate regions, but in the rainy tropics and along our own Southern coast many flowering plants live attached to the branches of trees. They usually have leathery leaves and a low transpiration rate. Many have water-storage tissue in fleshy stems or in thickened leaves. Others are called *tank epiphytes* because they catch water in the axils of the leaves or in pitcher-like leaves. Ephiphytes cling to the supporting tree by means of roots that act both as holdfasts and water-absorbing organs. They do not take their nourishment from the plants on which they grow. They carry on food syntheses as other green plants do, but depend for their water upon the evenly distributed rainfall and for their mineral substances upon dust and the decay of the bark on which they live.

Epiphytes are pronounced xerophytes, for there is probably no habitat in which it is more difficult to maintain a water balance than the one in which they live. It is not surprising, therefore, to find that among the epiphytic plants of the West Indies there are several species of cactus. Among epiphytes there are many species of ferns, and many species belong to two families of flowering plants, the bromelias and orchids. The bromelias are related to the pineapple and have leaves of the same form (Fig. 162). The orchids have flowers remarkable for their shapes and colors, and have the distinction of being the highest-priced of all flowering plants. The Spanish moss of Florida, a flowering plant, is perhaps the best known of American epiphytes. It is an extreme form and is devoid of roots. Spanish moss and some other members of the bromelia family have peculiar scale hairs through which water is absorbed directly by the leaves and stems. The roots of many epiphytes contain chlorophyll and assist in the manufacture of food.

Roots in relation to bacteria. The roots of many plants have bacteria or fungi growing about them or inside them. The best-known crop plants belonging to this group are the clover, cowpea, and alfalfa; their roots develop small nodules in which certain kinds of bacteria change nitrogen of the air into nitrogen compounds which may be used by the plants. More information about these bacteria will be found in Chapter XXXVIII, on "Bacteria and the Nitrogen Cycle" (page 396).

Mycorhiza. A root which has a fungus regularly associated with it is called a *mycorhiza* (Greek: *myco*, fungus, and *rhiza*, root). Many of our trees, like the oaks, maples, poplars, and conifers, have fungi surrounding their roots. The beech tree, for example,

Bureau of Science, P. I.

FIG. 112. Epiphytes on the branches of trees in wet mountain forests of the tropics. The plants in the picture are mostly ferns and orchids. (*Photo by H. N. Whitford.*)

flourishes only when it grows under such conditions. The difficulty in transplanting azaleas, laurels, and rhododendrons from the woods to our lawns lies largely in supplying conditions favorable to the fungi that infest the roots. It is easy to supply the proper shade and water conditions for the shrubs, but it is difficult to furnish soil conditions favorable to the life of the fungi. The transplanting of these shrubs is therefore most frequently successful when they are planted in large bodies of soil brought with them from their natural habitat. Such soils may be kept in their natural acid condition by the use of tan-bark extract and alum, but the addition of lime is harmful. In orchids and some ferns the fungi live inside the cortical cells of the roots. Just how the fungi aid the plant is not fully understood; that they are essential is very clear. A few of these fungi are known to furnish nitrogen to the roots, and they may also aid in the absorption of water and minerals from the soil.

PROBLEMS

1. Make a classification of roots and cite examples, on the basis of: (1) their origin, (2) their form, (3) their environment, and (4) their function.
2. Why cannot a dead root system absorb as much water as it did when alive?

CHAPTER TWENTY-TWO

THE PROCESSES OF ROOTS

THE absorption of water and mineral salts is the process most generally associated with roots. The three physical processes involved in absorption were defined in Chapter XI, and are here briefly summarized.

Diffusion is the movement of molecules and atoms from places of greater concentration to places of less concentration. When the diffusion of water into a substance, or body, results in swelling, the process is called *imbibiiion.* The diffusion of water through a differentially permeable membrane that separates a mass of water, or a dilute solution, from another is called *osmosis.* In osmosis water moves from the place of *its* greater concentration through the membrane to the place of *its* less concentration. If the solution is inclosed by walls, the movement of water into the solution produces a pressure known as *osmotic pressure.*

Absorbing mechanism of roots. The epidermis is the primary absorbing tissue of the root. We have seen that it consists of delicate walled cells, some of them prolonged outwardly as root hairs. The wall is composed inwardly of cellulose and outwardly of pectic material, the latter having a powerful imbibing capacity for water. The wall is permeable to both water and dissolved salts.

The cytoplasm of the epidermal cell is the differentially permeable membrane which separates the cell sap from the water in the soil. It prevents the outward diffusion of sugar and other soluble organic substances, and permits the inward diffusion of water and mineral salts. The epidermal cell, then, has a wall with a great capacity for imbibition of water; it forms an efficient osmotic cell for the taking up of water; and it affords a permeable medium for the inward diffusion of salts. Because of the presence of sugars and other substances made by the plant, the concentration of soluble substances in the cell sap is necessarily greater than in the soil solution in which a plant grows.

The cortical cells are essentially like those of the epidermis, and the concentration of their cell sap is progressively higher

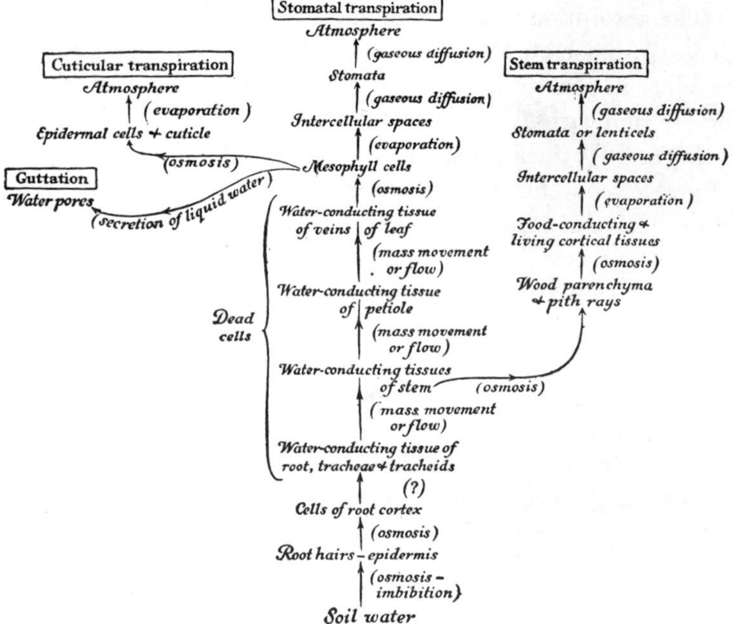

FIG. 113. Diagram of the path of water through a plant. The most important process involved in each step is indicated.

from the epidermis inward to the water-conducting tissue. Consequently water may pass by osmosis from cell to cell across the root to the tracheæ. Mineral salts may diffuse across the cortex into the tracheæ.

At this point the fourth factor in absorption enters. This is the tension, or pull, on the water in the tracheæ of the root caused by the evaporation of water from the above-ground shoot.[1]

[1] Large trees have been kept alive for days by placing the cut-off trunks in water. This shows that sufficient water to maintain the water balance of the plant for at least several days may be lifted in a plant by the pull of transpiration without the aid of roots. It is of practical interest to know that cut flowers will last much longer if the ends of the stems are bent over into a vessel and cut under the water.

The soil as a water-delivering mechanism. Soil is composed of various-sized solid particles massed together with small spaces between them. A root in its development pushes in among the particles and pushes some aside. The very small root hairs grow outward between the soil particles, and press against them on all sides. Due to rains, water enters the soil and spreads rapidly downward between the soil particles. Some of the minerals among the soil particles dissolve, and the result is a dilute solution, usually called "soil water." The water sinks, partly because of its weight; that is, it is pulled down by gravity. It is also pulled in all directions by capillarity, just as water is pulled into small tubes, or as it is pulled into blotting paper.

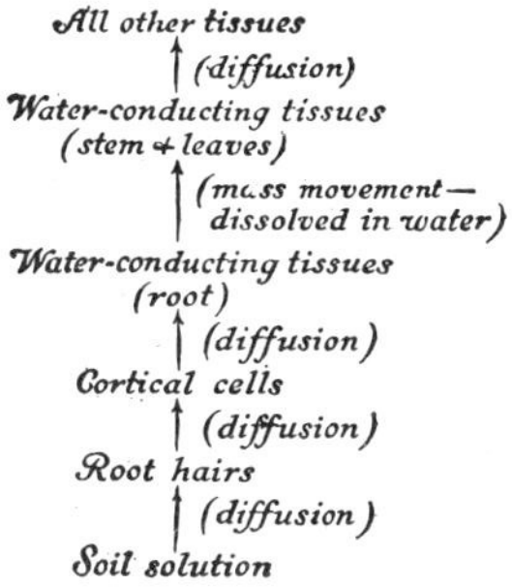

All other tissues
↑ *(diffusion)*
Water-conducting tissues (stem + leaves)
↑ *(mass movement— dissolved in water)*
Water-conducting tissues (root)
↑ *(diffusion)*
Cortical cells
↑ *(diffusion)*
Root hairs
↑ *(diffusion)*
Soil solution

FIG. 114. Diagram to show the processes involved in the movement of mineral salts into the tissues of a plant.

If a flower pot having a perforated bottom is filled with soil and placed in a pail of water, the air in the spaces between the soil particles will be gradually driven out and water will take its place. The soil is then saturated with water. When the pot is lifted out of the pail, a part of the water drains out and it will continue to drip for some hours. This is the water that is pulled down by gravity. The water that remains is held by capillarity and by the attraction of the soil particles. As water percolates out of the mass of soil, air is again drawn into a part of the soil spaces, but films of water surround every soil particle.

If a small plant is growing in the soil, its roots and root hairs

If cut in the air, air bubbles get into the water-conducting tubes and prevent the subsequent movement of water into them. Air bubbles already in stems that have been cut in the air may sometimes be removed by placing the lower ends of the stems in warm water and cutting off an inch or two. After standing in water for a day, the tracheæ may become clogged with bacteria and the rise of water prevented. Florists avoid this possibility by cutting off an inch or two from the stems of cut flowers each day.

are in contact with only a small part of the soil solution. Let us suppose that transpiration is active and that all the water in immediate contact with the roots passes into the plants. When this occurs, water moves from adjoining spaces to replace it and all the spaces again have about the same relative amount of water in them. This movement of water to the spaces near the roots continues in the soil as long as the capillary columns of water among the particles near the plant do not break. If this happens, the movement of water in the soil toward the root is stopped and the soil no longer delivers water to the root. If transpiration continues, the plant wilts.

The freedom with which water moves by capillarity varies greatly in different soils. Since it is the lateral and upward movement of water that determines the continued supply to the root, plants wilt in some soil sooner than in others. Wilting is determined by the availability of water, and this in turn depends upon the conditions in the soil that maintain continuous water movement toward the root. Consequently all kinds of plants show wilting when the water content of a given soil is reduced to a certain point.

Summary of absorption. The roots and root hairs form a mechanism into which water and mineral salts move readily by diffusion, or by those forms of diffusion known as *imbibition* and *osmosis*. Outside the root is another mechanism, the soil, in which water moves freely or with difficulty, according to its texture. If the water moves freely by capillarity, a large part of the water in a given body of soil will pass to the water-absorbing root, even though the root is in contact with only a small part of the soil water. If the soil spaces are very minute, the water is held more tightly by the soil particles, movement is impeded, the continuous capillary columns of water break, and the water ceases to move toward the root. Under these conditions absorption is stopped and the plant wilts, although there may still be a considerable amount of water in the body of soil as a whole.

Plants wilted in the daytime sometimes recover at night, because of the reëstablishment of the capillary water columns in the soil surrounding the root.

Before we leave the subject of absorption, attention must be called to the fact that the permeability of the epidermal cells determines what dissolved salts in the soil water will pass into the plant. These cells are more permeable to some salts than to others. Hence, some salts diffuse in more rapidly than others. But the root does not in any sense select the salts it *needs* and retard the salts it does not need. Neither does it prevent the entrance of poisonous substances. Salts of zinc, lead, copper, and arsenic readily pass into and accumulate in plants growing in the vicinity of smelters, and ultimately kill them.

Root pressure and sap pressure. If a number of well-watered plants are cut off just above the soil, some of them will exude water for a day or two. Experiments have shown that the sap may be forced out with pressure sufficient to raise water 30 to 40 feet. In most plants, however, a rise of only a few inches is obtained. This pressure is called *root pressure*. When such pressures exist in plants, they probably aid in the lifting of water in stems. Extensive experiments have shown, however, that root pressure is intermittent. It may exist at one time and not at another, and when transpiration is most active and the largest volumes of water are being raised in a plant, root pressure is wanting entirely. Because of all these facts, it is generally believed that root pressure is not a necessary, or important, factor in the raising of water in tall stems.

Imbibition and osmosis sometimes lead to the development of high sap pressure, and they are partly responsible for the flow of maple sap. Grapevines pruned in the spring exude water or "bleed" for days afterward. On a small scale the same thing may be seen when well-watered geraniums, begonias, and fuchsias, are cut off near the soil. There is evidence, however, that in "bleeding" only the cells near the cut surface are involved.

Roots and transplanting. Only a few years ago it was thought impossible to transplant large trees or even medium sized conifers. Today trees of large size are dug up, transported many miles, and replanted successfully. Even whole hedgerows several feet in height are transplanted without injury. This advance in the art of tree moving is a fine example of the application of a knowledge of root physiology to practical problems.

We have learned that the absorbing part of the roots is mostly in the root-hair zone near the root tips and that the older roots are largely organs of conduction. Formerly, when a tree was dug up for transplanting, all the roots were cut off 3 or 4 feet from the base of the stem. This operation destroyed practically all the absorbing organs, and the tree could not absorb water from the soil until a new set of roots had developed. Meanwhile transpiration went on and the plant cells lost so much water that they were injured and not infrequently killed.

Success in transplanting is attained by gradually trimming the roots months before the tree is moved, and by loosening the soil near the tree so as to develop a mass of absorbing roots near the base of the stem. When the tree is lifted, the roots are not cut off, but as many as possible of them are carefully removed from the soil. The small roots of trees are killed by drying, and for this reason they are protected from wilting by being bound up in wet moss. Sometimes the trees are loosened somewhat in the autumn and moved during the winter, together with much of the frozen soil surrounding the roots. Successful transplanting depends upon reducing temporarily the loss of water by trimming the top, preserving the absorbing roots, and exercising care in handling both roots and stems so that they may not be injured.

In transplanting smaller plants the greatest care should be taken to prevent the drying out of the youngest roots. Many of the roots are sure to be injured in the digging and resetting, and efficient absorption is thereby reduced. This reduced absorption may be balanced by trimming off a part of the stem or leaves.

Respiration in roots. Respiration goes on in the living cells of the roots, and this process requires a constant supply of oxygen. In obtaining oxygen, the division of the roots into numerous fine branches is an advantage, because a large surface is exposed to the soil air. Some plants are easily injured by the lack of oxygen in the soil; if water stands on the soil and excludes the air, the roots gradually suffocate. Suffocation of a part of the roots interferes with absorption and other root processes besides respiration, and the whole plant suffers. For example, corn becomes yellow and sickly in low fields where water has stood for some time. Such plants may recover if the soil is drained as soon as these symptoms appear; if delayed, the plants will never completely recover. Water plants and swamp plants can grow in poorly aërated soils either because the roots secure sufficient oxygen through the internal air spaces of the plants or because they have a low oxygen requirement.

The energy liberated in respiration is used in the chemical processes associated with food transformations in the cells of the root. If the aëration of the soil is poor, respiration is slowed down, and instead of carbon dioxide, poisonous substances, such as alcohols and organic acids, are formed. Hence the oxygen content of a soil is one of its important properties.

A soil is in its most favorable condition for plant growth when there is enough water to balance the loss by transpiration but not enough to interfere with the access of oxygen to the roots. Gardeners determine when it is in this condition by squeezing a handful of the soil. If it barely clings together in a ball when the hand is opened, it is about right. In this condition soil has its largest volume, best aëration, is mellowest, and is most easily penetrated by roots.

Carbon dioxide. A slight increase in the amount of carbon dioxide in the atmosphere increases the rate of photosynthesis and indirectly the growth of many plants.

When carbon dioxide accumulates in the soil, it becomes toxic

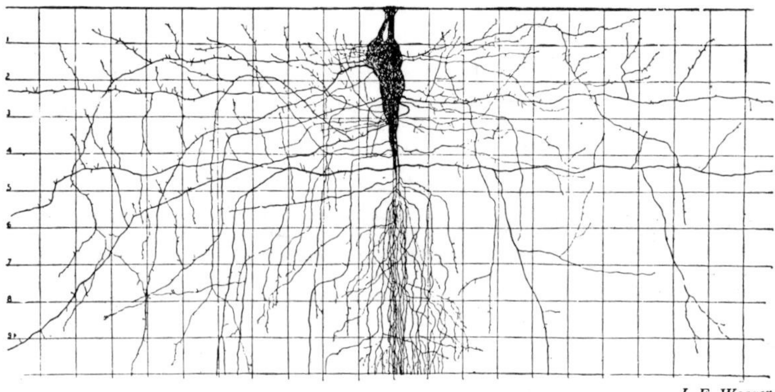

FIG. 115. Section showing part of the root system of the bush morning-glory (*Ipomœa leptophylla*), a widely distributed plant of the Great Plains region. The lateral and vertical branches start from a perennial fleshy root, one foot in diameter, and extend into more than 5000 cubic feet of soil. The section is divided into one-foot squares.

to some plants. Buckwheat, for example, can withstand a low content of oxygen in soils, but is killed by an accumulation of carbon dioxide. The black willow is indifferent to both low oxygen and high carbon dioxide content of soils.

The distribution of roots in the soil. In addition to the inherited root habits of plants, an important factor that determines the distribution of roots in soil is the oxygen supply. Various plants have different requirements, but all roots require oxygen for growth. Those who have seen stumps pulled from the land know that the roots go deep in upland sandy soils; that they do not go so deep in heavy clay soils; and that they are just beneath the surface in swamp and bog land. The principal reason why one finds the roots near the surface in swamps is that these roots were the only ones that continued to live and grow. The roots that in times of drought penetrated to greater depths were killed off (suffocated) when the water stood at higher levels. The distribution of roots in the soil, therefore, is determined principally by the combined influences of gravity, water, and oxygen. Water and gravity control the direction of growth, and the oxygen supply determines whether or not growth can take place or the roots survive.

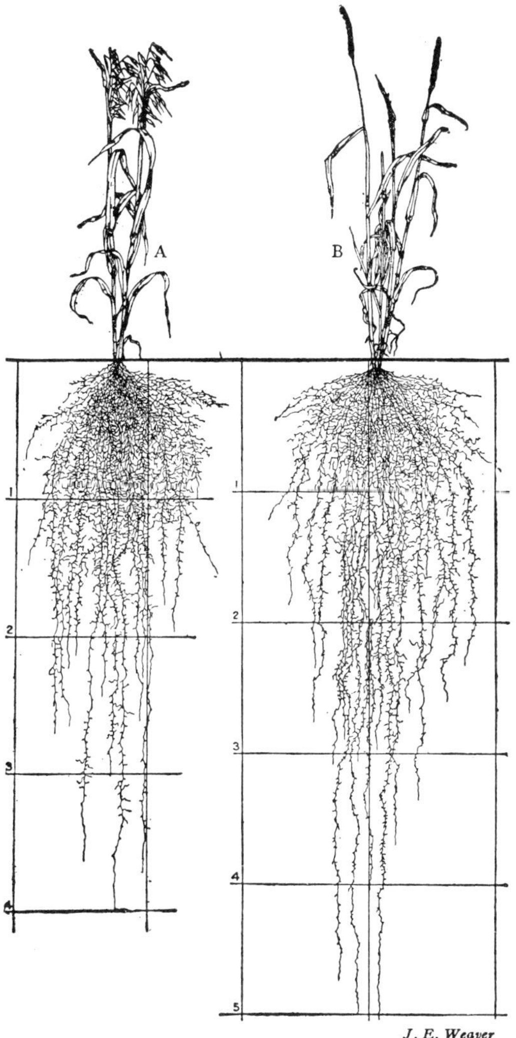

FIG. 116. Root systems of oats (*A*) and wheat (*B*) at time of blossoming; Lincoln, Nebraska. Section divided into one-foot squares.

In the plains of eastern Kansas, and California, the roots of plants such as alfalfa and sugar beet penetrate certain soils to depths of from 15 to 20 feet. The absorbing parts of these roots actually reach the water table; that is, they reach the level at which the soil is saturated, or the level to which water would rise in a well.

Two or more species of plants are sometimes found associated in dry regions, and locally in dry habitats, because their roots get their water at different levels and hence do not compete with each other. For example, in our Southwestern deserts the giant cactus often grows with the creosote bush. The former plant obtains its water from the superficial layer of the soil, while the latter obtains its water at deeper levels. The roots of lawn grass are very superficial, and lawn grass suffers from drought much sooner than do the deeper-rooted dandelion and English plantain that occur with it as weeds.

In dry regions where plants compete with one another, success comes mostly to those that secure a sufficient water supply. In moist regions success in competition between plants depends chiefly on ability to reach the light or withstand shade.

Temperature has a marked effect on root development. Many common annual weeds develop extensive root systems only at temperatures below 68° or 70° F. Poor root development means retarded vegetative growth and small plants. Wheat is an example of a plant whose root system develops best below 60° F. Its principal absorbing roots are those formed from the first node of the stem. These are formed down near the seed in cool soil. In warm soil the first node may be at, or above, the surface of the soil and the roots are then not advantageously placed in the soil.

The pressure of growth. In the growth of plant organs hundreds, or thousands, of cells are expanding simultaneously through the taking up of water. The pressure which these cells develop is called *growth pressure*. The pressure exerted by roots in penetrating the soil may be very great, amounting to hundreds of pounds to the square inch. This is readily appreciated when one sees cement sidewalks broken and large rocks moved by the

FIG. 117. Fern leaves pushing upward through a cement sidewalk. Growth pressure may amount to hundreds of pounds to the square inch. (*After G. E. Stone.*)

growth of roots under them. Growth pressure is just as powerful in stems and other growing parts. Fleshy roots like those of the radish and turnip sometimes force themselves partly out of the ground by the thickening of the upper portion. How is it possible for the cells of plants to withstand internal pressure of a thousand pounds to the square inch without bursting? They must have pressure equal to this amount or they could not move rocks or break cement walks, which they do. The explanation lies in the fact that they are so small that the pressure exerted by a single cell is trifling compared with the strength of its cellulose wall. In a cell mass expanding against a pressure of 1000 pounds to the square inch, if the cells are of average size, say .03 mm. in diameter, the wall of each cell will not have to resist a pressure of more than .13 of an ounce.

Food conduction. The transfer of food takes place in the food-conducting tissue of roots in the same way as in stems and leaves. Substances that are transferred are in a soluble form, and they are usually in a comparatively simple form.

The movement of a substance into or out of a cell depends upon the permeability of the cell protoplasm to that particular substance; if the cytoplasm is impermeable to a substance, it cannot enter or leave a cell. That the direction of the movement of foods may change from time to time is shown by the fact that sugar and soluble proteins may move down into the root during one season and up out of the root at another season. For example, in the turnip or beet the excess food made by the leaves during the first summer passes downward into the roots; the next year, food passes upward from the roots to the developing stems and leaves. This may be due to changes in the permeability of the cells, or to changes in the foods stored in the cells.

These changes in the behavior of organs, tissues, and cells are clear evidences of life. In physical apparatus the behavior is fixed and a process soon comes to a standstill. In living things changes are continually taking place in the living matter itself, and these bring about a continuation of the processes that are going on, or changes in these processes.

Accumulation of food in roots. Food accumulates in the roots of many plants, notably in those of biennials like the beet, carrot, turnip, and salsify. The sweet potato and the dahlia are examples of perennials with large storage roots. The most common forms in which carbohydrates accumulate in roots are starch and sugar. Starch as a storage material has the advantages of being insoluble and more concentrated than sugar. When growth begins anew, starch is readily converted (digested) into sugar. Most roots also accumulate proteins to the extent of 2 or 3 per cent. When these are digested, they change to soluble amides. Fats occur in still smaller amounts. They break up when digested into fatty acids and glycerin, both of which may be further modified by enzymes into simple sugars. In the sugar beet sucrose is formed in the leaves and accumulates to a large extent in the same form in the root, although some of it is changed to starch.

CHAPTER TWENTY-THREE

ENVIRONMENTAL FACTORS AFFECTING GROWTH AND REPRODUCTION

THE environment of a plant is made up of many factors. Moreover, the factors are more or less interdependent, and it is very difficult, and often impossible, to change one factor without altering related factors. Consequently, it is often difficult to determine the underlying cause of a change in the form of a plant that is undoubtedly produced by something in the environment.

Changes in the water content of a soil are complicated by the effects of decreased oxygen content. The addition of lime to a soil changes the permeability of the roots to water and mineral salts, modifies the rate of transpiration, as well as alters the chemical and physical qualities of the soil itself. The final effect upon the plant is the combined result of all these changes.

In the following paragraphs the more important factors of the environment and some of their effects upon plant growth and reproduction are discussed.

Light an important environmental factor. The amount of light available to a plant depends primarily upon the intensity of the sunshine. This is greatest in the tropics and least at the poles. The total amount of light is influenced also by the length of the day. At the equator the daylight lasts 12 hours; at the poles the light continues all summer. So tropical plants have intense light during half of each day, while arctic plants have weaker light continuously through the growing season.

Orange growers at the northern end of the Central Valley of California are able to ripen their fruit for market 3 to 6 weeks earlier than their competitors 400 miles farther south, due to the longer daily period of sunlight and the protection from cool night winds afforded by the surrounding high mountains.

As much as 1000 bushels of potatoes have been grown on an acre of land along the Mackenzie River, at the arctic circle.

FIG. 118. The effect of long and short days on the evening primrose. Both plants were brought into the greenhouse in November. The one at the left received, in addition to daylight, illumination from an electric light from sunset to midnight for about two months. The one at the right was kept under exactly the same conditions, except that it received only the natural winter daylight. This is a typical long-day plant, in nature flowering when the days are long. *(Garner and Allard, U.S.D.A.)*

Wheat has been ripened at 57° north latitude, and immense yields of hay are produced at 60° north latitude along the coast of Alaska. These large crops are made possible by the continuous or long daily period of sunlight during the growing season.

Medium light favors the growth of vegetative structures. Attention has already been called to the fact that leaves and many kinds of stems attain their largest size in partial shade.

Exposure to full sunlight increases the rate of transpiration to a point where the water content of the plant tissues is reduced below that necessary for the greatest amount of growth. Vegetative organs, in general, do not require intense illumination for their greatest development, because only a small fraction of the sunlight is used in photosynthesis, and their growth is favored by moisture both in the soil and the air. Just how much the sunlight should be reduced to promote growth varies with different plants and in different geographic regions. A reduction of 20 to 50 per cent is favorable to vegetative growth in many plants.

In moist regions, as for example along northern coasts, the

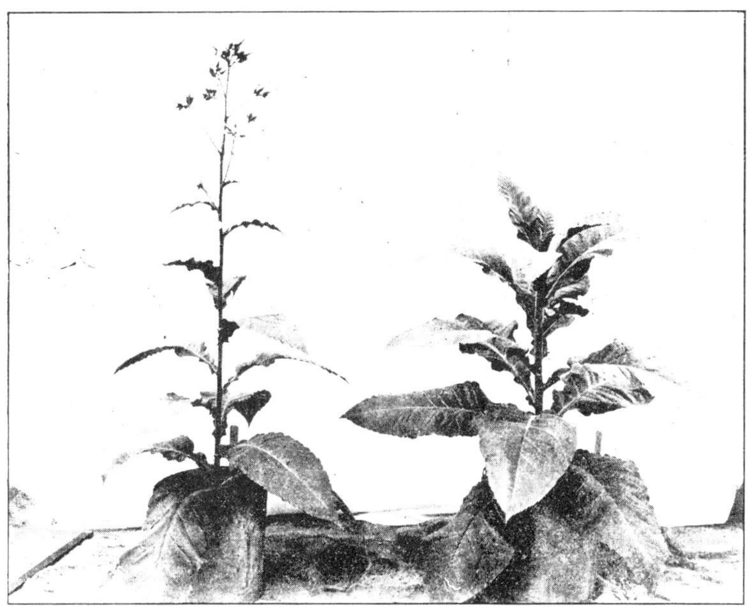

Short day Long day

FIG. 119. Effects of length of day on tobacco plant. Both plants were grown in a greenhouse during the winter. The plant at the right received, in addition to daylight, electric light from sunset to midnight, while the plant at the left received the natural daylight only. This is a typical short-day plant. When exposed to long days, this variety will grow 15 feet or more in height and produce upward of 100 leaves. (*Garner and Allard, U.S.D.A.*)

Short day Long day

Long day Short day

FIGS. 120 and 121. Like the evening primrose and in contrast to the tobacco shown in the preceding illustration, red clover (Fig. 120) is a long-day plant. The plants growing in the two pails were photographed June 28 and had each received the same treatment, except that those on the left were illuminated for only 10 hours daily, while those on the right received the light during the whole day. The ones exposed for the shorter period grew but 7 inches high and produced no flowers, while those illuminated during the whole day flowered abundantly and the tallest plants grew to a height of 33 inches.

On the other hand, the dahlia, shown in Figure 121, is, like the tobacco, a short-day plant. The plant on the right, beginning May 12, was exposed to 10 hours of light daily and flowered July 8, when the photograph was made. On the control plant, under the natural length of day, the first blossoms appeared September 27. *(Garner and Allard, U. S. D. A.)*

light intensity is often reduced locally by fogs and clouds. In the northern Lake states cloudy days may form a considerable part of the growing season, and the total light that reaches plants is much less than on the plains and deserts. Germany, France, Great Britain, and our own Northeastern states are noted for their high yields of potatoes, turnips, carrots, beets, and other vegetative crops.

The slope of the land, especially in mountain regions, may increase or decrease the intensity and the length of daylight. Finally, plants may have their light reduced or cut off by trees or other objects. Commercial growers of ginseng cover their gardens with slat frames so as to approximate the intensity of light found in the woods where ginseng grows wild. Tea, a leaf product, attains its best quality and yield in the shade of taller trees purposely planted in alternate rows with the tea plants.

The influence of sunlight and moisture may also be seen in the geographic distribution of the flax crop. Flax is grown for two distinct products: one, the bast fibers (a vegetative part), used in making linen thread; and the other, the seed (a reproductive structure), used in manufacturing linseed oil. The leading centers of *fiber production* are in northwestern Russia, northern

Short day Long day

FIGS. 122 and 123. Figure 122 shows, on the right, an apple seedling that grew more rapidly with 10 hours' daily illumination than the control plant on the left with a full day's illumination. In contrast, the maple seedlings (*Acer negundo*), shown on the left in Figure 123, were dwarfed and forced into dormancy by shortening the illumination period to 10 hours, while the plant exposed for the full length of day grew rapidly. The photograph of the apples was made July 13 and that of the maples September 22.

These photographs show clearly that light has effects other than those of photosynthesis on plants, and that its effects are different on different plants (*Garner and Allard, U. S. D. A.*)

France, Belgium, and Ireland, in regions of low light intensity and high humidity. The northern plains from Minnesota to Alberta, northern Argentina, Japan, Italy, and India are the centers of *seed production*, all of these being regions of low humidity and high light intensity during the season when the crop is grown.

Intense light favors the development of reproductive structures. The production of flowers, fruits, and seeds is promoted by bright sunshine, provided there is sufficient soil moisture to permit normal growth of the plant. Our greatest grain and fruit producing areas are in regions where these conditions prevail: the Middle Western states, Washington, California, and Colorado. In some of the areas the water supply is maintained by irrigation, and the intensity of the light is but slightly reduced by clouds or atmospheric moisture during the growing season.

It is a common observation that partial shade reduces flower production, and one of the difficulties in producing flowers in greenhouses in the winter time is the low light intensity. Keeping greenhouse glass clean at this season is as important as providing favorable temperatures.

In regions of intense light (Colorado, for example) many fruit trees blossom and set fruit every year, while the same varieties, in regions of less light, fruit only once in two or three years. In wet seasons in the Southeastern states the yield of cotton fiber and cotton seed per acre is greatly reduced, although the plants grow to more than normal size.

The most brilliant wild flowers occur in alpine meadows, where the light is intense and the moisture always sufficient. The flowers also are larger, although the plants are smaller than those of the same species growing at low altitudes.

Length of day. The length of the day is an important factor in determining the flowering and reproduction of some plants. For example, ragweeds given 7 hours of light daily blossom 2 months earlier than similar plants exposed for 14 hours daily.

The plants with a restricted period of sunlight grow to a height of 4 to 5 feet; the other plants grow to be 7 or 8 feet tall. Evidently long days favor vegetative growth in this plant; short days favor reproduction.

Under similar conditions radishes respond very differently. They continue to develop a thickened root throughout the growing season and do not form flowers when the daily period of illumination is shortened to 7 hours. With twice that amount of sunlight these plants bloom in about 1 month.

These two plants are each typical of many species whose vegetative development and reproduction are determined by the length of day. It is also probable that the length of day is the important factor that makes the beet a biennial in the latitude of Kansas and an annual in the latitude of Alaska.

Quality of light. The quality of light is also an important factor in growth. You have seen that when a beam of light is separated into its constituent rays, as in a rainbow, they form a series of colors running in order through red, orange, yellow, green, blue, indigo, and violet. The red rays have the longest wave lengths, while the violet rays have the shortest wave lengths. Ultra-violet light has still shorter wave lengths. Under natural conditions the longer light waves of the red end of the spectrum are most important in photosynthesis. The shorter wave lengths of the violet and ultra-violet rays are most important in inhibiting vegetative growth. Ultra-violet rays are sometimes used to kill bacteria, and they no doubt have similar deleterious effects upon the protoplasm of green plants. They are rapidly absorbed by the atmosphere and by clouds, and this probably has something to do with the difference of vegetative growth at low and high altitudes and in clear and foggy climates.

Indirect effects of light. The various effects of light upon growth and reproduction are a result of the physical and chemical effects of light upon the numerous physiological processes in the plant. The synthesis of carbohydrates and other organic compounds in the plant depends directly or indirectly upon sunlight. As we shall see later, the relative amount of carbohydrates produced in the plant determines in the main whether the plant continues vegetative growth or carries on reproduction. Light also greatly affects transpiration, directly through raising the temperature of the plant and indirectly by causing the stomata to open.

Atmospheric water. The water in the air affects plants directly in several ways. The moistness or dryness of the air determines whether less water or more is required for transpiration, and the amount of water precipitated from the air in the form of rain determines to a large extent the amount of water available in the soil. Atmospheric water condensed in the form of fog and cloud reduces transpiration and also lessens the amount of light that reaches the plant. Under conditions of high humidity and favorable temperature vegetative activity reaches its maximum. Drought greatly decreases vegetative growth and shortens the vegetative period of plants. A high rate of transpiration may not only prevent any increase in the size of a plant during the daytime but may actually bring about a decrease.

Distribution of rainfall. The *distribution* of rainfall through the year is of the greatest importance to vegetation. When the period of heaviest rainfall coincides with the hottest part of the year, the conditions are best for the rapid growth of plants. If the rainfall is scanty during the time of highest temperatures, plants are hindered in their growth, and only xerophytes may be able to withstand the conditions. In these regions irrigation is absolutely necessary for the growth of mesophytes. The greater amount of available sunlight in summer-dry regions accounts in part for the unusually large crops that can be raised on the irrigated lands of the Western states. This is one of the principal reasons why California has become an important center of the production of flower and vegetable seeds.

Methods of conserving soil water. — In dry regions there are two methods by which the soil water is conserved. By cultivating

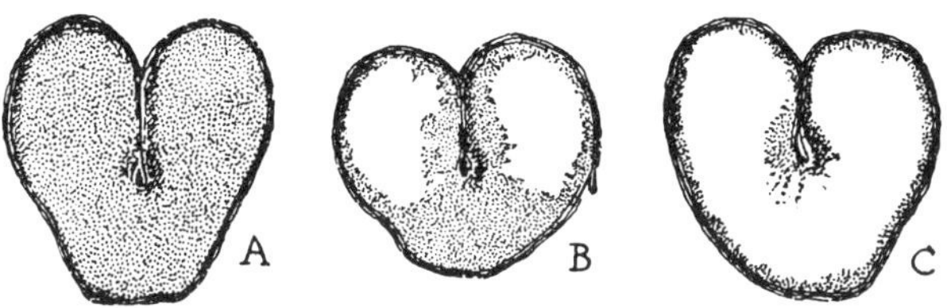

FIG. 124. Cross-sections of kernels of hard or macaroni wheat. This wheat is grown in dry regions and is valued because of its large content of protein. In the figures the flinty or high-protein parts are shaded and the soft or starchy parts are white. When the wheat is grown under the conditions of dry farming, the protein content is highest (*A*); when regularly irrigated, the same wheat produces soft, starchy grains (*C*). An intermediate condition is shown by *B*. This exemplifies the effect of the water balance on the composition of a grain.

the soil several times during the growing season, the soil structure is broken a few inches below the surface, where evaporation takes place. Consequently the capillary water columns are broken and water rises only to the top of the undisturbed soil layer. The cultivated layer soon dries out and forms a blanket that reduces water loss by evaporation. The rough, loose surface is of further advantage when it rains, in that the water settles into the soil quickly and there is little run-off. This method of conserving water is called the "dust mulch."

A second method of making land suitable for crop production in regions of slight rainfall is to plant crops only in alternate years. By plowing the land so that it will take up water as fast as it falls, and especially by destroying weeds which would otherwise remove large amounts of water, each crop has a large part of two years' rainfall available. These two methods of conserving water form the basis of the so-called " dry farming."

Effect of temperature. As one goes north or south from the equator, the temperatures of the soil and the air decrease. Increasing altitude in mountains brings about the same effects. Temperature directly influences the rate of all plant processes, and most plants grow best under certain rather fixed temperature conditions. For tropical plants, air temperatures above 90° F. are most favorable. Temperate plants develop best at between 60° and 90° F. Arctic and alpine plants grow at temperatures but little above the freezing point.

The time during which the temperature remains above the freezing point is the *growing season.* In the tropics this extends throughout the year. In arctic and alpine regions it may be reduced to 2 or 3 months. The temperature of the air and the length of the growing season are important factors in determining the amount of food a plant may manufacture, and consequently the amount of growth. Rice and peanuts, for example, require high temperatures for their best growth, while cotton must have a long season in which to mature its seeds. None of these crops, consequently, is profitable north of Tennessee.

Air temperatures influenced by air drainage. Cold air is heavier than warm air; consequently it accumulates in low grounds and reduces the temperature there. In low places frost occurs later in the spring and earlier in the autumn than on hills. Crop plants like beans, that are easily injured by frost, can be planted earlier and grown later on uplands. Peach orchards are more profitable on uplands than in valley bottoms, because on the uplands they are more likely to escape late spring frosts.

Soil temperatures also are important. Dark-colored soils are warmer than light-colored soils of the same texture, because they absorb the sun's rays more readily. Well-drained soils are warmer than wet soils, (1) because less heat is required to raise their temperatures, and (2) because the temperature of a wet soil is lowered by the constant evaporation of water. The most valuable farm lands are those with dark-colored, well-drained soils. On north slopes, soils do not warm up so rapidly in the spring, and plants growing there start their growth later than do those on the south slopes of the same hills. Peach growers prefer not only uplands but north slopes.

Recent investigations indicate that much of importance may be learned from a more careful study of the effect of soil temperatures upon plant form and behavior. Very low soil temperatures have been reported to shorten the vegetative period of beets, with the result that only slender roots are obtained, while very high temperatures shorten the vegetative period of kohl-rabi and prevent the formation of the thickened stems for which the plant is cultivated. Wheat germinated at low temperature produces its adventitious root system at a favorable depth beneath the soil surface. If germinated at higher temperature the adventitious root system is produced at or above the surface of the soil, with the result that a very weak plant results even if the conditions after germination are the best obtainable.

Disease-producing organisms are much more destructive at one soil temperature than at another. If the best temperature for the growth of the organisms differs from that of the host plant, the disease may often be avoided by planting when the soil temperatures are favorable to the crop and unfavorable to the disease-producing organism.

Freezing. When plant tissues freeze, the formation of ice takes place in the intercellular spaces. As the ice forms, water is withdrawn from the cells, just as it is when a plant wilts. The result of this withdrawal of water is a greater concentration of the salts inside the cell and a higher osmotic pressure. Water-imbibing substances and the osmotic pressure resist the outward movement of water. In general it has been found that those plants and tissues that have the highest water-holding power are also most resistant to freezing injuries. They are likewise least affected by drought. One of the sources of injury to the cell when a large proportion of its water is removed is the precipitation of proteins. Young growing tissues usually have a high water content and the cells contain but little of the water-holding substances; consequently they are very susceptible to freezing injuries.

Hardening of plants. "Hardening" is a term applied by gardeners to the practice of rendering young plants immune to drought and frost injuries. Seedlings grown in hotbeds and greenhouses in early spring, if set out directly into the open ground, are very susceptible to drought and freezing temperatures. If kept, however, in cold-frames for a few days at temperatures several degrees above the freezing point, they increase in hardiness, and if set out will withstand frost.

Investigation shows that hardened plants differ from tender plants in having (1) more water-imbibing substance in the cells and (2) in having more soluble proteins. The former prevents the withdrawal of too much water from the cells when freezing or drought occurs. The latter prevents the precipitation of the proteins when the cell sap becomes more concentrated by the partial withdrawal of water. Some plants may be hardened by subjecting them to drought before freezing weather, and the changes in the cells are quite similar to those that occur when hardening is brought about by the low temperature of cold-frames.

The hardiness of certain varieties of peach is due to the slowness with which they take up water in early spring. The cell sap in the buds is consequently very concentrated, and light frosts are not sufficient to freeze the water in the tissues.

If cultivated perennials are kept rather dry in the autumn, they are much less likely to be winter killed than if they are kept wet and green up to the time of killing frosts.

Plants differ greatly in their ability to synthesize the water-holding substances that produce hardiness. Wheat, for example, hardens readily and can withstand drought and extremely low temperatures. Oats, on the contrary, seems to lack these water-holding substances and is resistant neither to drought nor to low temperature.

Winds. Winds and air currents are of importance, as they affect the rate of transpiration or modify the temperature. Pre-

W. S. Cooper

FIG. 125. Limber pine (*Pinus flexilis*) at timber line on Long's Peak, Colorado, showing effects of environmental factors on growth. When exposed to the wind, snow, and low temperature of the mountain peak, the tree has the scraggy, much-branched form shown in the illustration. At lower altitudes it is a single stemmed timber tree.

vailing winds increase transpiration and slow down growth on the windward side of trees to such an extent that a larger part of the crown of any tree standing in the open is on the leeward side of the trunk. This is so generally true that one can tell the direction of the prevailing winds of a region by a careful examination of the trees. Occasionally violent winds may destroy large areas of timber and crops, and along exposed coasts and mountain tops bring about the development of stunted and gnarled trees.

Gravity. The direction of growth of many plant organs is determined by gravity. The downward growth of primary roots, the upward growth of stems, and the direction of growth of lateral branches are responses to gravity acting as a stimulus.

The peculiar shapes of York Imperial apples are the results of gravity stimulating growth in a vertical direction, no matter what the position of the axis, or core, of the apple. If the apple hangs vertically downward during growth, the mature fruit will have a long, sheep-nosed form. If it extends horizontally from the branch, the apple will be short, flat, and vertically elliptical instead of round in cross-section. If it hangs obliquely from the branch, the fruit will be obliquely elongated. In all cases the apple is longest in the direction of the pull of gravity.

The annual rings in the horizontal branches of trees are often thicker on the lower side than on the upper. When corn and other grasses are blown down by wind, they again become upright, because gravity not only stimulates growth of the nodes, but causes the lower half to grow faster than the upper half, until the stem is brought again to a vertical position.

Chemical elements essential to plants. We have already learned that in order to have a plant grow the soil must furnish it with sufficient water for transpiration and for the manufacture of food. At the same time, the soil must not be so filled with water as to exclude oxygen from the roots. Carbon dioxide and water supply the plant with the three elements, *carbon*, *hydrogen*, and *oxygen*, needed in building carbohydrates.

From the soil solution plants obtain other essential chemical elements used directly, or indirectly, in the manufacture of food and in the development of their tissues. These elements are *potassium*, *calcium*, *magnesium*, *nitrogen*, *phosphorus*, *sulfur*, *iron*, and possibly *manganese*. The growth of plants is hindered, and certain plants are excluded from soils that contain insufficient amounts of any of these substances. It should be borne in mind, however, that from 60 to 95 per cent of a plant is water and that most of the remainder is organic matter. When plants are burned, the water and organic matter pass into the air and only the mineral matter remains as ash. The ash seldom amounts to more than 3 per cent of the green weight, and sometimes it is as low as .3 of 1 per cent. It is evident, then, that each of these essential elements has one, or several, special uses in the plant, and no other can be successfully substituted for it. There are other elements, like silicon, aluminium, and sodium, that accumulate in plants but take no necessary part in either their processes or structures. These non-essential elements, however, may greatly affect the growth of plants. When present in small amounts they may be favorable to the plant, but when present in larger amounts they may be injurious.

Nitrogen enters into the composition of all proteins and of many related but less complex compounds, like chlorophyll, amino acids, and alkaloids. Carbohydrates furnish the basis of these compounds, and both carbohydrates and proteins enter largely into the making of protoplasm. Hence, when there is an abundance of carbohydrates and nitrogen, vegetative growth is greatly increased. This condition may be seen especially in potatoes. If too much nitrogen is available, the plants develop enormous tops but produce almost no tubers, because as fast as carbohydrates are made nitrogen is available for the production of proteins and protoplasm and further growth of the shoots ensues. If the amount of nitrogen is just sufficient for the growth of an average potato plant, there will be an excess of carbohydrates formed, and these will accumulate in the tuber as starch. This example illustrates what is meant by the proper balance of carbohydrates and nitrogen.

Another example of the *carbohydrate-nitrogen balance* may be seen in wheat. If the proportion of nitrogen is too large, the wheat grows tall and the straw is so weak that it falls over, and the grains fail to accumulate the usual amount of starch. Carbohydrates form the cell walls, and if they are all consumed in extending the stems and leaves, there are none left for thickening the cell walls, upon which the stiffness of stems depends, and none for filling the grain. Too much nitrogen added to orchard soils leads to great vegetative growth and very little fruit. Insufficient nitrogen leads to poor growth and few blossoms, and the fruit is small and woody, because the excess carbohydrates accumulate as starch and cellulose.

Calcium occurs in many plant cells in the form of calcium oxalate crystals. These may be large, rounded masses occupying most of the cell, or bundles of microscopic needle crystals. If you happen to have bitten into the corm of the common Jack-in-the-pulpit and felt the stinging sensation in your mouth, you have come in contact with the needle crystals, even if you have not seen them. The crystals pierce the soft tissues of the mouth and continue to irritate until they are dissolved. Oxalic acid is produced in plant cells under certain conditions, and in the presence of calcium it is precipitated as calcium oxalate (Fig. 28).

The middle lamella, which holds plant tissues together, is composed of calcium pectate, a compound of calcium and pectic acid. The use of calcium in the formation of this compound is probably the most important rôle of calcium in the plant, since no other element can be successfully substituted for it in the building of the cell wall. The presence of calcium in the soil also affects the permeability of the cell membranes and thus facilitates the absorption and retention of other salts by the roots.

Calcium is also important in soils because it neutralizes acidity. Red clover, alfalfa, and blue grass, for example, cannot withstand acid soil conditions. This explains why lime and wood ashes are recommended for improving lawns. Lime improves the texture of many soils, and this in turn improves its drainage, water-supplying power, and its aëration.

Potassium is essential to the growth of plants, although we know of few potassium compounds in plant tissues. Cell division does not occur in its absence, and it plays an important rôle in the chemical transformations that are continually being made in the living cells among carbohydrates, organic acids, fats, proteins, and other less familiar substances. Weak-stemmed plants that occur in the absence of sufficient potassium appear to be due to the need of potassium in the synthesis and translocation of carbohydrates necessary for the formation of thick cell walls.

Magnesium forms a part of the chlorophyll molecule and is therefore indispensable to all green plants. It is also necessary to the growth of non-green plants.

Sulfur forms a part of all plant proteins. It also occurs in certain compounds common in the mustard family, that give

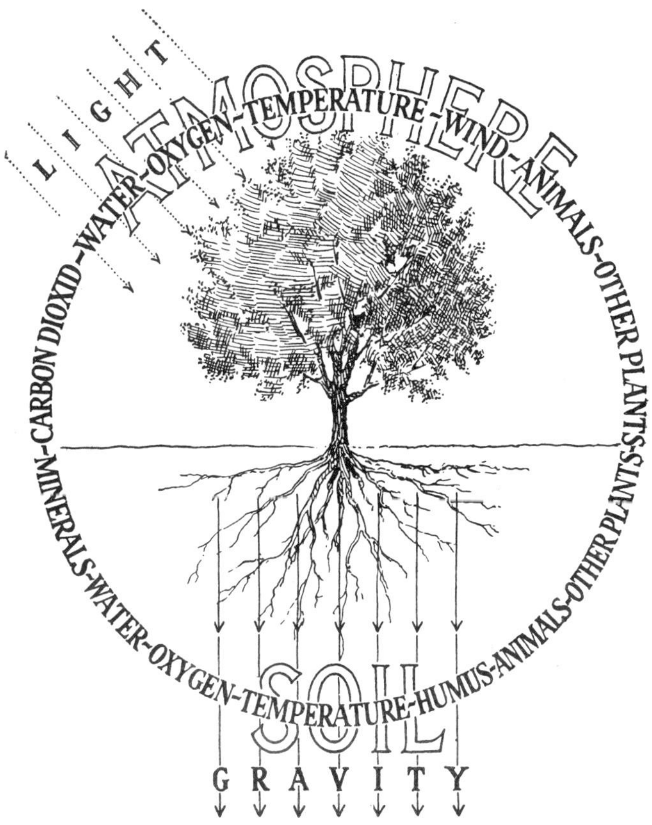

FIG. 126. Diagram showing the principal factors in the environment of land plants.

them their pungent odor and taste. Onions, garlic, and leeks, members of the lily family, owe their flavors to sulfur compounds.

Iron is essential for the development of chlorophyll, although it forms no part of the chlorophyll molecule. It appears to function chiefly as a catalyzer in the plant.

Phosphorus is a necessary element in certain compounds found in the nuclei of cells. It is essential for cell division and many enzyme activities.

Manganese, like iron, is a catalyzer and is said to be associated with iron in the formation of chlorophyll.

Fertilizers. Fertilizers are added to agricultural soils, (1) to increase the supply of the essential mineral elements, (2) to improve the texture of the soil and its water-supplying qualities, (3) to liberate other mineral elements by breaking up insoluble compounds in the soil, and (4) to correct acidity.

Of all the elements needed for plant growth, phosphorus, potassium, and nitrogen are most frequently found in quantities insufficient for the best yields of agricultural products. Phosphorus may be added in the form of crushed phosphate rock, or as "acid phosphate" (rock phosphate treated with sulfuric acid), which is more soluble and contains sulfur in addition. Potassium may be supplied by the use of potassium chloride.

Nitrogen may be added to soils in the form of sodium nitrate or ammonium sulfate, but these salts are very expensive and can only be profitably used on truck gardens where the value of the crops amounts to hundreds of dollars per acre. In agricultural soils generally, the nitrates are best secured and maintained by making the soil favorable for the growth of the nitrogen-fixing bacteria. Calcium is usually added to soils in the form of crushed, or burned, limestone, not only to furnish this element to the plants, but also to improve the physical condition of the soil. Lime also helps to liberate potassium from insoluble compounds.

Acidity of soils. Most plants grow best in soils that are nearly neutral. The use of lime in neutralizing acid soils has been mentioned previously. There are some plants, however, that are favored by acid soils. Cranberries, blueberries, azaleas, laurels, and rhododendrons flourish only under these conditions. In growing these plants in cultivation the acidity of the soil is sometimes maintained by adding ammonium sulfate, by watering occasionally with tanbark extract, and by adding alum.

Bog soils, which are naturally acid, must be neutralized when reclaimed for the growing of celery, onions, cabbage, and mint. These soils are also deficient in potassium, and this element must be supplied in some form to obtain the best yields.

Alkalinity of soils. In arid regions, the evaporation of water may cause salts to accumulate in the surface layers of the soil to such an extent that most or all plants are excluded. About many of the lakes in the Great Basin region are alkali lands of this kind. Various salts have been leached from the rocks and minerals of the mountains, and washed down into these lake basins, which have no outlets. This has been going on for thousands of years and the water that carried them has evaporated, leaving the salts behind. Some of these salts, like sodium chloride (common table salt), sodium carbonate (washing soda), sodium sulfate (black alkali), and borax, are poisonous to plants. Others are not poisonous, but when present in considerable amounts interfere with the absorption of water by roots. When the concentration of salt is slight and relatively pure water is available, these lands may be irrigated and drained and a part of the alkali removed. These lands then become valuable for agriculture. The cultivated sugar-beet and alfalfa lands near Salt Lake City are of this character.

Humus. Another soil factor of great importance is *humus.* This material, which gives the brown and black colors to rich agricultural land, is composed of the partially decayed remains of plants. Leaves and other plant organs that fall to the ground are slowly changed and broken up by bacteria, fungi, and other agencies until only the brown, powdery humus remains. Moist or wet grasslands accumulate more humus than forested lands, because so large a part of the plant is underground where the decay is slower, and because these lands are covered with water during a part of the year so that there is less oxygen available for completely oxidizing the plant remains.

Humus favors plant growth by increasing the water-holding

capacity of the soil and so rendering the water supply more uniform throughout the growing season. It improves the physical properties of the soil by making it mellow. Humus also makes it possible for bacteria and other organisms that increase fertility to live within the soil.

Loam. Soils containing a large percentage of humus are called *loams*. Some of the prairies of Illinois, Iowa, and southern Minnesota were originally poorly drained areas largely covered with water during late winter and spring. During the summer they dried off and were covered with tall grasses that died down to the ground in the late autumn. During the winter they became matted together, forming a thick layer of plant materials. In time these partially decayed, and each year a new layer was added, until after hundreds and thousands of years humus accumulated to a depth of from 1 to 5 feet. Later, when the settlers broke the prairie-grass turf and the land was drained by tiles and ditches, these areas became the most productive lands in the United States and the center of production of corn and wheat.

Animals as a factor in plant environment. Leaf-eating insects, such as the potato beetle, injure the plant by destroying the chlorenchyma and thus preventing food manufacture. It is estimated that grasshoppers and other insects often eat as much of the grass in a pasture as do the farm animals. Plant lice, leaf hoppers, and scale insects remove the sap from the cells of the tender growing parts and may kill the entire plants. Plant lice and leaf hoppers may also carry disease-producing organisms from one plant to another. Other animals, like the earthworm, favor the growth of plants by loosening the soil and promote the formation of humus by eating and by pulling bits of leaves into their burrows. Herbivorous (Latin: *herba*, herb, and *vorare*, to eat) wild animals, like the rabbits, squirrels, and deer, markedly affect natural vegetation, while the domesticated cattle, sheep, and hogs to a large extent determine what plants can survive in pastures and grazing lands.

W. A. Orton, U. S. D. A.

FIGS. 127 and 128. A healthy potato plant and one showing the mosaic disease. The organisms that produce disease are a most important factor in the environment of both wild and cultivated plants.

Man, more than all other animals put together, has modified the natural vegetation of the earth. In some cases he has destroyed it; in other cases he has encouraged and protected it. Most of all, he has selected certain plants and made of them the food supply of the world. If he understands the interrelations of the processes occurring in plants and how these processes are affected by the various factors of the environment, he may secure desirable modifications of both the vegetative and reproductive structures of the plant.

Other plants as an environmental factor. Other plants, such as weeds growing among cultivated crops, may modify the environment of plants by shading them and by removing water and soluble salts from the soil. Or a plant may directly affect another plant by growing on it and taking its nourishment from it. For example, the mistletoe grows on trees and injures them. Corn smut and wheat rust live on corn and wheat, and decrease or prevent the production of grain. These are only two out of many disease-producing organisms that injure and destroy wild and cultivated plants.

Importance of further study of the environment. From this brief survey of the more important factors it must be evident that the plant lives in a highly complex environment, that these factors vary from one season to another, and that they are closely interrelated. For these reasons it is difficult fully to explain the effects produced by changing one of these factors. Certainly the day has passed when offhand answers can be given to the many questions arising from intelligent observation of plants in nature or in cultivation. These questions can only be answered correctly by experiments carried on by men who have made a special study of plants in relation to environmental factors. Furthermore, only well-trained men can make investigations that will advance our knowledge in this important field. Yet this more than any other is the field of botany that will contribute information of fundamental importance to the farmer, the gardener, and the forester. Every advance in our knowledge of the relation between a plant process and a definite environmental factor can be advantageously applied to improve cultural practices.

CHAPTER TWENTY-FOUR

VEGETATIVE MULTIPLICATION AND PLANT PROPAGATION

THE development of new plants from roots, stems, and leaves is called vegetative multiplication in distinction from reproduction by seeds. Growers of plants make use of these natural methods of multiplying plants, and in addition have devised methods of artificially multiplying them by grafting, budding, and the growing of cuttings. In the plant-growing arts these methods are grouped under plant propagation.

Vegetative multiplication. In the discussion of stems attention was called to the fact that one of the advantages in underground stems lies in the facility with which the plant may be multiplied. From rootstocks arise new terminal and lateral buds that later form new aërial shoots, and through the death of the older parts of the underground stems these branches become separate plants. Bulbs, corms, and tubers bring about vegetative propagation in a similar way.

Plants may multiply from the aërial vegetative parts also. The stems of the black raspberry commonly bend over, and where they touch the ground they form buds from which adventitious roots and new upright stems develop. A grapevine will take root where a node comes in contact with the soil. In the walking fern the tips of the leaves (Fig. 291) develop buds, roots, and new plants when in contact with the soil. The strawberry is an example of certain plants, including many grasses, which have horizontal branches (runners) on the soil surface that take root at intervals and produce new plants. In Bryophyllum, a persistent weed in cultivated fields of the West Indies, the leaves when they fall to the ground develop new plants from the notches in their margins.

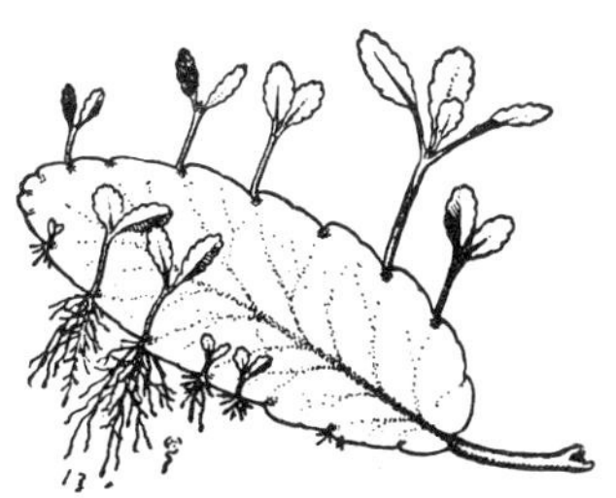

FIG. 129. Bryophyllum leaf, with young plants starting from the notches in the margin.

These illustrations, which might be indefinitely multiplied, show the importance of vegetative propagation in the increase and spread of plants. In nature it is probable that vegetative multiplication is as effective in spreading plants as is reproduction by seeds. By the former method the young plant is able to start more vigorously than a seedling, because it is able to draw water and food materials from the parent plant until its own root and leaf systems are well developed.

On this account, among wild herbaceous plants vegetative multiplication frequently determines which species shall dominate a habitat. Vegetative multiplication, for example, gives blue-grass the advantage over other plants in our lawns. Cat-tails and water lilies frequently exclude all other plants from their habitats by this means. Grasslands the world over are dominated by perennial grasses with underground stems. Denuded soil areas near cities are at first populated by annual weeds, but in a few years are occupied by perennials, which have gradually crowded out the annuals.

Office of Farm Management (J. S. Cates)

FIG. 130. Underground bulb of wild garlic, showing vegetative multiplication by the formation of three bulbs from the one planted. The terminal bud of the original bulb developed a flowering shoot, and three of its lateral buds formed the new bulbs.

Vegetative propagation of cultivated plants. In agriculture and horticulture, vegetative propagation is relied upon for starting many cultivated plants. Especially with plants that do not usually produce seed, and desirable hybrids and horticultural varieties that do not come true from seeds, is this method of propagation used. Potatoes, mint, horse-radish, sugar cane, sweet potatoes, and cer-

Office of Farm Management (J. S. Cates)

FIG. 131. "Sets" of wild garlic, showing the flowering heads, in some instances entirely made up of small bulbs. These small bulbs are very effective in spreading the plant, which often becomes a serious weed in pastures.

tain varieties of onion are examples of crop plants started in this way. Propagation by bulbs, corms, or rootstocks is the method commonly employed in starting lilies, tulips, hyacinths, irises, cannas, caladiums, and chrysanthemums. Most of our fruit trees are multiplied by budding and grafting, which are specialized methods of vegetative propagation. Geraniums, coleus, willows, currants, grapes, and most ornamental shrubs are grown from cuttings. These cuttings are pieces of a stem usually containing several nodes. Cuttings with a single node may be used when it is desired to propagate from a very limited supply of stock. It is obvious that a cutting must either have a small leaf surface or sufficient stored food to carry on growth until a leaf surface is developed.

Hardwood cuttings. Cuttings from woody plants are usually made when the wood is dormant in the fall or early winter. They are immediately tied in bundles of twenty-five or more, and buried in a trench with the uppermost buds turned downward. Sand or light soil is then added until the basal ends are covered 2 or 3 inches. This method of storage lessens the freezing and thawing of the upper buds, and in the spring the basal ends are warmed first and start developing roots. Cuttings may also be kept over winter in cool cellars in sand. In the spring the bundles are taken up and the cuttings set about 3 to 4 inches apart in trenches, with only the topmost end above the soil. The object of the winter treatment is to allow the formation of a *callus* at the lower end of the cutting. From this callus the first roots develop.

Softwood cuttings. Cuttings or roses, geraniums, chrysanthemums, coleus, and begonias are commonly termed "slips." These are propagated in greenhouses and hotbeds under glass. Most of the leaves are removed to prevent excessive transpiration and wilting, and the cuttings are placed in rows in sand kept constantly moist. Leaf cuttings are often used in propagating begonias. Parts of leaves, or complete leaves, may be used, and are simply laid on the surface of clean, moist sand. New plants develop at the base of the leaf, or at the lower end of the principal vein of a leaf segment. When the cuttings are transplanted, they should be placed immediately in moist soil before the root hairs are killed by drying.

Grafting and budding. As pointed out in Chapter XVIII, there are methods of propagating desirable varieties of a plant by growing them on the roots of a less desirable variety. Seedling apple and pear trees occur everywhere and are almost invariably worthless, but can be made to bear choice fruit by grafting twigs from desirable trees on them when young. This is so easily accomplished that it seems a pity to find hundreds of these trees along roadsides and fence rows bearing worthless fruit.

By grafting standard apple, pear, cherry, peach, and apricot cions on certain slow-growing stocks, with small root systems, dwarf trees are produced. Apples, for example, may be dwarfed by grafting them on " Paradise " stocks, pears on quince roots, and standard cherries on native shrubby plums. The root systems of these stocks, being small, decrease the water content of the cion and use a smaller part of the food manufactured in

FIG. 132. The banana, a perennial herb propagated by planting the "suckers" that develop from the thick underground stem. The aërial stem, because of the many layers of overlapping leaf sheaths, appears to be nearly a foot in diameter, but in reality it is only about as thick as that seen on a bunch of bananas.

growth. Consequently the plants accumulate food more rapidly and come into bearing earlier.

Plants may be grown in a region where they would otherwise perish by grafting them on other stocks. For example, the vineyards of France were threatened with destruction by the ravages of a root louse (*Phylloxera*). American grape roots were found to be immune to attacks of this insect, and the grape industry of France was saved by grafting the French vines on roots of American grapes. In Florida it was found possible to extend the cultivation of oranges farther north by growing the edible orange on the roots of the Japanese bitter orange, which is quite hardy.

In testing apple seedlings for their possibilities as new varieties, plant breeders take 1-year-old stems and graft them into the branches of a large, thrifty tree. As the tree has a large store of carbohydrate food at hand, fruit may be developed and its value determined on these cions the second or third year. To test the seedlings on their own roots would require perhaps from 10 to 15 years. In this instance grafting is used to hasten reproduction.

When two varieties of apple are grafted together and the cion does not make a perfect union with the stock, food may not pass freely from the cion to the stock. This results in accumulation of food above the point where the cion was set. The accumulated food leads to increased growth and the formation of a thicker trunk above than below the union.

Sprout forests. With the exception of the California redwood, cypress, and pitch pine, most conifers reproduce only by seed. Redwood, poplars, oaks, chestnut, and many other broad-leafed trees develop sprouts from stumps. Sprout forests, or coppice, as foresters call them, grow more quickly because the sprouts have a root system already established in the soil, while a seedling must first manufacture food and grow one. Chestnut coppice will grow large enough to furnish railroad ties in 25 to 35 years, or in about half the time required by seedlings. Sprout forests

Bureau of Science, P. I.

FIG. 133. Cuttings of sugar cane. This plant is not propagated by seeds, but by pieces of the stalk placed in furrows in the field and partially covered with earth.

do not grow as tall as forests developed from seed, and they are more subject to disease, because the trees become infected through the decay of the stump. Nevertheless, coppice is a rapid and efficient method of growing small timbers, posts, and pulp wood.

Advantages of vegetative propagation. Vegetative propagation has been found advantageous in crop plants wherever its use is possible, (1) because desirable varieties which do not come true from seeds may be perpetuated, (2) because some plants, like the sugar cane, banana, and horse-radish, do not produce seeds, (3) because it saves time in securing the product, as a longer period is required for the maturing of plants started from seeds, (4) because by grafting and budding plants may be grown in regions where they could not survive, and standard plants may be dwarfed to fit special conditions.

CHAPTER TWENTY-FIVE

FLOWERS AND FLOWER CLUSTERS

THE flower is a specialized shoot in which the reproductive processes, pollination and fertilization, lead to the production of

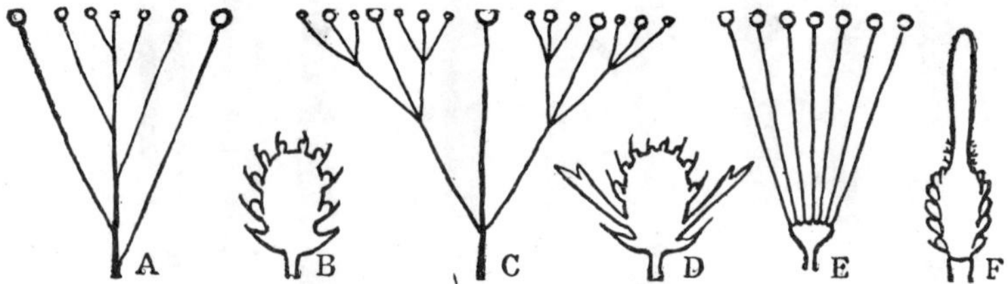

FIG. 134. Diagrams illustrating terms applied to flower clusters: *A*, corymb; *B*, head; *C*, compound umbel; *D*, head with disk and ray flowers; *E*, umbel; and *F*, spadix.

seed. Commonly the word "flower" is associated with the brightly colored parts that make many of our garden and house plants so attractive. But here we shall include under the term the simple structures associated with seed production in plants like the grasses, poplars, and birches, that have merely scale-like leaves and bracts inclosing the reproductive parts. In the conifers

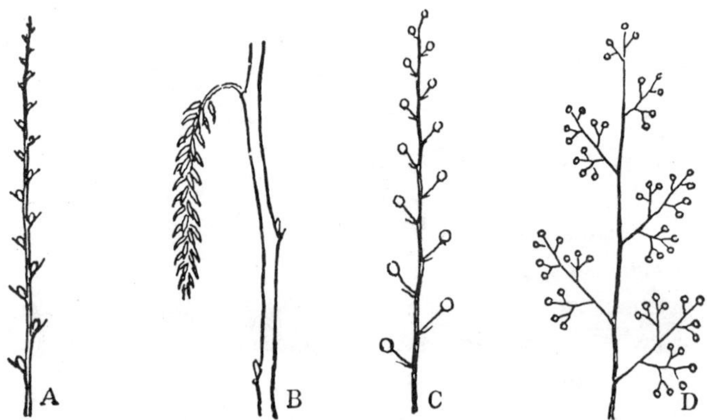

FIG. 135. Diagrams illustrating terms applied to flower clusters: *A*, spike; *B*, catkin; *C*, raceme; *D*, panicle.

the seeds are produced on scale leaves arranged spirally in cones. These cones may be looked upon as a lower type of flower, structurally very different from the flowers of the monocots and dicots.

Flower clusters. The arrangements of flowers on stems are very varied in different plants, and many descriptive terms have

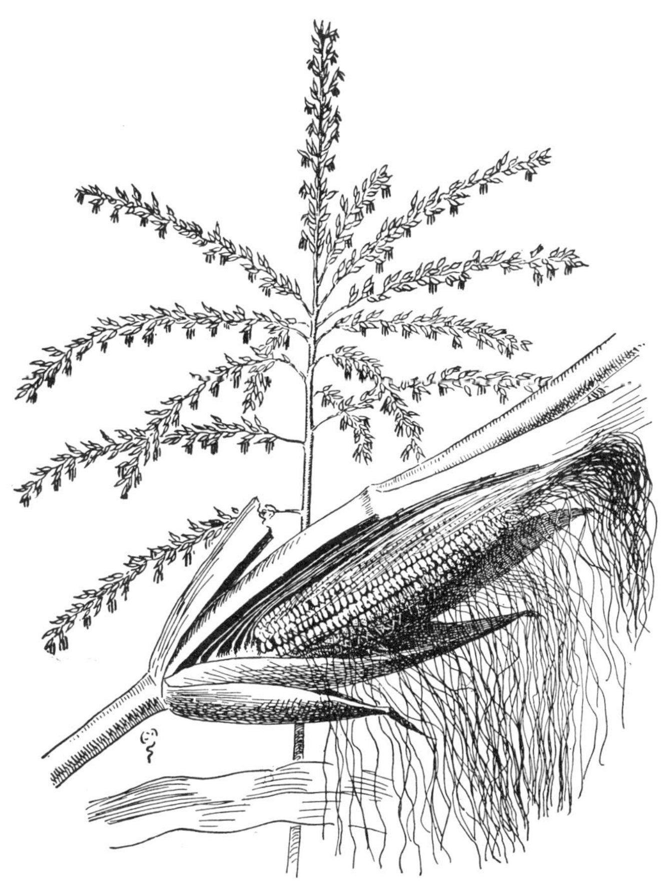

FIG. 136. Flowers of the corn plant. The panicle of staminate flowers (tassel) is shown above. Below are the pistillate flowers arranged in a spike (ear), inclosed by sheathing leaves. The only part of the pistillate flowers exposed to the air is the long style (silk).

FIG. 137. Fan palm with panicles of flowers. The photograph was made in Cuba.

FIG. 138. Catkins of staminate flowers of red oak (*Quercus rubra*).

been invented to describe them. In many plants the flowers occur singly at the ends of stems or lateral branches, as in the tulip and in some varieties of roses. The stem which bears a flower or flower cluster is called a *peduncle.* In flower clusters the small branches which bear the individual flowers are called *pedicels.*

In the *spike* of the common plantain, cat-tail, and timothy, the flowers are arranged along the sides of the upper part of the peduncle. The *catkin* of the willow, poplar, alder, and oak differs only in that it droops. The *raceme* of the garden currant differs from the spike in the fact that the flowers are borne on long pedicels at some distance from the peduncle. In the *umbel* of the onion, milkweed, carrot, and cherry, the pedicels all arise at the top of the peduncle and are of about the same length, so that the cluster is more or less flat topped. The *corymb* of the hawthorn

FIG. 139. Staminate inflorescence and opening bud of the white ash (*Fraxinus americana*).

is a flat-topped cluster in which the pedicels vary in length, the outer being the larger, and arise from different nodes of the peduncle. In a *panicle* the peduncle is repeatedly branched and the branches are wide-spreading. Yucca, hydrangea, the "corn tassel," and many other large grasses will exemplify the panicle. The *head* is a flower cluster in which the flowers are all crowded together at the end of a peduncle as in the red and white clover. The head of flowers of daisies, asters, sunflowers, and chrysanthemums are often mistaken for simple flowers, because the larger ray flowers have the appearance of petals and beneath them is a cycle of *bracts* that might be mistaken for the parts of a calyx.

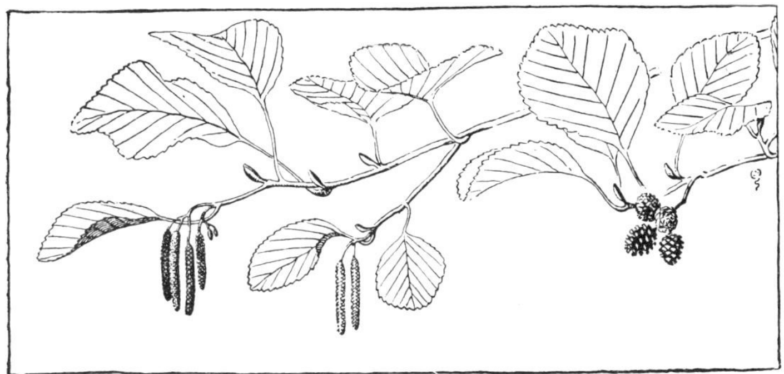

FIG. 140. Flower spikes of the alder. The two clusters on the left are staminate spikes; on the right the mature pistillate spikes are shown.

The parts of the flower. The apex of the flower stalk is called the *receptacle*. It is often enlarged and serves as a place of attach-

FIG. 141. Spike and flower of wheat (*Triticum vulgare*), showing two bracts, lemma (at right) and palea (at left), inclosing three stamens and a pistil with two plumose stigmas. At the base of the flower inside the lemma are two minute scales, the lodicules.

ment of the various floral organs. The outer whorl of scales or leaf-like organs is the *calyx*. It usually is green in color, and in

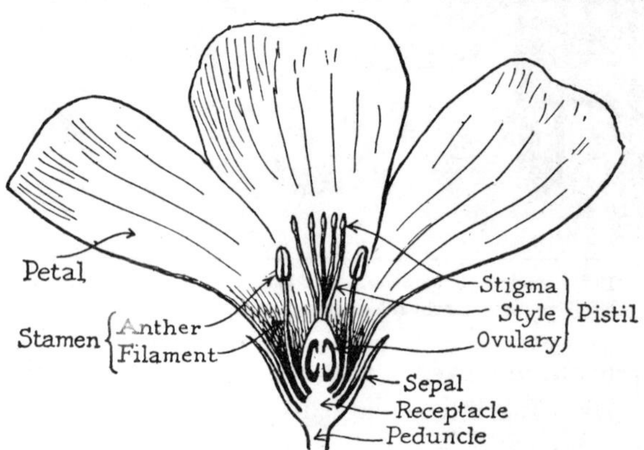

FIG. 142. Flower of flax sectioned to show the several parts of a typical flower.

the bud stage it completely incloses the flower. The individual parts of the calyx are called *sepals*. Next inside the calyx is a

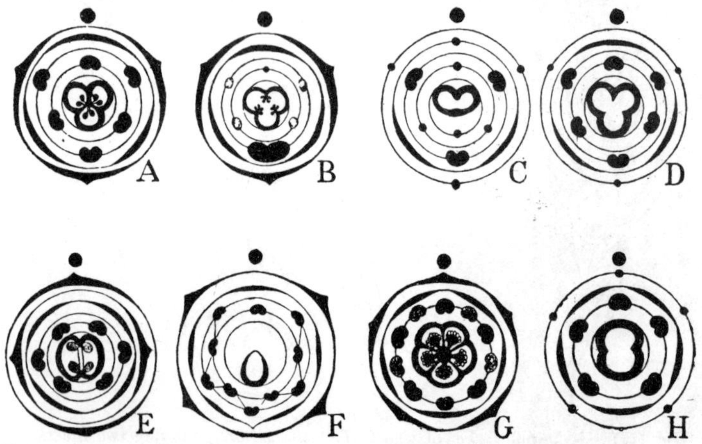

FIG. 143. Floral plans of several families of plants. The large dot above each figure represents the position of the axis; small dots represent missing members of a cycle. Unshaded stamens indicate presence of stamens without anthers. *A*, lily family; *B*, orchid family; *C*, most grasses; *D*, bamboo; *E*, mustard family; *F*, legume family; *G*, heath family; and *H*, composites. (*After Frank.*)

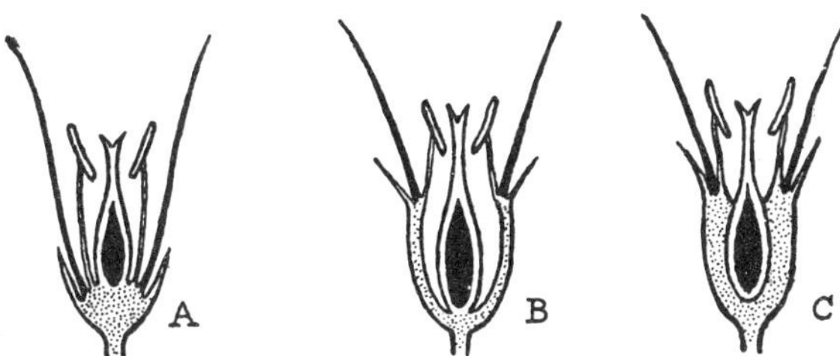

FIG. 144. Terms used in describing flowers. When the stamens, petals, and sepals are inserted on the receptacle below the pistil (*A*), they are said to be *hypogynous*, when united around the ovulary (*B*), they are *perigynous;* and when united above the ovulary (*C*), they are described as *epigynous*. The position of the ovulary in *A* is *superior* — that is, above the insertion of the stamens and perianth; in *C* the ovulary is *inferior*—that is, below the insertion of the perianth.

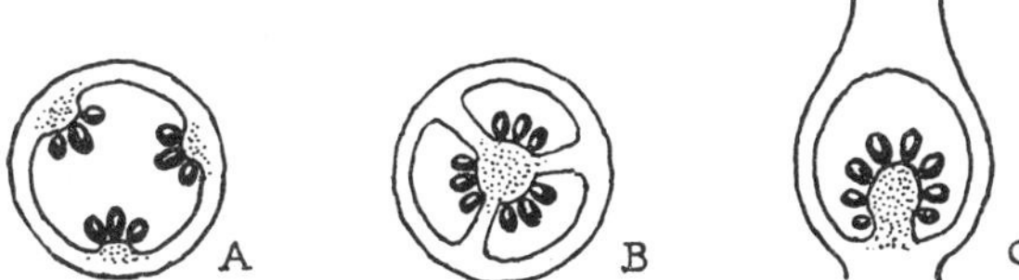

FIG. 145. Terms used in describing ovularies: *A*, one-celled, with ovules on three parietal placentas; *B*, three-celled, with ovules on three central placentas; *C*, one-celled ovulary with free central placenta. The surface to which the ovules are attached is the placenta.

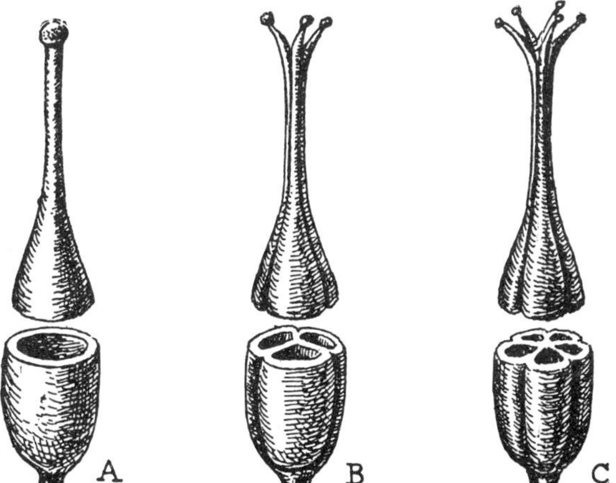

FIG. 146. Pistils formed of one, three, and five carpels, respectively.

FIG. 147. A tropical orchid (*Lælia*). The perianth consists of three sepals and three petals, one of which is greatly modified.

whorl of white or brightly colored leaves that make up the *corolla*. The several parts of the corolla are called *petals*. The corolla is usually the attactively colored part of the flower, but in some flowers, as in the tulip and clematis, the sepals have the same coloring as the petals. The calyx and corolla are often spoken of as the *floral envelopes*, because in the bud they form a wrapping, or envelope, for the inner parts of the flower.

Inside the corolla is a group of *stamens*, each composed of a stalk-like *filament* and an *anther*, that contains the pollen. The center of the flower is occupied by one or more *pistils*, each made up of an *ovulary*, *style*, and *stigma*. The ovulary is the enlarged part of the pistil that contains the *ovules*, which develop into the seeds. The style is the stalk above the ovulary that bears at its summit the stigma. The stigma is usually an enlarged surface, which secretes a sticky, sugary solution in which the pollen grains are caught and in which they germinate. The pistils and stamens are called the "essential organs" of the flower, because they produce the ovules and pollen which are the two elements necessary in the production of seed.

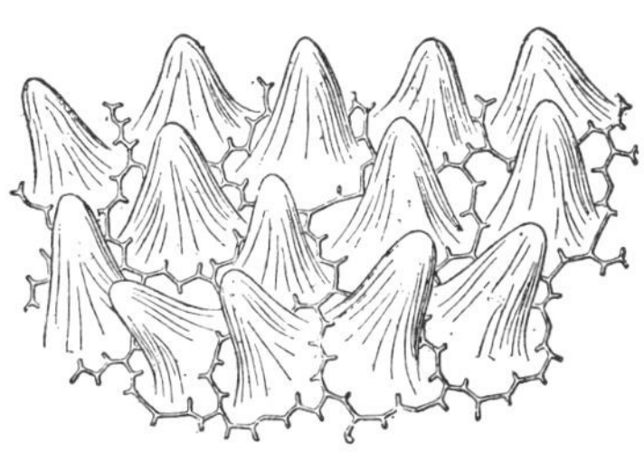

FIG. 148. Epidermis from the petal of a geranium. The velvety appearance of many leaves and flowers is due to similar projections of the epidermal cells.

Pistils are variously constructed out of one or more leaf-like parts called *carpels*. For example, the pod of the bean or pea is composed of a single carpel. It may be compared to a simple leaf folded at the midrib, with the margin united. The fruit of the yucca, tulip, and lily is composed of three carpels. The apple, pear, and quince pistils are made up of five carpels, which constitute the hard papery walls of the seed cavities in the "core."

The variety of floral structures. The above is a description of a typical flower; but in the plant world we find an almost endless variation in the number, form, size, color, and arrangement of

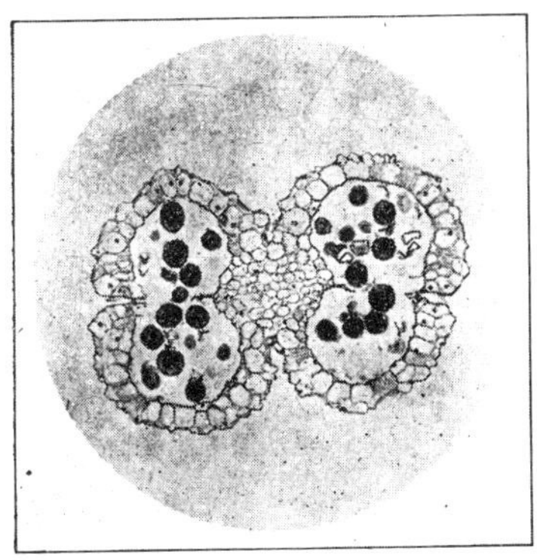

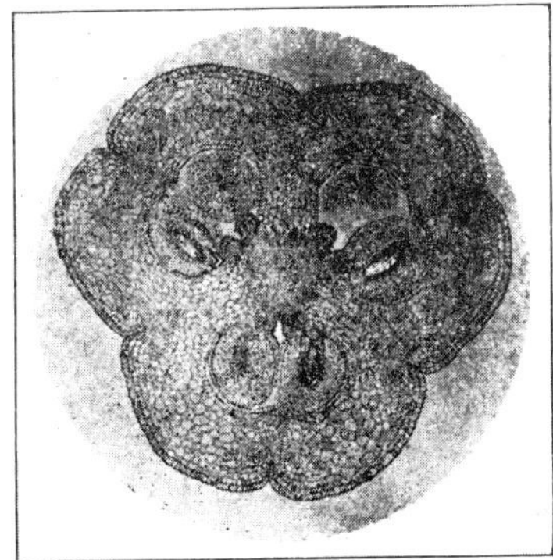

FIGS. 149 and 150. Cross-sections of the "essential organs" of a flower. At the left, anther of a lily, showing the four microsporangia and the contained pollen (microspores); at the right, ovulary of a lily, showing six of the ovules, arranged in pairs within the three carpels.

these parts. In some flowers the calyx, or the corolla, may have its parts united into a tube, or one or both may be wanting. Or the flowers may lack either pistils or stamens. For example, the red maples and the cottonwood bear pistillate flowers on some trees and staminate flowers on others, and the corn has staminate flowers on the tassel and pistillate flowers on the lateral branches or ears. It is not our purpose to name and describe here the many different variations in floral structure; a visit to a conservatory or a tramp through the near-by fields and woods is the most effective way of securing an idea of the great diversity of flowers.

CHAPTER TWENTY-SIX

SEXUAL REPRODUCTION IN FLOWERING PLANTS

THE most important fact that should be associated with the flower is that in it occurs the sexual reproduction of the plants. Sexual reproduction takes place by the union of two specialized cells, the *gametes*. One of these cells is called the *male gamete*, or *sperm*, and the other the *female gamete*, or *egg*. When these cells unite they form a single cell called a *zygote*, or *fertilized egg*. This process of sexual union, or fertilization, is the first step in the development of nearly all plants, and every individual plant normally starts as a single cell, no matter how complicated it may be at maturity. The flower is a complicated structure in which the development of the gametes, fertilization, the formation of the zygote, and its further development takes place. These several steps ending with the seed are described in the following paragraphs.

Pollination. If the anthers of a lily or nasturtium are examined, a fine yellow powder will be found which under a microscope appears as a multitude of small grains. This is *pollen*. In the production of seed it is necessary that the pollen grains be carried to the stigma, and this transfer is called *pollination*. In some plants the pollen merely falls by gravity on the stigma. Wheat and oats are examples of plants that are pollinated in this way. In other plants, like the pines, elms, birches, oaks, rye, and corn, the pollen is carried by the wind. It is an interesting fact that the stigma of wind-pollinated flowers are usually roughened by hairs, which probably make them more effective in holding the pollen.

The pollen of most plants with conspicuous flowers is carried by bees, flies, butterflies, or moths. As the body of one of these insects is rough or hairy, pollen grains become attached to it when the insect enters a flower. Then when the insect passes to another flower, some of the pollen from the first flower is brushed off on the stigma of the second. Thus pollination is brought about by the insects in the course of their visits to successive flowers. It is an advantage to the plant to have its pollen carried by insects directly from flower to flower instead of having it blown about and reaching a stigma by mere chance. If the amounts of pollen produced by the pine, corn, ragweed, and other wind-pollinated plants are compared with the amounts produced by plants that are pollinated by insects, it will be seen that insect-pollinated plants generally produce less pollen and are no less effectively pollinated.

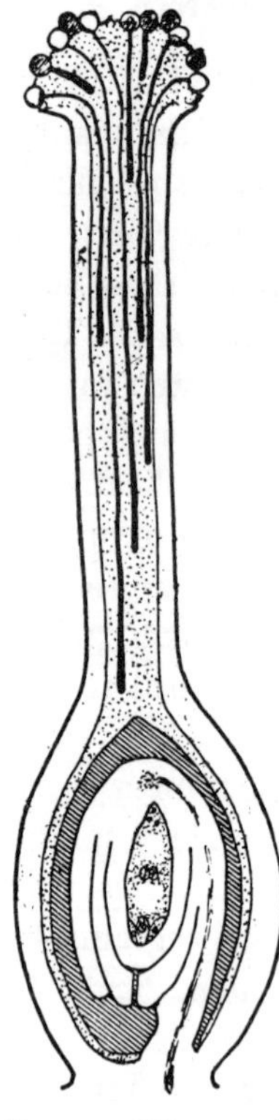

FIG. 151. Diagram of a pistil with germinating pollen grains and pollen tubes of various lengths. The embryo sac is in the seven-celled stage, with a central fusion nucleus and an egg (below). Fertilization occurs when one of these pollen tubes reaches the egg. (*After Buchholz.*)

Why insects visit flowers. Insects do not visit flowers to carry pollen for the plants. They eat the pollen or feed their young on it, and they also secure nectar from the flowers. The nectar is a watery solution containing sugar, which is secreted by glands called *nectaries*. One or more of these nectaries is usually located near the base of the corolla, inside the flower. The insects visit the flowers and secure food for themselves, but as they make their visits they brush against the anthers and become covered with pollen. Later they come in contact with the stigmas of other flowers and leave pollen adhering to the stigmatic surface. In this way they perform a service for the plants. The perfumes of flowers assist the insects in finding them, and conspicuous white or brightly colored parts of flowers may aid in the same way. The massing of many small flowers in clusters and heads certainly makes them more conspicuous.

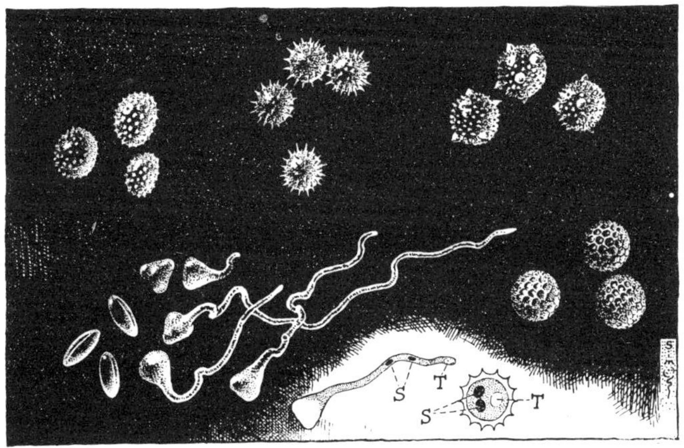

FIG. 152. Pollen grains and pollen tubes. *S* is the two sperms or male cells, and *T* the tube nucleus.

Cross-pollination. When a flower is pollinated with its own pollen or with that from another flower on the same plant, it is said to be *self-pollinated.* If the pollen comes from another plant, a flower is said to be *cross-pollinated.* In many plants it makes no difference whether the pollen comes from the stamens of the same plant or from those of another plant. In the common tobacco plant the pollen may be transferred to the stigma of the same flower, and seeds will be produced. In some plants, however, it is only when the flowers are cross-pollinated that seeds are formed. The sunflower is a good example of this kind of plant. In still other plants, seeds that are formed after self-pollination are less vigorous than those formed after cross-pollination.

In some species of plants that are self-sterile, the pollen from plants started by cuttings will not fertilize the egg cells on other plants derived from cuttings of the same plant. This is a matter of practical importance in cultivating blueberries, which are propagated by cuttings. Unless fertilization takes place, perfect fruits are not formed. Hence, cuttings from different sources must be alternated in the field in order to secure abundant production.

Certain varieties of strawberries, which are usually propagated by runners, must be alternated in culture in order to secure fruit, because they are either self-sterile or produce no pollen.

From the above statements it will be seen that cross-pollination is an advantage to some plants. Many flowers have arrangements that make self-pollination impossible. Often the anthers do not shed their pollen at the time when the adjoining stigma is in condition to receive it. The pollen may be shed either before or after the ripening of the stigma. In such plants there is little possibility of the stigma's being pollinated from the stamens of the same flower. So, as insects go from one flower to another they transfer pollen from flowers in which the pollen is ripe to flowers in which the stigmas are ripe. This favors cross-pollination.

It is exceedingly interesting to study the various other mechanisms that favor cross-pollination, but it should be done in the field or with the flowers in hand. In the white lily the stigma is out of reach of the insects when the pollen is shed. In other plants the pistillate and staminate flowers may occur on different individuals, or on different branches of the same plant. In primroses and bluets the stigmas and stamens each have two different lengths; the flowers on one plant have long styles and short stamens, while the flowers on another plant have short styles and long stamens.

The most remarkable cases of cross-pollination by insects are those in which a particular species of insect is necessary for the pollination of a plant. Such relations exist in the yuccas and in some orchids. In the absence of the particular insect, pollination and seed production fail. Yuccas may be grown in our Northern states, but in certain localities they fail to produce seeds because the moth (*Pronuba*) needed to pollinate the flowers does not live there. The Pronuba moth collects pollen from the anthers of the yucca flower, and carries it to the top of the pistil and pushes it down into the tubular stigma. The eggs of the moths are then laid in the ovulary by piercing the pistil wall. As a result of pollination the ovules develop and furnish food for the young Pronuba larvæ. But the larvæ eat only a small percentage of the ovules. So the larvæ of the moth are fed on the ovules that resulted from pollination. The yucca matures many undisturbed seeds in every pod where none are produced in the absence of the moth. How the relation became established we do not know, for the Pronuba moth never sees her offspring and they never see her.

Germination of the pollen. Further steps in the production of seed are the germination of the pollen, the formation of the pollen tube, and the fertilization of the egg. The details of these processes vary in different plant groups, but the account here given is representative of what is found in many flowering plants. At the time of shedding, the pollen grains of many plants contain three cells. One of the three cells is active in the formation of the pollen tube; the other two are the sperms, or male gametes. The stigma, as we have learned, secretes a sticky fluid containing sugars, acids, and other substances. Pollen germinates best in fluid secreted by the stigmas of the same kind of plant, and it usually germinates imperfectly, or not at all, on the stigmas of other kinds of plants. Germination results in the formation of a microscopic tube that grows downward among the cells of the stigma and style into the ovulary and into an ovule. Usually this is but a short distance. In corn, however, the silk is the style and stigma, and the pollen tube must grow several inches, or a foot, down the silk before reaching the ovules below. As it grows downward, the two sperms move along near the end of the tube.

To summarize the steps in the formation and movement of the male gametes preceding fertilization, there is (1) formation of pollen in the pollen sacs, (2) opening of the pollen sacs and shed-

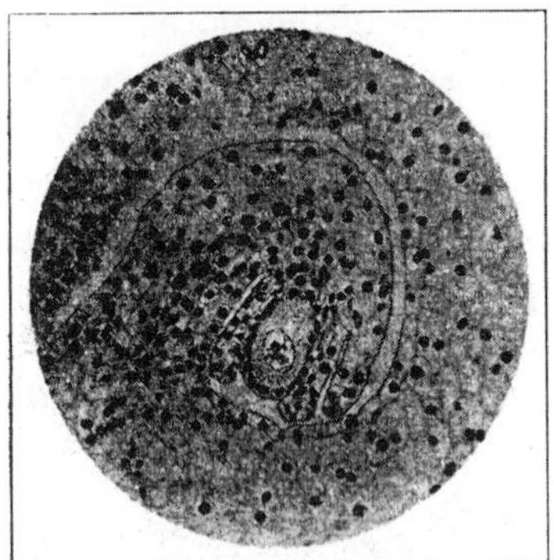

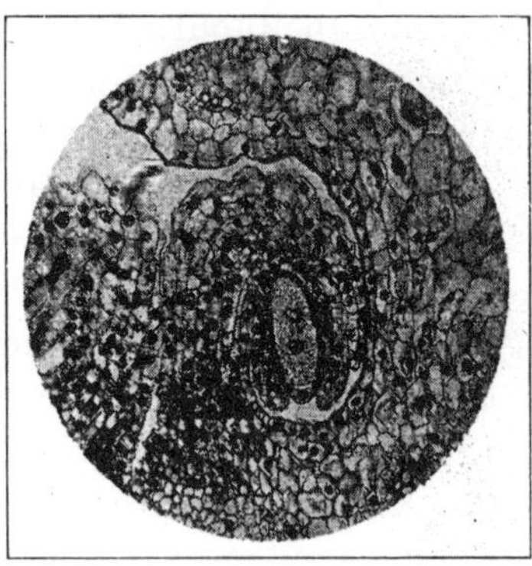

FIGS. 153 and 154. Cross-sections of ovules. At the left, megaspore of lily; note the surrounding nucellus and two integuments making up the ovule. At the right, the first division in the megaspore (embryo sac), resulting in two nuclei.

ding of the pollen, (3) pollination, or transfer of pollen to the stigma, (4) germination and formation of the pollen tube, (5) growth of the pollen tube into the ovule, and (6) movement of the sperms to the end of the pollen tube.

Development of the egg in the ovule. Inside the ovule a parallel series of cell activities is going on which results in the production of the embryo sac and the egg, or female gamete.

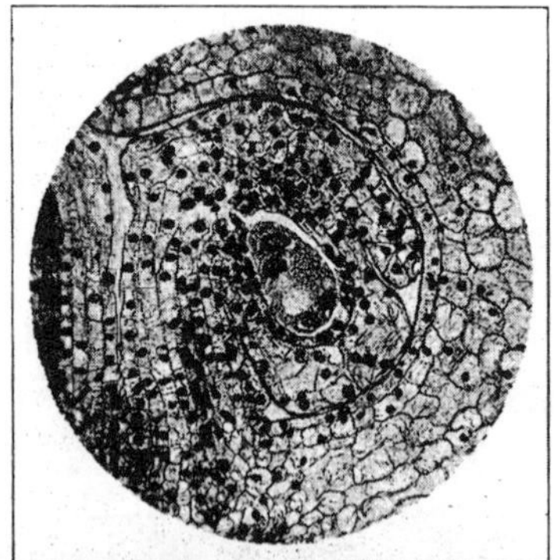

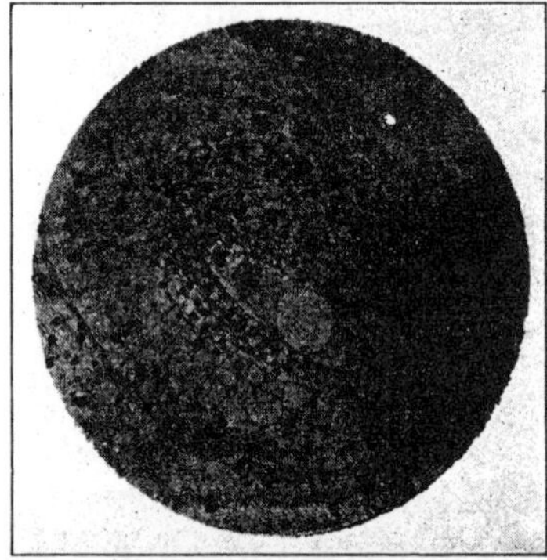

FIGS. 155 and 156. Cross-sections of ovules. At the left the second division in the embryo sac, resulting in four nuclei. At the right, the eight-celled stage of the embryo sac; two nuclei are about to unite to form the fusion nucleus.

At an early stage the center of the ovule is occupied by a single large cell. This cell continues to enlarge as the ovule grows, and its nucleus divides, forming two nuclei, each of which divides a second and third time, forming all together eight nuclei. The large cell we will now call the embryo sac. There are four nuclei near each end of it. Three nuclei out of each group move nearer the ends of the sac and form cells. The two remaining move to the center of the sac and fuse, forming one large fusion nucleus. One of the cells at the outer end of the embryo sac is the egg. This completes the development of the female gamete.

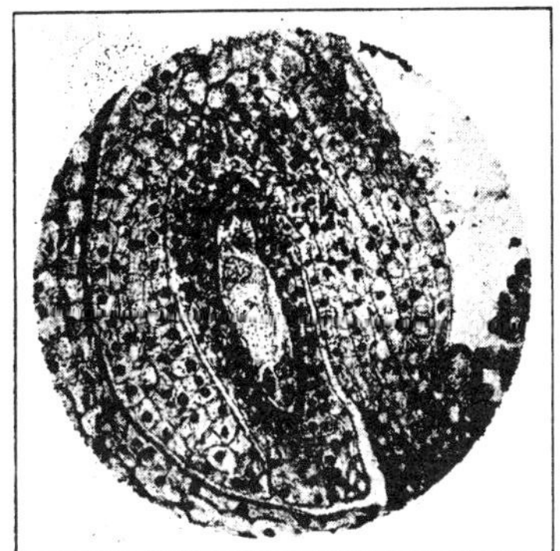

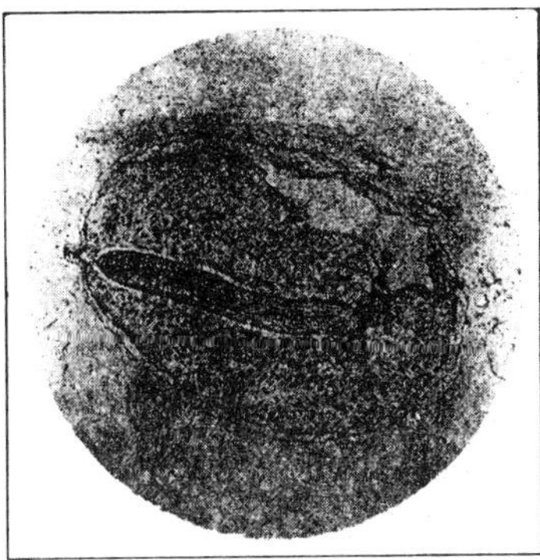

FIGS. 157 and 158. Cross-sections of ovules. At the left, the seven-celled stage of the embryo sac just before fertilization; note the large fusion nucleus and just above it the egg. At the ight, the embryo sac after enlargement through the development of the embryo and the surrounding endosperm.

The series of events inside the ovule preceding fertilization begins with (1) the formation of a large cell, (2) three successive divisions of the nucleus, forming the eight-celled embryo sac, (3) the formation of three cells in each end of the embryo sac, (4) the fusion of two of the nuclei at the center of the sac, and (5) the changes connected with the development of one of the three cells at the outer end of the sac into the egg.

Fertilization. At the beginning of fertilization the pollen tube grows into the embryo sac and the two sperms are liberated. One of the sperms moves to the egg and fuses with it, forming the fertilized egg. In more general terms, the male gamete unites with the female gamete and forms a *zygote*. This is the beginning, or first cell, of the embryo. Fertilization is the essential part of sexual reproduction both in plants and animals, and it marks the actual beginning of a new individual. After fertilization the zygote may develop into a new plant or animal of the same kind as its parents. It should be clear that the sperms from one pollen tube fertilize the egg in only one ovule, and that to fertilize all the eggs in a pistil, as many pollen tubes must grow down through the style as there are ovules in the ovulary below.

The endosperm. At the same time that the egg is fertilized, the second sperm from the pollen tube unites with the fusion nucleus at the center of the embryo sac and forms the *endosperm nucleus*. The endosperm nucleus is therefore a nucleus made up of three nuclei, and on this account the process by which it is formed is called *triple fusion*. This fusion is followed by rapid cell division and the formation of a soft tissue filling the rapidly enlarging embryo sac. The tissue thus formed is the endosperm, and into it pass large amounts of food from the plant. In the grains and some other kinds of seeds the endosperm occupies most of the space within the seed coat.

The embryo. The zygote, or fertilized egg, starts growth and cell division at once. As the mass of cells enlarges, it grows farther and farther into the endosperm, from which its food materials are derived. It finally takes on the form of the embryo, or young plant, that we find inside the seed. As development proceeds, the growth is slowed down and finally ceases until the seed germinates. Sometimes, as in the bean, all the contents of the endosperm are consumed and all that remains of it in the mature seed is a thin layer of cells around the embryo. In other seeds the endosperm partly or wholly surrounds the embryo with a thick layer of cells, as in the castor bean, corn, and lotus.

The perisperm. In some seeds the tissue immediately surrounding the embryo sac becomes much enlarged and accumu-

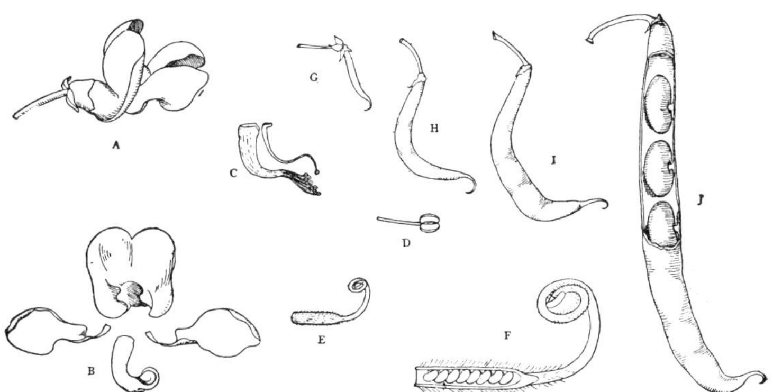

FIG. 159. Floral organs and development of fruit of bean: *A*, side view of floral envelopes; *B*, petals; *C*, stamens in two groups; *D*, stamen with anther; *E*, young pistil; *F*, young pistil enlarged to show ovules; *G*, *H*, *I*, and *J*, stages in the development of the fruit.

lates food materials. In the mature seed this tissue resembles the endosperm. Physiologically, endosperm and perisperm are alike, in that they supply food to the growing embryo during the germination of the seed. Corn cockle, spinach, and pepper seeds have the food supply in the perisperm tissue.

The seed. The final product of pollination and fertilization is the seed. Its complete development ends the rôle of the flower. During the development of the endosperm and embryo the wall of the ovule, commonly called the *integument*, enlarges and may change in various ways, sometimes forming a hard outer coat of stone cells and an inner soft coat. The primary parts of a seed, then, are (1) the seed coats, (2) the embryo, and (3) the endosperm (or perisperm). When we plant a seed, we are placing a small, partly developed plant, with a limited supply of food, under conditions in which it may continue its growth. Seeds are discussed in more detail in the next chapter.

CHAPTER TWENTY-SEVEN

FRUITS AND SEEDS

THE term "fruit" is commonly used to designate a great variety of organs that are developed as a result of the flowering

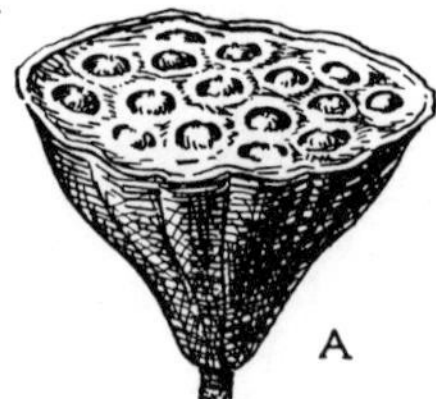

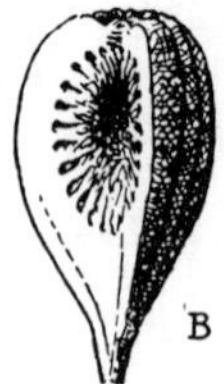

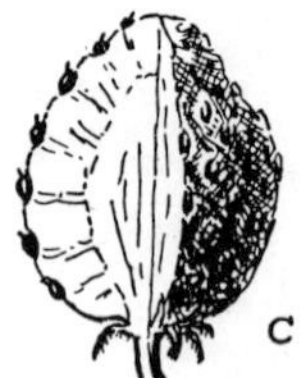

FIG. 160. Types of fruits: *A*, enlarged receptacle with imbedded nut-like pistils (water lotus); *B*, fleshy stem tip with a central cavity containing many minute flowers (fig); *C*, enlarged fleshy receptacle with pistils attached to surface (strawberry); *D*, fleshy urn-shaped calyx-tube with pistils inserted on the inner surface (rose).

of plants. The direct result of pollination and fertilization is the production of the seed. The indirect effect of pollination is the further development of adjacent structures. The pistil, or at least the ovulary wall, enlarges and sometimes becomes greatly thickened. Primarily the fruit is the enlarged pistil or ovulary, but in many cases the calyx and the receptacle also enlarge and form a part of the fruit, sometimes most of it. The pineapple is a fruit in which an entire flower cluster has become fleshy, and this fruit is formed without fertilization. Like the common banana, it is seedless.

In some fruits the enlarged pistil forms a thin wall inclosing the seeds. At maturity the pistil wall dries out, forming a *dry fruit*. In others the pistil wall, or some of the adjacent structures, become enormously enlarged by the formation of soft parenchyma tissue in which sugars, fats, acids, and other substances accumulate. These are distinguished as *fleshy fruits*. Among the dry fruits the most familiar are the *grains*, illustrated by wheat, barley, and corn. The outer coat is the ovulary wall; the embryo is small, and most of the seed is made up of the starchy endosperm.

Very similar is the *akene*, a small dry fruit inclosing a single seed, as in the buckwheat, buttercup, and sunflower. The pod, or *legume*, is a dry fruit of one carpel which splits down the sides when mature, freeing the several seeds, as in the bean, pea, clover, peanut, and alfalfa.

Among fleshy fruits the commonest is the *drupe*, or *stone fruit*, illustrated by the plum, cherry, olive, and peach, in which a single seed is surrounded by an inner stony layer and an outer fleshy layer. The *pome* is a fleshy fruit in which the receptacle enlarges and surrounds the pistil (core), which is composed of five carpels each containing several seeds. Pomes include apples, pears, and quinces. The fruits of tomatoes, potatoes, currants, grapes, cranberries, and blueberries are true *berries* having a fleshy wall inclosing several seeds. The *pepo*, or *gourd* fruit, is exemplified by the cucumber, watermelon, and cantaloupe. It is a greatly enlarged and fleshy ovulary containing numerous seeds.

There are many other kinds of fruit distinguished, but they are too numerous to describe here. In the strawberry the fruit is the greatly enlarged receptacle bearing numerous little akenes on its surface. The rose fruit is similarly an enlarged cup-like recep-

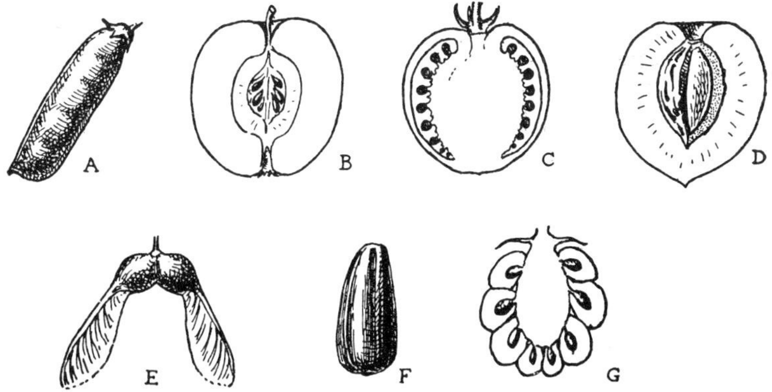

FIG. 161. Types of fruits: *A*, legume (pea); *B*, pome (apple); *C*, berry (tomato); *D*, drupe, or stone fruit (peach); *E*, samara, or key fruit (maple); *F*, akene (sunflower); *G*, aggregate fruit (blackberry).

FIG. 162. Pineapples growing in Porto Rico. The entire flower cluster becomes fleshy and forms the fruit. Like most of the members of the Bromelia family, the leaves are leathery, rigid, and arranged in a rosette. Pineapples are propagated by planting cuttings of small lateral branches.

tacle, with small akenes on the inner surface. The fruit of the fig is a greatly enlarged and hollow peduncle, with numerous flowers lining the inside. Blackberries and dewberries are clusters of fleshy pistils held together by the inclosed receptacle. Raspberries differ in that the cluster of fleshy pistils separate from the receptacle when ripe.

The development and ripening of fleshy fruits. The process of development and ripening may be illustrated by the changes that occur in an apple. As soon as fertilization occurs, the tissues that finally make up the fruit begin to enlarge by cell division. Food materials from the stem pass into this tissue and accumulate as starch, acids, fats, and proteins. In the young green apple the cells are very dense and gorged with starch. The sourness is due to malic (Latin: *malum*, an apple) acid. During the process of ripening, great chemical changes occur. The starch is changed to sugar, — sucrose, glucose, and fructose. The water content increases, and the acid gradually becomes less and less. The middle lamella of the cell walls is partly dissolved and the cells separate more or less, thus producing intercellular spaces, and making the fruit softer and more "mealy." The ripening process begins at the core and gradually extends outward, until all the tissues are affected.

U. S. Dept. of Agriculture

FIG. 163. Fruit of mango, now being successfully grown in southern Florida.

The middle lamella, composed of calcium pectate and pectose, makes up part of the cell walls. The changing of these substances to pectin and pectic acid, jelly-like substances, contributes to the softening of the fruit. The pectic compounds are important in jelly making, and those fruits that contain large amounts of them form jellies readily when they are mixed with sugar, boiled, and allowed to cool. Fruits like the quince, apple, and currant are plentifully supplied with pectic compounds. In elderberries and grapes the pectic compounds are less abundant, and juices of apple or quince are commonly added to them in jelly making to make them jell more readily. When fruit juices and sugar are boiled too long, they may not jell. This is because the pectins have been chemically changed to other substances which do not have this property.

Recently it has been found possible to remove pectin from carrot roots, which contain large amounts, and the pectin may be added to fruit juices to insure jelling.

U. S. Dept. of Agriculture

FIG. 164. Persimmon fruits. The persimmon grows wild over a large part of the Southeastern states, and improved varieties are now cultivated.

Economic importance of flowers and fruits. The economic value of flowers lies chiefly in their use for decorative purposes, but certain flower clusters like the artichoke, pineapple, and cauliflower are used as fruits or vegetables. The fruit industry needs only to be mentioned to call to mind the vast scale upon which plants are grown for their fleshy edible fruits. It should be noted that ripe fruits are made up largely of water pleasantly flavored with sugar, dilute acids, and aromatic substances. The amount of food actually present is not large. Most fruits contain from 10 to 15 per cent of carbohydrates, 1 to 2 per cent of proteins and fat. Persimmons and bananas run somewhat higher in carbohydrates; olives and avocados may contain as much as 10 per cent of oil. Fruits are valuable chiefly for the variety which they add to our diet. Through canning, preserving, and drying they are made available at all seasons of the year.

The coconut fruit when mature consists of a thick, fibrous husk surrounding the seed. This fruit is an important source of coarse fibers, both in the tropics and in temperate regions. Several million pounds are annually imported into America and manufactured into door mats, floor mats, and coarse brushes.

The olive is the source of olive oil, which is extensively produced in Spain, Italy, and California. In recent years it has been partly replaced as a salad oil by cottonseed and corn oils, or mixtures of them.

The structure of seeds. Although seeds vary as much in form as do other plant organs, the different arrangements of the three essential parts may be illustrated by a castor bean, a lima bean, and a grain of corn.

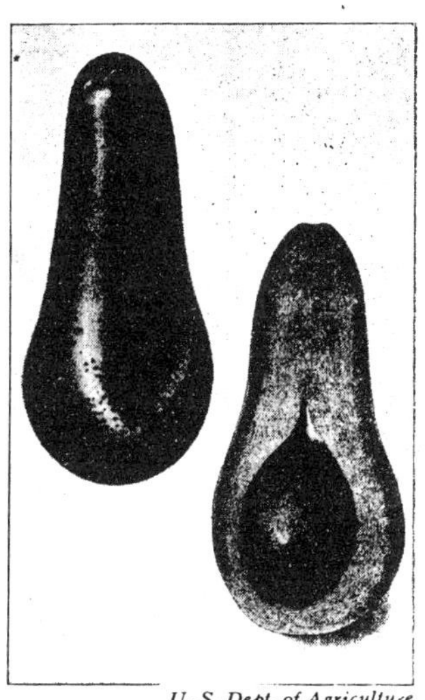

U. S. Dept. of Agriculture

FIG. 165. Avocado, or alligator pear, a salad fruit now being grown in southern Florida and California.

In the castor bean the seed coats consist of a hard outer layer and a thin inner membrane. These inclose an endosperm, which is a mass of cells containing food in the form of starch, oil, and protein. Within the endosperm lies the embryo, ready to grow when favorable external conditions for germination come. The embryo consists of the hypocotyl and two very thin cotyledons, with a small bud between the cotyledons, called the *plumule*. The cotyledons are the first leaf-like organs of the plant. The hypocotyl is the first stem, and the plumule is the first bud. No root is found in the embryo; but when the seed germinates the hypocotyl elongates, and from its basal tip the primary root develops. The cotyledons at first absorb food from the endosperm, expand, and when exposed to the light turn green and carry on photosynthesis. The plumule grows upward and forms the stem. These early stages of growth use up most of the food in the endosperm.

The lima-bean seed consists merely of the embryo, with a seed coat inclosing it. The food in this seed has already been absorbed into the embryo and stored in the greatly thickened cotyledons; that is, the young embryo continued its growth in the ovule and absorbed all the food from the endosperm. The parts of the embryo are the same as in the castor bean, but the cotyledons are

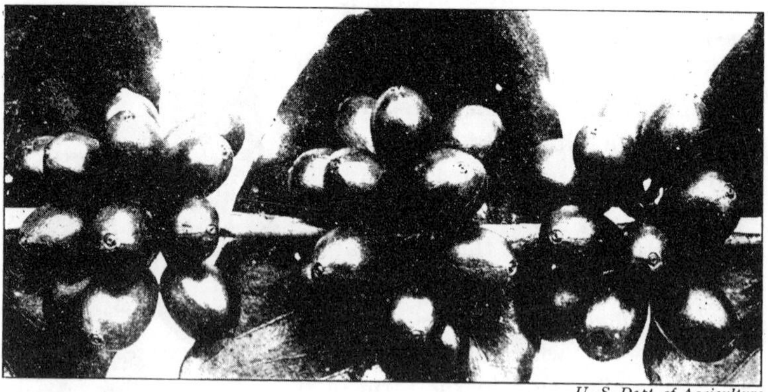

U. S. Dept. of Agriculture

FIG. 166. Coffee berries, natural size. Each contains two seeds.

thick and contain a great supply of food. The bean is an example of a large group of plants, including the pea, squash, apple, and pumpkin, in the mature seeds of which the endosperm is lacking.

A grain of corn is an example of a third kind of seed. In it there is a large endosperm, with a small embryo near one end of the seed. The embryo differs from the embryos of the bean and the castor bean in that it has only a single cotyledon, wrapped more or less around the hypocotyl and plumule. The plumule grows upward and forms the aërial shoot. As in the castor bean, the cotyledon is the absorbing organ through which the foods in the endosperm enter the young plant, but in the corn the cotyledon is not forced out of the soil by the elongation of the hypocotyl.

The flower and embryo in monocots and dicots. In discussing the subject of stems, attention was called to the fact that flowering plants are divided into two great groups, the monocots and dicots. The monocots have parallel-veined leaves; the bundles of the stem are closed (have no cambium) and are not arranged in a circle.

The terms "monocot" and "dicot" (or, as they are frequently written, "monocotyledon" and "dicotyledon") are based on the apparent number of cotyledons in the embryo, whether there are one or two. Any one who has watched plants beginning to grow in a garden will recall the two cotyledons of the bean, pumpkin, sunflower, and radish, raised above the soil. Seeds, of fruits, of our broad-leafed forest trees — maple, ash, tulip, linden — may be readily secured and germinated, and they too will be seen to be dicots. It will also be recalled that these plants have net-veined leaves.

The cotyledon of a monocot is usually an absorbing organ that remains below the ground in contact with the endosperm, and in wheat, corn, and other grasses it is the first leaf that appears above the ground — not the cotyledon. In other monocots, like the onion, the cotyledon is leaf-like and rises above ground.

The two groups differ in their flowers also. In the monocots the number of parts of the calyx and corolla is usually three, and the stamens and divisions of the pistil are three or some multiple of three. In the dicots the parts of the flower are typically in fives or fours, or in a multiple of these.

Thus the names "monocot" and "dicot" relate to the form of the embryo; but the two groups are further distinguished by differences in leaf venation, bundle structure, bundle arrangement, and flower plan.

The gymnosperms and angiosperms. We have previously learned that the conifers bear their seeds on scale leaves arranged in cones (page 232). A study of one of these cones shows that the seeds are formed on the upper surface of the scales and are not inclosed in capsules. When the scales mature and become dry, the cone opens and the seeds fall out. The word "gymnosperm" means "naked seed," and this is the group name for the conifers and all other plants whose seeds are not inclosed in an ovulary.

The angiosperms are what we usually call the flowering plants, although some of them, like the grasses and many forest trees, produce small, inconspicuous flowers without colored parts The seeds of an angiosperm, in contrast to those of the gymnosperm, are inclosed in an ovulary commonly called a pod or capsule, as in the bean, horsechestnut, hickory nut, and watermelon.

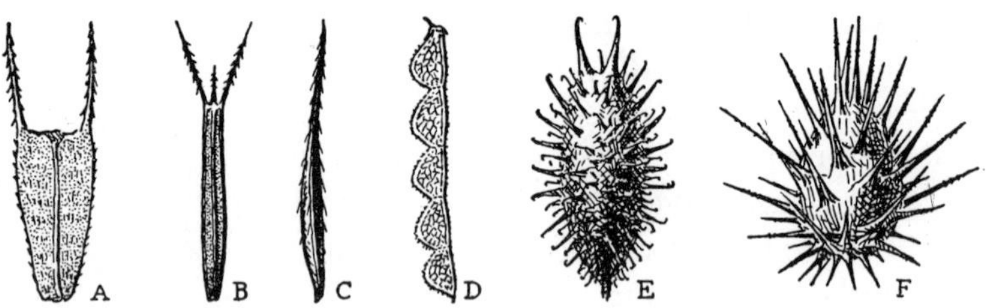

FIG. 167. Fruits frequently transported by animals: *A*, beggar-ticks (*Bidens*); *B*, Spanish needles (*Bidens*); *C*, sweet cicely (*Washingtonia*); *D*, tick trefoil (*Desmodium*); *E*, cocklebur (*Xanthium*); and *F*, sand bur (*Solanum*).

The term "angiosperm" means "hidden or covered seed."

The gymnosperm seed consists of an embryo, surrounded by an endosperm (rich in carbohydrate, fat, and protein material), and two seed coats. The embryo has several distinct cotyledons.

Separation of the seed. Seeds become free of the fruit or the parent plant in various ways. Fruits of the akene type (sunflower, Spanish needle) are dry, one-seeded fruits, and are set free at maturity by abscission from the receptacle. In the case of some legumes, like the bean, the pods split open and the two halves curl and twist, forcibly expelling the seeds. In the witch-hazel the pod dries out at maturity and the outer wall shrinks more than the inner, thus producing a tension. When a certain tension is reached, the four-seeded capsule suddenly springs apart, throwing the seeds several feet. In the walnut, coconut, and many fleshy fruits (e.g., the apple) the seeds are set free only by the decay of the fruit. The hard, resistant fruit wall of the coconut is of advantage, since this palm is a common seashore plant. The seeds may be carried for weeks by ocean currents without injury. Orange and lemon seeds sometimes germinate inside the ripe fruits.

The dissemination of seeds. The wind is probably the most important agent in transporting seeds. How far a seed may be carried by the air depends on the amount of surface it exposes in proportion to its weight. The greater the surface in proportion

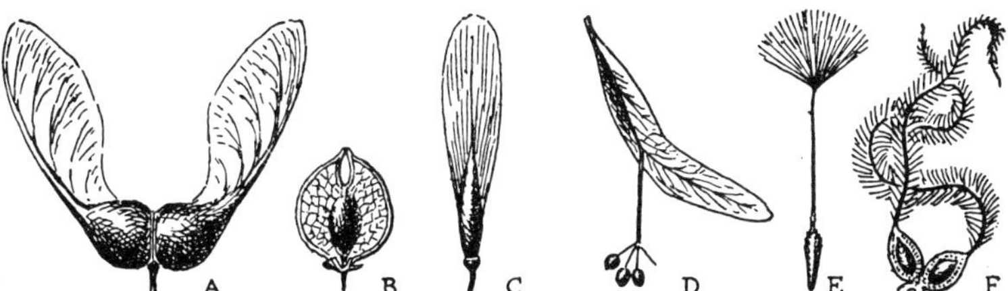

FIG. 168. Fruits frequently transported by wind: *A*, maple; *B*, elm; *C*, ash; *D*, basswood; *E*, dandelion; and *F*, clematis.

to the weight, the more the resistance it offers to falling through the air and the farther it may be carried by the wind. The plumes of the milkweed, thistle, dandelion, willow, and cottonwood increase the surface tens, hundreds, or thousands of times without materially increasing the weight. Consequently, these seeds may be carried many miles if they get well up into the air at the start. Maple, elm, ash, and catalpa have relatively large surfaces (wings) for their mass and they are easily blown about. The chief disadvantage of wind as a disseminator of seeds is that so many seeds are carried to habitats where germination will not take place.

Many seeds are transported by streams and lake currents. During spring freshets enormous numbers of seeds are picked up from the overflowed lands and transported downstream. After floods in the Mississippi and other large rivers one may find numerous rows of seedlings extending along the sides of the valley at definite levels, marking the height of a rather prolonged stage of high water, during which the seeds were washed ashore. Similar lines of seedings are not uncommon along the shores of the Great Lakes at certain times of the year. Here they are soon destroyed by storm waves. The seeds most commonly transported by water are those of water, shore, and bottom-land plants.

Seeds are transported by animals in several ways. They may be inclosed in fruits like the burdock, cocklebur, and Spanish needle, and become entangled in the fur coat of the animal. They may be eaten and, due to impervious seed coats, survive the digestive juices of the animals. They may be carried and buried by squirrels, ground squirrels, and gophers. The walnut, butternut, and hickory nut have no other means of being carried away from the parent plants. Small seeds that have fallen on the muddy banks of ponds and streams may be carried by water birds in the mud that clings to their feet. The mistletoe produces seeds with an outside sticky coat. These also are said to be spread to other trees by adhering to the feet of birds. Finally, the greatest of all transporters of seed are human beings. Wherever man goes there follows shortly in his trail a host of weeds. His ships carry them across the oceans, and his railroad trains scatter them over the land. The continual shipment of agricultural and horticultural products of necessity leads to the spread of seeds of various other plants that grew with them. More than half of our weeds have been introduced from Europe in this way.

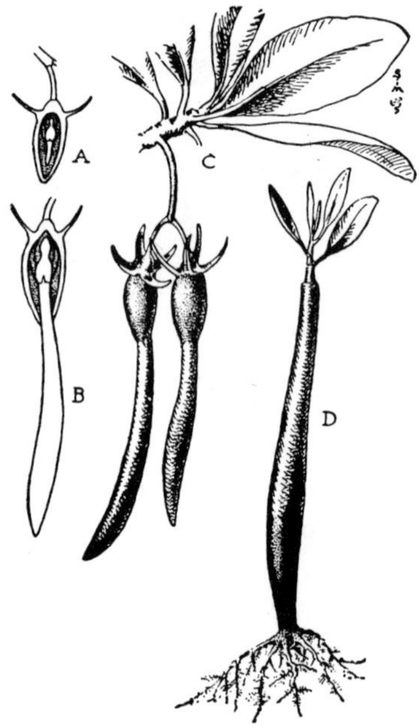

FIG. 169. Development of mangrove seedlings. This small tree grows on soft mud flats in the tropics and semi-tropics. The seed (*A* and *C*) germinates while still attached to the tree and forms an embryo a foot or more in length. The embryo finally drops endwise like an arrow into the mud below and starts a seedling (*D*).

Economic importance of seeds. Seeds and grains supply the most concentrated foods derived from plants. They provide the larger part of the food of all human beings. Seeds of cotton, corn, and coconut furnish enormous quantities of oils used in the manufacture of various fats and soap, and nuts of various kinds are coming to be used more and more extensively as foods. Corn oil is widely used in the making of a rubber substitute. Flax-seed is the source of linseed oil, which is used in the manufacture of paints. From the grains we derive also starch, glucose, alcohol, ether, and many related organic substances. The seeds of the coffee and cacao plants supply pleasant and mildly stimulating drinks. The hairy covering of the cottonseed is the most important fiber used in the manufacture of cloth.

PROBLEMS

1. Why does a corn plant growing alone produce imperfect ears?
2. When cucumbers are grown in commercial greenhouses, how is pollination accomplished?
3. Why are large, heavy seeds of agricultural plants more desirable for planting than small, light seeds?
4. What common market "vegetables" are included in the botanical term "fruit"?

CHAPTER TWENTY-EIGHT

DORMANCY AND GERMINATION OF SEEDS

Although the embryo is made up almost wholly of meristematic tissue, seeds do not usually germinate as soon as they are mature. In nature only a very small percentage of embryos germinate and grow to maturity; most of them either fail to germinate, or they perish in the seedling stage.

Dormancy. When a seed does not germinate immediately after leaving the parent plant, it is spoken of as being in a state of *dormancy*. The seeds of most wild plants do not germinate immediately after ripening, even though external conditions are favorable. The seeds of many cultivated plants, on the other hand, have little or no dormancy and will germinate during wet weather even before they leave the parent plant. Corn germinating in the ear, and wheat and oats sprouting in the shock, are common occurrences in extremely wet weather. The lack of dormancy in seeds of cultivated plants is due, in part, to the fact that man has unconsciously selected those seeds that germinate readily. Obviously, only those seeds that germinate soon after sowing produce mature seeds at harvest time. In the long history of agriculture those individuals and races of plants whose seeds did not germinate readily were largely eliminated.

An interesting example of the difference in dormancy between wild and cultivated species may be seen in the Dakotas. The common oats and wild European oats both occur and produce seeds in summer. The cultivated oats, however, germinate in early autumn and the seedlings freeze and die during the cold winter. The seeds of wild oats do not germinate until spring, and consequently they become troublesome weeds in grain fields.

Sometimes the period of dormancy is very short, as in the seeds of willow, elm, cottonwood, and other spring-fruiting trees. Seeds of this type do not withstand drying and consequently never germinate unless they fall in a moist habitat. The soft maple is another seed of this kind. When its water content falls below 30 per cent, it dies.

Dormancy is generally more pronounced in seeds produced in late summer and autumn. Even when the seeds are kept in conditions usually favorable for germination, they will not germinate for several months — sometimes not for several years.

External causes of dormancy. Dormancy may be due to various causes. Some of these causes are environmental, others lie within the seed itself. In late autumn and winter the temperature may be too low for germination. This condition often prevails when the seeds of many late flowering plants are mature. On the other hand, the temperature may be too high, as it often is when the seeds of cool temperature species of the desert mature.

The seeds may fall on ground that is too dry, or they may fall to the bottom of a pond where the oxygen content is very low or insufficient. They may be covered with earth, under conditions that exclude oxygen entirely. Some seeds will not germinate in the absence of light. Unfavorable temperature, too little water, the insufficiency of oxygen, are the most common external causes that prevent germination in nature.

Internal causes of dormancy. The failure to germinate, even when external conditions are favorable, often depends on certain characteristics of the seed coats or of the embryos. It is usually possible to break dormancy when the cause is known, a fact that is of great importance to horticulturists in their work of growing seedlings. Many of these specific causes have been determined; some of the most important are given below.

1. *Seed coats impermeable to water.* One of the commonest causes of dormancy is the exclusion of water by hard seed coats. The seeds of many water plants, and also of plants belonging to the legume family (clover, alfalfa, and lupine), have coats of this character. When placed in water no absorption takes place, and germination is therefore impossible. In some species all the seeds are hard and impervious to water; in others, only some of

them. In either case the germination of a given crop of seeds, under natural conditions, would take place over a series of years. Red clover seeds, for example, when placed in water do not all swell at once, but only a few at a time; some may remain hard, yet alive, at the end of ten years. It has been found that all these seeds will germinate at once if the seed coats are scratched, broken, or removed by being shaken together with sharp sand. Ninety per cent of these clover seeds may be germinated immediately also by immersing them in boiling water for 50 to 60 seconds, or by pouring strong sulfuric acid over them and then after 30 to 50 minutes carefully washing them with a 5 per cent solution of sodium bicarbonate (baking soda) and water. In this kind of dormancy the embryo is in a condition to grow when mature, but it cannot do so until the seed coat allows water to enter. In nature, freezing and thawing, bacteria, and fungi are important agencies in breaking the coats of this class of seeds.

2. *Seed coats mechanically resistant.* In many seeds the seed coat is a membrane strong enough to prevent the expansion of the embryo by the absorption of water. Although such seed coats are permeable, when put in water only a small amount enters. The seeds of the common pigweed, water plantain, black mustard, shepherd's-purse, and peppergrass furnish examples of this kind of dormancy. In nature the dormancy of such seeds is broken by chemical changes which weaken the seed coats as they lie on the soil. Dry storage promotes changes of the same kind, but if the seeds are stored in water or wet soil they remain dormant for many years. This explains why they are such persistent weeds in gardens, even in gardens that are kept clean for a number of years. So long as the seeds lie deep in the soil they will not germinate; but when brought to the surface by cultivation, the coat dries out, and as soon as they are again moistened germination follows.

3. *Seed coats impermeable to oxygen.* In a third class of seeds dormancy is brought about by the exclusion of oxygen. The

seed coats or fruit coats inclosing these seeds are either impermeable to oxygen or at least retard the entrance of oxygen to such an extent that germination cannot take place. The best-known example of this class of seeds is the cocklebur, a weed very difficult to eradicate in low grounds. The dry fruit of the cocklebur contains two seeds, an upper and a lower. In nature the lower seed usually germinates in the spring following maturity, while the upper seed does not grow until later, or not until the second spring. Both seeds require considerable oxygen to germinate, but the upper requires about one third more than the lower. So long as the seed coats remain intact, no germination is possible; but if the embryos are removed, they germinate at once. In nature, freezing and thawing and the action of the bacteria and fungi of the soil alter the seed, so that after a few months sufficient oxygen penetrates the lower seed to start growth in the embryo; but it requires more than a year for the seed coats of the upper seed to decay, or to become altered sufficiently for enough oxygen to reach the embryo to induce growth. Many grasses — for example, the Johnson grass — and many composites have seeds whose behavior is similar to that of the cocklebur. Seeds of this kind have been found to germinate rather readily when kept for a time at a temperature above 120° F.

4. *Embryos requiring acidity.* The seeds of the peach, hawthorn, red cedar, hard maple, basswood, ragweed, and to a less extent apple seeds, will not germinate when first matured, even if the seed coats are removed. In these seeds dormancy is due to the condition of the embryo. Experiments have shown that the stem-part (hypocotyl) of the embryo is neutral or alkaline in reaction in the mature seed, and that dormancy lasts as long as this condition lasts. As soon as the whole embryo becomes acid in reaction, the seed will germinate.

Seeds of this class may be germinated at once by removing the seed coat and immersing the embryos in weak acids. It has long been the practice of horticulturists to "layer" such seeds;

that is, place them under a thin layer of soil over winter. In the spring they germinate readily. The same seeds kept at room temperature and then planted will not germinate for a much longer period. A low temperature (about 40° F.) is most favorable for the development of acidity in the growing embryo. Acidity favors growth because it increases the ability of the cells to take up water, and it favors the production and activities of enzymes that are necessary for digestion, assimilation, and respiration.

5. *Imperfectly developed embryos.* A fifth form of dormancy occurs in many seeds in which the embryo is only partly developed when the seed matures. Sometimes it is little more than a fertilized egg; but one may find all gradations between these few-celled embryos and those that are nearly or completely grown. They occur among a wide range of plants, but most of these plants are not commonly known, except the ginkgo, holly, buttercup, dogtooth violet, and many orchids. Orchid seed, formerly thought to germinate only in the presence of certain fungi, may be germinated in a sugar solution, which furnishes both the water and the food material necessary to complete the growth of the embryo.

Longevity of seeds. How long do seeds live in a dormant condition? This question is frequently asked, and in connection with it there are many stories told that are either based on wild guesses or are merely fiction. Unscrupulous individuals have at times taken advantage of people by these fictitious stories and made large profits by selling small vials of ordinary seeds at high prices. " Miracle " wheat, purporting to have been taken from mummy cases in Egypt, had a wide sale among unsuspecting farmers until investigation showed the true source and the worthlessness of the seeds.

There are few unquestionable records of seeds germinating after 100 years of dormancy; even those germinating after periods much shorter than 100 years are rare. The seeds of a few legumes have germinated after storage for 80 years. Experiments have shown that the seeds of a number of our common weeds will withstand burial deep in the soil for more than 30 years. The seeds of water plants will remain alive under aërated water for the same length of time. The seeds of several land plants have remained alive under water for periods of from 4 to 12 years. It is therefore safe to say that the seeds of most cultivated plants deteriorate rapidly after 2 years; that the seeds of many wild plants remain alive for 5 to 10 years; and that a few may live under favorable conditions 25 to 50 years. No seeds are known to have remained alive 200 years.

Storage of seeds. One of the important conditions for the storage of most seeds is that they be kept dry. When seeds are stored for long periods in soil, the absence of germination seems to depend on lack of sufficient oxygen. The same is probably true for storage under water. Seeds like those of the soft maple live longest when kept cool and moist.

In general, seeds may be kept longer and show greater vitality if they are thoroughly mature when harvested. Corn and wheat seeds lose their vitality rapidly when not mature and well dried out. The seeds produced in wet seasons usually show poorer germination than those produced in dry seasons. Moreover, corn matures late in the autumn, and unless its water content falls below 20 per cent before killing frosts come it is sure to be injured. Corn that is to be used for seed the following year should be gathered as soon as mature and placed in racks, so that it will dry out rapidly.

The changes that take place during storage that lead to the death of the embryo have been much studied. One important fact discovered is that during prolonged storage the proteins are gradually coagulated or changed into insoluble forms, so that when the seeds are planted the proteins do not become soluble and the protoplasm dies. Seeds will not remain alive, therefore, after their proteins have coagulated. Since this takes place more rapidly at a high temperature than at a low, with other conditions favorable, seeds will keep longer when the temperature is low.

External conditions necessary for germination. We have already discussed some of the internal conditions necessary for the germination of seeds. The seed coats must be permeable to water and oxygen, and they must allow complete swelling of the embryo. The embryo must be fully grown and in some cases in an acid condition.

The first external condition necessary for germination is abundant moisture, but there should not be enough moisture to interfere with the access of oxygen. Water is needed to bring about the swelling of the cells and tissues; to dissolve various salts, sugars, and other organic substances in the cells; and to facilitate chemical changes in the cells.

Oxygen is needed for respiration. Oxidation liberates energy for chemical changes in the cells. The respiration of germinating seeds goes on at a very high rate; when compared with that of human beings the rate is several times as great. Human beings give off carbon dioxide equivalent to 2.5 per cent of their dry body weight in 24 hours; germinating seeds may give off from 5 to 20 per cent of their weight of carbon dioxide in a day.

Seeds are planted as near the surface of the soil as possible to insure an adequate oxygen supply. They are planted below the surface to insure a sufficient supply of water. As the relation of water and oxygen to soil particles varies greatly in different soils, it is evident that to obtain a sufficient amount of both oxygen and water, seeds must be planted deeper in some soils than in others. We may plant them deeper in loose sandy soil, for example, than in tight clay.

The third important external factor for germination is temperature. Temperatures favorable for germination are usually lower than those for the subsequent development of the plant. But as few seeds germinate much below 50° F., the temperature of the soil should be at least as high as this when seeds are planted. If the temperature is much lower, the vitality of the seedling is reduced, and the plant is then more readily attacked and injured or destroyed by bacteria and fungi. On the other hand, when germination occurs in soil that is higher than 70° F., many plants form very poor root systems; consequently the growth of the plant is retarded.

The germination of most seeds takes place equally well in light or in darkness. Light retards the germination of some seeds, while others, like those of bluegrass, certain varieties of tobacco, mullein, and mistletoe, germinate better in the light.

Seedlings. Recent experiments have definitely proved that large, vigorous plants develop only from vigorous seedlings. In many plants large seeds produce better seedlings than small seeds. Therefore, in order to produce the best plants we must start with seeds of good quality and we must make sure that the seedlings are not interfered with during their early development. Planting many seeds in a row and then removing all but the most vigorous of the seedlings is, therefore, good agricultural practice. Removing weeds and keeping the soil porous and fairly moist keep up the water and oxygen supplies for the roots and prevent any interference with the light that the seedlings should receive for food manufacture.

CHAPTER TWENTY-NINE

PLANT BREEDING

THE origin of our most important cultivated plants is in most instances shrouded in mystery, for they were brought into cultivation by prehistoric peoples. When Columbus discovered America, the Indians of the New World were cultivating corn, potatoes, cotton, kidney and lima beans, arrowroot, peppers, peanuts, pineapples, tomatoes, tobacco, sweet potatoes, squash, pumpkin, and a number of tropical food and fiber plants. The other important food plants, like wheat, rice, barley, rye, and oats, were mostly selected from wild species by the prehistoric races of Asia and Africa. It is a singular fact that within historic times no important additions have been made to the food plants of the world except through borrowing from the so-called primitive races. During historic times, however, these food plants have been greatly modified and innumerable superior varieties have been developed. Among plants that produce edible fruits, berries, and nuts, some additional species have been brought into cultivation during recent centuries — notably grapes, cranberries, raspberries, dewberries, cherries, plums, and pecans.

Objectives in plant breeding. Plant breeding is concerned with the improvement of economic plants, with the discovery of new varieties, and with the production of new plants of economic

A. B. Stout

FIG. 170. An ear of white sweet corn partly pollinated by pollen from black Mexican sweet corn. The color is in the endosperm (xeniophyte) and was produced by the factor for color carried by the sperm nucleus which furnishes one of the three sets of chromosomes in the endosperm nucleus.

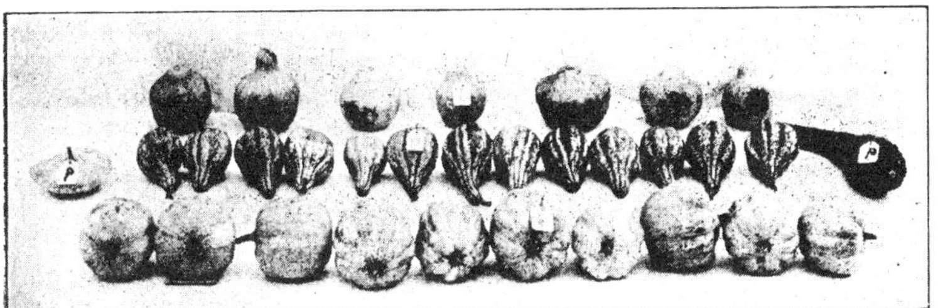

New Jersey Expt. Sta.

FIG. 171. Three new varieties of squashes produced by crossing a white scallop summer squash (*P*, at the left side of picture) with a warty, yellow-colored summer crookneck (*P*, at right side). The photograph shows three new varieties that have been produced. The upper row shows a type of short-necked "jug" fruit of medium size with a smooth, cream-colored surface. The middle row shows a longer-necked type of "jug" fruit, somewhat like the crookneck in shape, but *green-striped* and *not warty*. In the lower row the fruits are very thin-fleshed, nearly spherical, cream-colored, and not warty. After the first crossing, the plants were selected and self-bred for five generations, after which some of the new kinds would breed true enough to make new varieties.

value. Plant breeding is actively carried on at the State Agricultural Experiment Stations, by the United States Department of Agriculture, by experimenters at several of the larger universities, by seedsmen, and by breeders of nursery stock. The activities of plant breeders are being directed toward four principal objectives: (1) the breeding of plants with more desirable products, as flowers, fruits, leaves, and fibers; (2) the breeding of new varieties which will increase the yield per acre; (3) the securing of varieties better fitted to particular climates and soils; and (4) the producing of varieties capable of greater resistance to diseases.

Breeding for more desirable products. The first object of plant breeding may be illustrated by recent improvements of the pecan. Among the hundreds of thousands of pecan trees scattered through the Southern states, a few trees have been discovered that produce nuts of large size and good flavor and with thin shells. Breeders have found that these types may be preserved by budding pecan seedlings with buds from the most desirable trees. The best paper-shell pecans on the market are now grown in orchards started in this way from perhaps a single tree.

The most valuable fiber plant known is a new variety of "long-staple" upland cotton. It was produced by hybridizing or breeding together two well-known varieties of Egyptian cotton. Among the numerous varieties obtained from this cross was one whose seeds were covered with hairs an inch to an inch and a half in length. This variety has been propagated by the United States Department of Agriculture and is now widely grown.

The Concord grape is now grown in most temperate regions of the earth. It was produced by Ephraim Bull in New York by crossing two wild species, and was one plant selected in 1853 from among 22,000 seedlings tested.

Breeding for increased yield. The way in which the yield per acre may be increased is strikingly illustrated by a tobacco discovered and distributed by the Connecticut Experiment Station. The usual varieties of tobacco develop about twenty leaves and

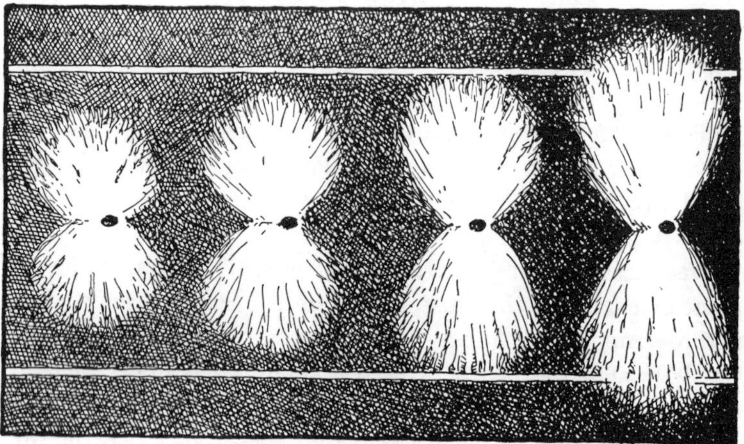

Fig. 172. Fiber from new varieties of long-fibered cotton at the right, obtained by hybridizing and selecting progeny from the two forms producing the shorter fibers at the left. The hybrid offspring excel both parents in the length of fiber produced.

then produce a flower cluster. The new varieties found by the plant breeder occurred as a few scattered plants, among the hundreds of acres grown in the state. These plants had indeterminate growth; that is, the stem continued to produce leaves until the end of the growing season. Seeds were secured by transferring the plant to the greenhouse. The average number of leaves on these plants is seventy, and the yield of tobacco per acre has been increased 90 per cent. Since the cost of growing a crop is nearly the same in both cases, the increased yield is largely added profit. (See also page 205.)

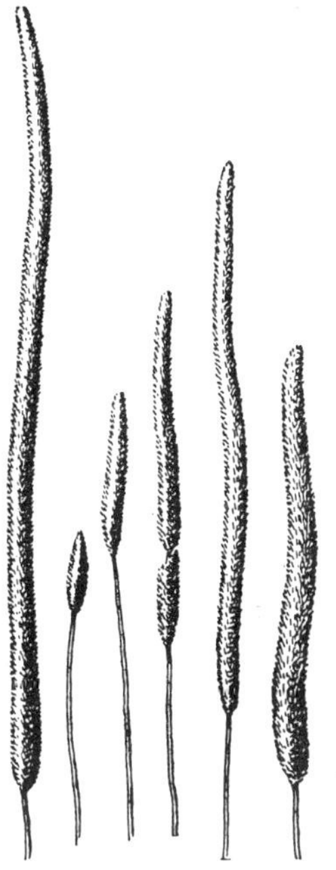

FIG. 173. Variations in length of timothy spike.

By the selection and propagation of timothy plants of large size for a few years, the Cornell Experiment Station was able to furnish seed to growers of hay which increased the yield 36 per cent over ordinary timothy.

At the Maine Experiment Station experiments in the breeding of oats led to the separation of varieties which gave a yield of 80 bushels per acre, where the best commercial varieties produced but 75 bushels.

Better varieties for particular climates. The Florida Velvet Bean was formerly confined to Florida and the Gulf Coast. By the selection of early varieties which suddenly appeared at several different places, the plant can now be grown throughout the cotton belt, and there are in our Southern states more than six million acres devoted to this crop.

Plums suited to the Northern Plains region have been produced by crossing the Japanese and European plums with the wild plum of the region, and selecting the best of the resulting hybrids.

Wheat, barley, oats, and rye have all had their areas of cultivation extended by the discovery of, or production of, new varieties with qualities which enabled them to be grown in other climatic regions.

Greater resistance to diseases. A striking example of breeding for resistance to a plant disease is the successful production of watermelons resistant to " wilt." All the edible varieties were highly susceptible to this disease. By crossing the Eden variety of watermelon with an inedible citron which was highly resistant, hybrids of great vigor and productiveness were produced. After 8 years of selection and trial a uniform edible variety was isolated which possessed the good qualities of the Eden watermelon and the wilt resistance of the citron. Curiously enough, this resistance is maintained throughout the eastern United States, but in California the new melon is susceptible. This emphasizes the importance of breeding plants for particular regions.

Univ. of Wis. Agric. Expt. Sta.

FIG. 174. In a field of cabbage that was almost entirely destroyed by yellows, a plant that had formed a good head was found. This plant was saved for seed.

Univ. of Wis. Agric. Expt. Sta.

FIG. 175. The rows of cabbage at the right were grown from seed from resistant stock. They have inherited the power of the parent plants to resist the disease. The plants on the left are from ordinary seed.

Successful disease-resistant plants have been discovered and bred among cabbage, tomatoes, asparagus, potatoes, cowpeas, flax, wheat, and cotton.

The basis of breeding. Plant breeding with these purposes in view is possible and profitable because (1) variations naturally occur among plants; (2) some variations are inherited and may be preserved by selection and propagation; and (3) different varieties and species may be crossed, producing hybrids having a still larger range of variations than the parent plants, or possessing new combinations of desirable qualities which may be selected and preserved.

The methods used by the plant breeder depend upon the reproductive structures and habits of the particular plants with which he is working. For example, the means by which a plant is naturally pollinated will determine how it must be handled at the time of flowering to secure self-pollinated seed or cross-pollinated seed.

Plants like wheat self-pollinate naturally. A particular variety may, therefore, be readily kept pure; and the seeds produced come true when planted. Seeds from self-pollinated corn do not produce the most vigorous plants and the largest yields. The best corn seed is obtained by planting in rows and removing the tassels (the staminate flowers) from alternate rows. The seed for the next year is then collected only from the detasseled plants. In this way the breeder is assured that the pollen came from another plant and that the plants grown from these seeds will have the increased vigor that is characteristic of cross-pollinated corn. Sunflowers are not self-fertile. That is, even though pollen does fall on a stigma of the same flower cluster, fertilization does not occur and no embryo is formed. In plants of this type pollen *must* be carried from one plant to another. These three examples illustrate some of the details of reproduction with which the plant breeder must be familiar before he can intelligently engage in plant-breeding work.

Methods of vegetative propagation may be used to multiply

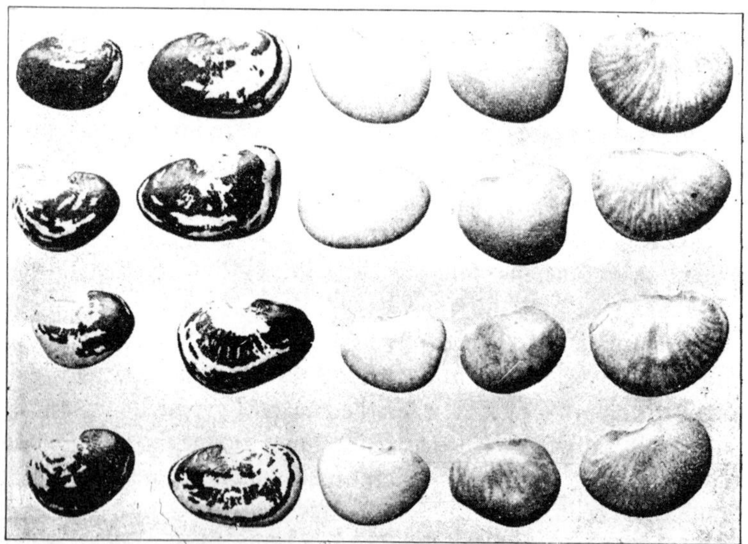

Bureau of Agriculture, P.I.

FIG. 176. Varieties of lima beans, showing differences in size, shape, and color.

Bureau of Agriculture, P.I.

FIG. 177. Four types of corn (*Zea mays*): sweet, dent, flint, and pop — mutant varieties that have been selected, each for a particular quality, since prehistoric times.

perennial plants after a desirable variety has been produced. In this way the plant grower avoids cross-pollination and the variations that appear when many cultivated crops are grown from seed. The best ways of discovering, selecting, hybridizing, and propagating particular crop plants may be found described in recent publications devoted to plant breeding. Much literature on the subject may be secured from your state experiment station and from the United States Department of Agriculture.

REFERENCE

BABCOCK AND CLAUSEN. *Genetics in Relation to Agriculture.* McGraw-Hill Book Company.

CHAPTER THIRTY

VARIATIONS AND MUTATIONS

No two fruits, flowers, or other plant organs are exactly alike. The variations may be small or large, and there may be every gradation between the extremes of any character. The several thousand sunflowers that might be grown from a pound of seed would vary in height of stem, amount of branching, and size of flowers. Not only may there be variations in the structures of plants, but there may also be variations in the composition of the plant organs. For example, the great variety of colors, flavors, and other qualities of apples is due to variations in the chemical composition of this fruit. The variation in each of these characters is quite independent of variations in other characters. A thorough knowledge of the possible kinds of variations that occur in plants is a necessary preliminary to progress in discovering their causes and in utilizing them in plant breeding.

FIG. 178. Two plants of sweet corn of the same variety, one grown in poor soil and one in soil to which fertilizer was added. The differences in the plants are due to the environment.

Variations due to environment. Many variations are due to the environment. When a crop plant like corn is grown in rich soil and in poor soil, there are great differences in the size of the plants and in the yield per acre. These variations in size and yield are due to the environment. Even though the seed planted in each kind of soil is exactly the same, there will be wide differences in the plants. Examples of these variations were discussed in the chapters on ecological variations of stems and leaves and in the chapter dealing with environmental factors.

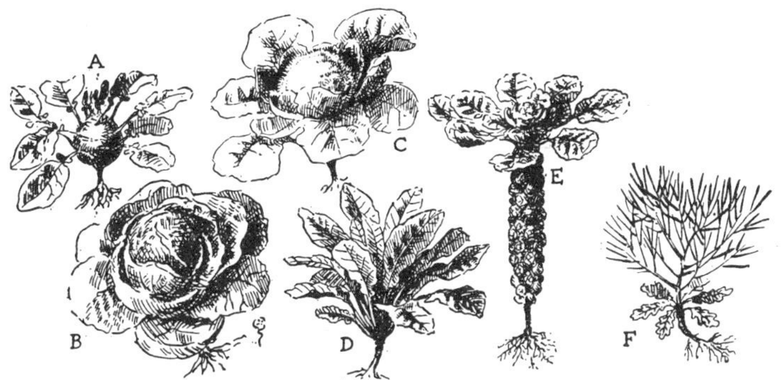

FIG. 179. Varieties derived from the wild cabbage (*F*), a native plant of Europe. *A* is kohl-rabi, *B* cabbage, *C* cauliflower, *D* kale, and *E* Brussels sprouts.

Temperature, moisture, light, mineral salts of the soil, all have effects upon the plant. Consequently, when plants having the same hereditary qualities are grown in dissimilar habitats, the different environment brings about marked variances in the expression of those qualities. Variations of this kind are not inherited, although when the vigor is decreased by the conditions in which the plant is grown, the plants of the next generation may get a poor start by less vigorous seedlings and show at maturity some of the effects of unfavorable conditions to which the previous generation was exposed.

Fluctuations. Variations due to differences of the environment are often called *fluctuations*, and we can now associate many of them with the particular external factors which produce them. Another class of fluctuations are those which appear to be due to unknown internal causes. The leaves that occur on a mulberry tree, for example, may vary from leaves that are almost perfectly heart-shaped to those with several lobes. The number of leaflets that makes up a compound leaf of the horse chestnut, walnut, ailanthus, and sumac varies somewhat widely. In a California privet hedge one finds branches usually with leaves opposite,

U. S. Dept. of Agriculture

FIG. 180. Tobacco plants of the same variety grown from large, medium, and small seeds, showing the relation between the size of the seed and the size and vigor of the seedling. Is the difference in size in the plants due to environment or to differences in the plants themselves?

but an occasional branch has three leaves at a node, and another has only one leaf at each node. Variations of this kind are not heritable. Perhaps they are mere accidents of development.

Heritable variations. These variations are the result of hereditary differences in the nature of the plant. For example, dwarf nasturtiums were variations that occurred among the common tall forms. The tall nasturtiums showed fluctuations in height, but all the fluctuations were near a certain size. The dwarfs were very different in size at the start, and when bred among themselves it was found that the small size is heritable. Similar variations have resulted in the production of dwarf peas from tall peas, bush beans from the pole beans, bush squashes from squashes with long, trailing stems. The differences between variants are not always so great as in the examples just mentioned. Indeed, the amount of variation may be very small. The several varieties of lima beans, for example, show only slight differences in shape and size. The varieties of mock orange show small differences in the form of the flower and leaf. The important point about these variations is that the particular characteristics of each are inherited and appears in successive generations. Heritable variations are the result either of mutation or of hybridization.

It is possible to get all kinds of combinations of different characters, and by careful hybridizing and selection to combine many desirable qualities in a single plant. The Shasta daisy, for example, was made by breeding together the English, American, and Japanese daisies, and combining in one plant the pleasing foliage of the English species, the free-blooming habit of the American daisy, and the waxy luster of the petals of the Japanese plant.

Variations of all kinds are of interest to the plant breeder, because he must learn to distinguish between the two kinds. His attempts to develop a new variety from plants having certain qualities will be futile unless he is dealing with heritable variations. It is not always a simple matter to discover the nature of a particular variation. It may require careful breeding experiments carried through several generations to determine whether a variation is due to environment, to heritable causes, or to internal non-heritable causes.

Mutations. Sometimes, among many thousands of individuals a single plant appears which is markedly different from all the others. For example, a few years ago a sunflower was discovered that had some red pigment near the base of the otherwise yellow corollas. Among the millions of sunflowers that have been seen, this was the first one in which a red color was noticed. In some unknown way there was produced in this plant a red pigment not formed in other sunflowers. From the seeds of this plant there were developed other plants having red pigment in their flowers. Evidently the new character is inherited and these sunflowers have a chemical constitution which enables red pigment as well as yellow to be formed. The sudden appearance of the sunflower with the red pigment is an example of *mutation*. Individuals that first show new characters are called *mutants* (Latin: *mutare*, to change).

T. D. A. Cockerell

FIG. 181. The red sunflower, a color mutant from the common sunflower of the plains (*Helianthus annuus*).

What the plant breeders have long known as " sports " are the rare mutants in which notable changes have occurred. They show new characters, and these characters are inherited. Consequently, their discovery is of the greatest importance. However, mutations are not necessarily large, and the term *mutant* is applied to any variant showing a distinct heritable character.

The many modern varieties of tomato have been developed from mutations that occurred among the currant tomatoes or love apples grown for ornament in our great-grandmothers' gardens. The original fruits resembled large red currants.

Today single tomato berries may weigh a pound. In color they may be red, yellow, or pink, and in shape they may be spherical, plum-shaped, or flattened. They exhibit at least three types of leaves and two types of stems. The characteristics due to mutation are inherited, no matter what the soil and climatic conditions may be.

Bud mutations. Mutations occur not only among plants grown from seed but also among plants, or plant parts, developed from buds. These are called *bud mutations* or *bud sports.* On fruit trees one branch will occasionally produce fruit that is of different quality from the fruit produced on other branches. If the quality of the fruit is superior, these branches may be used in budding and grafting to preserve the variety. Known bud sports are comparatively rare, but it is estimated that at least several hundred horticultural varieties have originated from them. In this country the improved varieties of seedless or navel oranges have been secured entirely by this method. The Boston fern and its forty or more varieties originated in bud mutations from a wild tropical fern. In the potato and some other plants that are usually propagated vegetatively, bud variations are known to occur; but they are so difficult to discover in plants of this kind that they have not been of much practical value.

CHAPTER THIRTY-ONE

HYBRIDIZATION AND SELECTION

THE crossing of two species or varieties of plants is known as *hybridization*. It is brought about by transferring the pollen from one plant to the stigma of the other plant, which ultimately results in cross-fertilization. The plants grown from seed produced in this way are called *hybrids*. Hybrids may resemble one of the parents, or they may have some characters of both parents. In the second generation derived from crosses some plants show a wide range of variations, with all possible combinations of the characteristics of the parent plants. Successful hybridization, therefore, increases the number of variations available for selection by the plant breeder.

Hybrid vigor. Hybridizing often has a physiological effect which is of importance, for in many plants it increases the vigor of the offspring. This may result in increased yield of grain, or in greater resistance to disease organisms, to drought, and to the effects of high or low temperature. Sometimes the vigor is expressed merely in the size attained by the hybrids. For example, the hybrids secured by crossing the American sunflower and the Russian sunflower, neither of which is over 10 feet in height, grow to a height of 15 feet. In a series of experiments with corn, hybrids gave an average yield more than 50 per cent greater than the average of their parents.

Mendel's experiments. To see more definitely what may happen in hybridization, we may review an experiment performed by Gregor Mendel about the middle of the last century. Mendel crossed a tall variety of the common garden pea with a dwarf variety. In this way he secured hybrids which had received some characters from each of two different parent plants. One of the most important discoveries he made was that in plant breeding every feature of a plant must be studied separately. Considering height growth alone, he found that when he planted the seeds secured from the cross all the plants grew tall. This first hybrid generation is called among breeders the "first filial" (Latin: *filialis*, related as a child), or F_1 generation.

Mendel concluded from this and many similar experiments that in any pair of contrasting characters one usually appears to be unmodified by the other. In other words, one of the two characters shows in the F_1 plants, and the other does not. In the experiment with peas cited above, tallness dominates over dwarfness. Mendel, therefore, called that character which appeared in the F_1 generation *dominant*, and the contrasting character which did not appear *recessive*. Working further with peas, Mendel discovered that purple flowers are dominant over white flowers, smooth seeds over wrinkled seeds, and yellow seeds over green.

The second important fact brought out by Mendel is that when the F_1 plants are self-pollinated and the resulting seeds planted, an F_2 generation is obtained in which both tall and dwarf plants occur. Furthermore, there is a *definite* ratio of 3 to 1 between the number of tall and dwarf plants. The results of these experiments may be expressed in the following diagram:

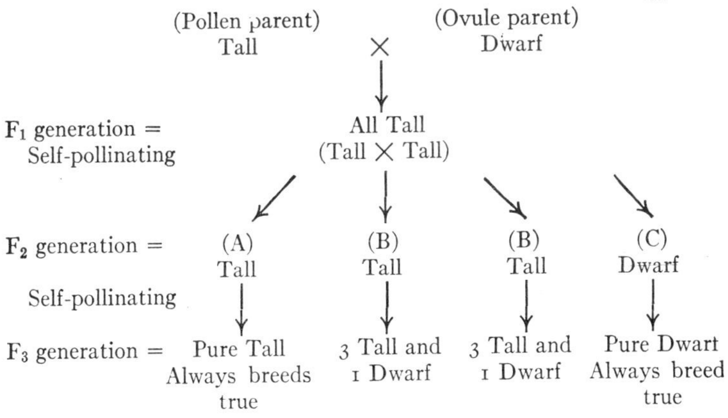

Evidently the factor for dwarfness is carried along with the factor for tallness in the F_1 generation, but the tallness dominates it and the dwarfness does not show. In the F_2 generation there are two kinds of tall plants: (*A*) those that breed true and (*B*) those that, like the hybrid talls of the F_1 generation, have the dwarfness latent in them. These latter plants produce three tall and one dwarf in the F_3 generation. The recessive dwarfs that come out in the F_2 generation always breed true. Hence, in the F_2 generation one fourth of the plants are pure talls, two fourths are hybrid talls, and one fourth are pure dwarfs.

A plant that breeds true for any character when self-pollinated is said by breeders to be *homozygous* with respect to that character. This implies that it carries only one kind of factor out of any pair of contrasting characters. In the A-group of tall plants above, both the pollen and the ovule (or more definitely the sperm and the egg) carry only the factor for tallness. Hence all the offspring are tall. The other two tall plants, those of the B-group, that appear among every four in the F_2 generation, carry two different factors and are called *heterozygous*. Half of the number of their eggs and sperms will carry the factor for tallness, the other half carry the factor for dwarfness. They produce, when self-pollinated, plants with the dominant character and plants with the recessive character in the proportion of 3 to 1.

With these facts in mind we can rewrite the above diagram, using T for the dominant character, tallness, and d for the recessive character, dwarfness:

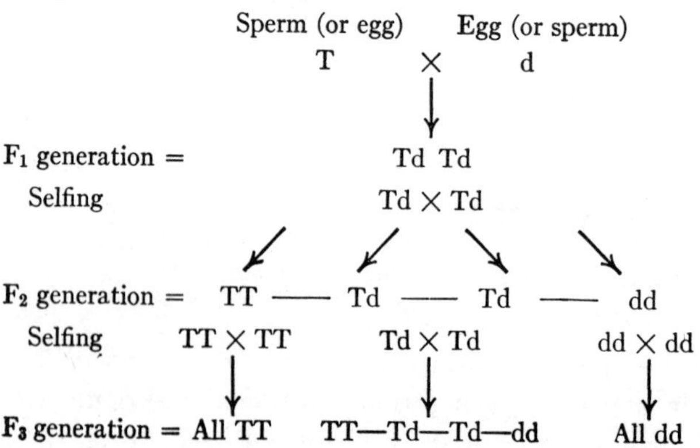

Mendel concluded from these and other experiments that most characters that make up a plant are each inherited independently of others. He also concluded that each egg and sperm carried but one of two contrasting characters, while the zygote which is formed by the union of the sperm and egg will, therefore, be either homozygous (containing two similar factors) or heterozygous (containing two contrasting factors). In the latter case the plant that develops will show the *dominant* character but not the *recessive.* Its offspring, however, will show both characters, three fourths being like the dominant and one fourth being pure recessives.

The combinations of characters that will appear in the progeny in the F_2 generation above may be shown more clearly by a diagram.

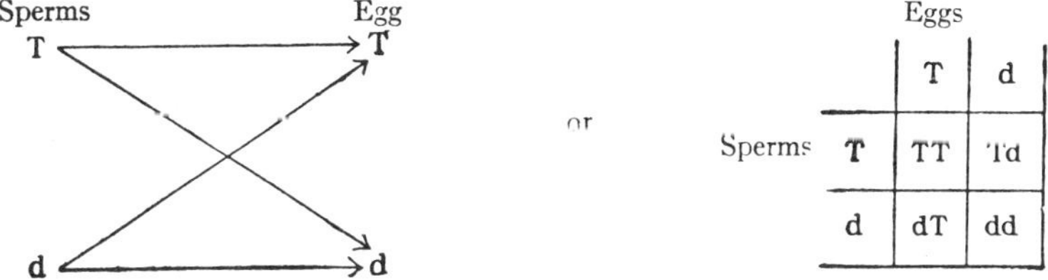

		Eggs	
		T	d
Sperms	T	TT	Td
	d	dT	dd

Application of Mendel's laws. Just how Mendel's laws may be applied to a particular breeding problem will be apparent if we follow the steps in the making of a hybrid with two pairs of contrasting characters. Suppose we have a tall pea with smooth seeds and a dwarf pea with wrinkled seeds. Tall peas require a longer time to develop and produce more peas per plant. Dwarf peas are, therefore, earlier. Wrinkled peas are generally sweeter than smooth peas because they contain more sugar. For these reasons it is decided that a tall pea with wrinkled seeds would be a desirable variety to produce. How can the plant breeder secure this combination from the two plants in his possession? Knowing that tallness (T) is dominant over dwarfness (d) and that smoothness (S) is dominant over wrinkledness (w), the problem is a rather simple one.

The plant breeder starts by cross-pollinating the two plants, a process which may be represented as follows:

T S × d w

He wishes to secure a plant that will have the composition T w.

	Sperm *Egg*	
	T S × d w	
All plants have	TdSw	in F_1 generation

When the plants of the F_1 generation are mature, they will all be tall and will have smooth seeds.

He now allows these hybrids to self-pollinate and plants the resulting seeds. Only one of each pair of contrasting characters will be contained in each sperm and each egg from these plants. It will be either the factor for tallness or for dwarfness, but not both. It will be either the factor for smoothness or wrinkledness, but not both. Hence, there will be four kinds of sperms and four kinds of eggs, whose composition may be indicated as follows:

Sperms	*Eggs*
TS	TS
Tw	Tw
dS	dS
dw	dw

Any one of these sperms may unite with any one of the eggs, and if hundreds of zygotes are formed the several kinds of sperms will unite with each of the several kinds of eggs in about equal numbers, and there will be sixteen possible combinations, which may be seen by writing out the following checkerboard diagram. Write first the four kinds of eggs above, then the four kinds of sperms on the left side. Then put in each square the combination of the factors carried by the egg (indicated above) and the sperm (indicated to the left). At the right of this diagram we may make a second diagram giving the characteristics of each of the plants represented by the sixteen squares:

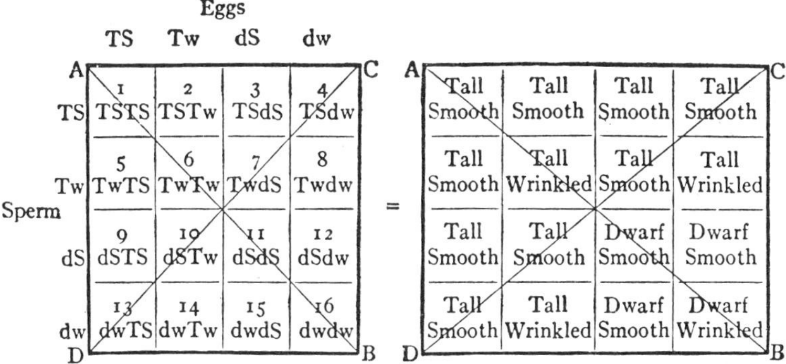

Note that in square 6 there is represented a plant that has the factors desired and no others. It is homozygous for both the desired characters, tall plants and wrinkled seeds. There will be twelve seeds out of every sixteen that will produce tall plants in the third generation, and four that will produce dwarfs (ratio 3 : 1). Which one of the twelve will produce only tall plants with wrinkled seeds can be determined only by planting the seeds of the F_2 generation and watching the characteristics and proportion of tall and dwarf and smooth and wrinkled seeds developed from each individual plant. If this is accurately done, the plant breeder will be able to select the plants having the composition represented by square 6 and his problem will be solved.

Referring now to the diagram at the right, all the plants represented by the squares along the axis AB are homozygous and will breed true if self-pollinated. All the plants represented by the squares on the axis CD are exact duplicates of the hybrid or F_1 generation and will produce all the sixteen possible combinations when self-pollinated. These four squares represent plants that are heterozygous for both pairs of characters. The remaining eight squares represent plants that are homozygous for one character but heterozygous for the other. By means of similar diagrams it is possible to calculate the chances of securing a

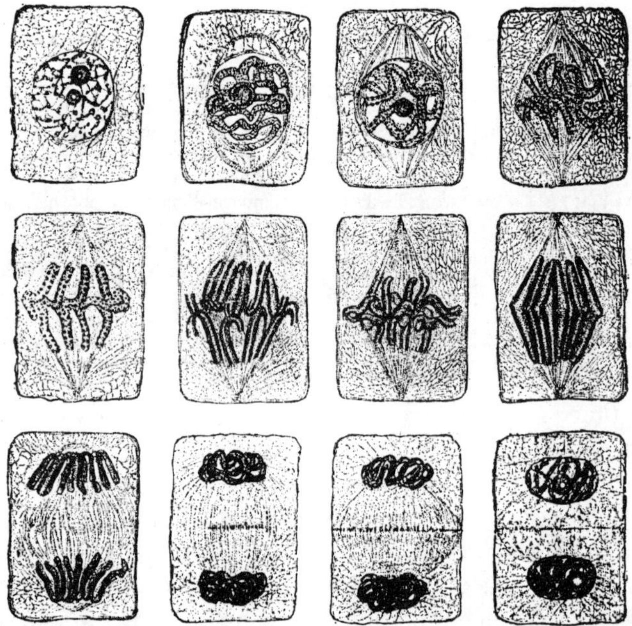

FIG. 182. Diagram to illustrate the behavior of the chromosomes during vegetative cell division. (*After Strasburger.*)

certain combination and to plan a definite set of crosses that will bring the desired result in a few generations.

Importance of Mendel's laws. With the above example in mind, it is not difficult to see that the use of Mendel's methods has greatly simplified the problems of the plant breeder and made possible the early attainment of desirable combinations of characters by hybridizing different varieties and closely related species of plants. It has made it possible to breed for definite combinations and to determine when the combinations have been secured.

Before the discovery of Mendel's laws of inheritance, all breeding was merely a matter of chance, and much of it is still being carried on in that way because we are still ignorant of the factors involved in many characters of numerous plants and animals. On the other hand, many important food and ornamental plants have been extensively studied and their hereditary behavior discovered. With these studies as a foundation further progress in improving, modifying, and combining qualities in new ways is rapidly being made.

Cell structures and Mendel's laws. Since Mendel's time much has been learned concerning the physiological explanation of these laws. It is now well established that the explanation rests on the behavior of certain nuclear structures, the chromosomes, during cell division, during the formation of pollen and embryo sac, and when fertilization occurs.

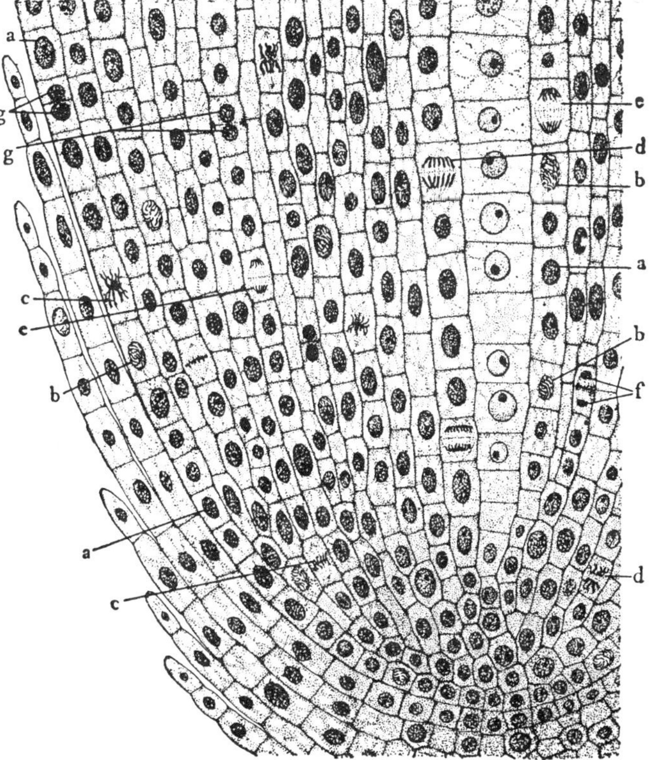

Fig. 183. Root tip of onion, showing cell division and enlargement. A series of stages showing behavior of the chromosomes during vegetative cell division is labeled *a* to *g*.

Chromosomes and vegetative cell division. In ordinary cell division in a growing tissue a cell near the growing point is more

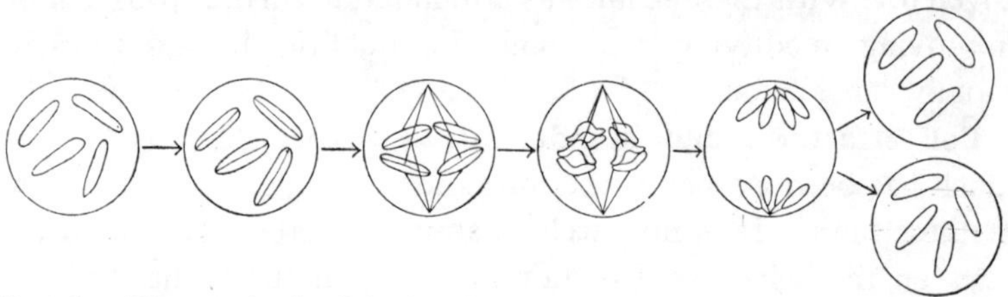

FIG. 184. Diagram showing behavior of chromosomes in division of a vegetative cell. The daughter cells have the same number and kind of chromosomes as the mother cell. (*After Sharp.*)

or less cubical. Cytoplasm and nucleus occupy all the space within the cell wall. The nucleus is very large in proportion to the volume of the cell. Inside the nucleus when properly stained may be seen a tangled network of material called *chromatin* (Greek: *chroma*, color) because it may be stained deeply by certain dyes. When the cell is about to divide, the chromatin becomes aggregated into a single much-twisted thread. The thread then shortens and thickens and becomes arranged in a number of loops, and a little later it divides into a definite number of segments, the *chromosomes* (Greek: *chroma*, color; *soma*, body). The chromosomes are U-shaped and collect in the equatorial region of the nucleus. By this time lines have appeared in the nucleus radiating from the two opposite poles of the nucleus, forming the so-called *spindle*. Each chromosome splits longitudinally, and a half moves toward each of the two poles. A little later the chromosomes become merged at each pole, granules appear at the equator of the spindle, and the first layer of a transverse wall is laid down. At the former poles of the spindle two new nuclei are organized about the chromatin. The dividing wall between the cells is thickened and we have two *daughter cells* in place of the original or *mother cell*. This process is repeated as often as new vegetative cells are formed in roots, stems, leaves, or other organs.

The number of chromosomes in the vegetative cells of any species of plant is definite, and in the cell divisions that occur

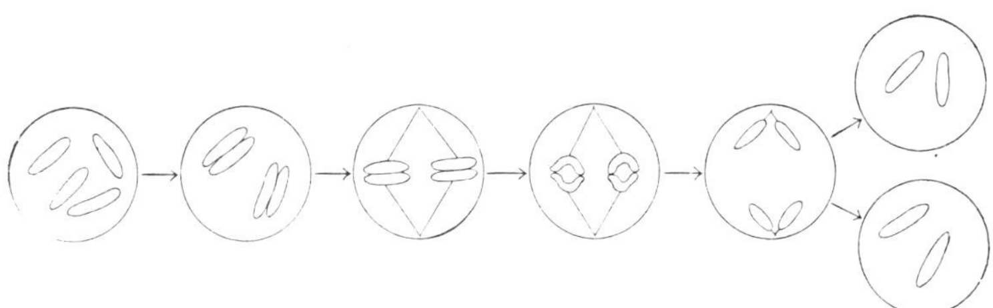

FIG. 185. Diagram showing behavior of chromosomes in the reduction division. The daughter cells have half as many chromosomes as the mother cell. (*After Sharp.*)

during vegetative growth each chromosome splits longitudinally and one half goes to each of the daughter cells. So every cell has a set of chromosomes similar to that of every other cell. Each chromosome is the bearer of certain hereditary factors. In vegetative multiplication the chromosome complement of cuttings, cions, and vegetative offshoots is the same as that of the parent plant. Hence, the hereditary qualities of vegetative propagating shoots are similar to those of the parent plant.

Behavior of chromosomes in sexual reproduction. In Chapter XXVI (page 249) the process of fertilization is discussed and attention is called to the fact that fertilization consists of the union of a sperm and egg, forming a zygote. This fusion would lead to a doubling of the number of chromosomes in each successive generation if it were not for the fact that in the formation of the pollen and the embryo sac which precedes fertilization a cell division takes place in which the behavior of the chromosomes is different from that described for vegetative growth.

In all the complex plants the mother cells, which give rise to the pollen and embryo sac, and ultimately the sperms and eggs (gametes) divide by a method called the *reduction division*. The word "reduction" refers to the fact that in this division the number of chromosomes is reduced to half the number that occur in the vegetative cells.

Reduction division. In the reduction division the chromosomes become arranged in pairs in the early stages of nuclear division and group themselves at the equator of the spindle, as double chromosomes. *Then, instead of splitting, as in vegetative cell division*, the two chromosomes of each pair separate and migrate to the opposite poles of the spindle, forming the chromosomes of the new daughter nuclei. As a result each daughter nucleus contains one half the number of chromosomes contained in the vegetative cells of the plant.

For example, the vegetative cells of the white lily contain twenty-four chromosomes. When the mother cells of the pollen and the embryo sac divide, the daughter cells each contain twelve chromosomes. These daughter cells divide again and again, until the sperms and eggs are finally formed and each nucleus contains twelve, the reduced number of chromosomes. When fertilization takes place and the sperm and egg unite, each carries twelve chromosomes. The resulting zygote nucleus contains twenty-four. The zygote is the first cell of the plant of the next generation, and in the further development of the embryo and plant all the cells carry twenty-four chromosomes.

Explanation of Mendel's laws. There are many reasons for concluding that the chromosomes are the principal carriers of the factors of heredity, and that each individual chromosome carries certain particular factors. We can explain the Mendelian behavior of hybrids if we assume that in the reduction division each of the two chromosomes that are paired in the early stages contains one of two contrasting factors. Then, when they separate, one of these factors is carried to one of the daughter nuclei and the contrasting factor to the other daughter nucleus. In this way each sperm and each egg will contain only one member of each pair of contrasting factors, and the plant that develops from each zygote has its characteristics determined by the factors contained in the chromosomes of the sperm and egg that united to produce it.

FIG. 186. Vegetable trial grounds of the Office of Seed and Plant Introduction, United States Department of Agriculture, Washington, D.C.

Mode of inheritance the same in plants and animals. One of the most remarkable facts of heredity is the similarity of behavior of hereditary factors in both plants and animals. Not only can we predict from the behavior of one plant what will occur in another plant, but we can predict in many instances what will happen in animals. This may be taken as further evidence of the essential similarity and close relationship of plants and animals.

Selection. The heritable variations or mutations and the products of hybridization furnish the materials from which valuable new varieties of animals and plants may be isolated. The plant breeder grows large numbers of plants derived in these several ways in gardens which he calls "plots," in order to determine what the plants do or how they will behave under a certain set of conditions. Then, from among the hundreds o thousands of individuals he selects those most nearly approacl

U. S. Dept. of Agriculture

FIG. 187. A field of upland cotton, in South Carolina, attacked by the wilt disease. The fungus that causes the wilt remains in the soil for many years.

ing the ideal or standard he has in mind for further study and testing. This process may be repeated year after year until he has secured the desired qualities.

Mass selection. For many years selection was carried on, particularly in the case of cereals, by what is termed *mass selection.* This method consists in selecting seed from all those plants that most nearly approach the breeder's ideal. The next year these seeds are planted and the process repeated. In this way yields of many crop plants were improved. But the method is one that may require a long time to accomplish results, and the results are usually uncertain, because the end product is still a mixture of plants with a great variety of hereditary qualities and, therefore, lacks uniformity.

Selection of individual plants. In recent years, since the general recognition of the importance of mutations and the Mendelian behavior of hybrids, mass selection has been superseded by the selection of individual plants. Instead of taking seeds from

U. S. Dept. of Agriculture

FIG. 188. The same field shown in Figure 187, planted with seeds from plants that survived the attack of the disease.

the best plants found in a field and sowing them together the following year, the seeds from each of the best plants are kept separate. These are then planted separately in short rows and a record is kept of the performance of the progeny of each of the selected plants, and the most desirable varieties are speedily isolated.

Since these trial rows are short, the amount of seed produced is small; and when a desirable strain or variety is discovered it must be grown in larger plots ("increase plots") until sufficient seed is obtained for distributing or marketing.

Marquis wheat. As an example of what may be accomplished in plant breeding by individual plant selection attention may be called to Marquis wheat, which is perhaps the most valuable variety of wheat, and perhaps the most valuable variety of a food plant, thus far discovered. In 1903 it was found by Charles Saunders as a single plant derived from a cross made several years previously between two varieties of wheat, known as "Red Fife" and "Calcutta." In 1904 there were just twelve plants. In 1909 sufficient seed had been grown to distribute four hundred samples to farmers in various parts of Canada. So successful did it prove to be that its cultivation spread rapidly, and in 1918 North Dakota and Minnesota alone produced nearly 150 million bushels of Marquis wheat. It is now grown in all the states from Ohio to Nebraska and Washington.

Marquis wheat has short straw, a rather short spike, and short, broad kernels. Its straw is stiff and remains erect under unfavorable weather conditions. It ripens on an average about 115 days after sowing, and has repeatedly won the International Prize as the best spring wheat.

Summary. The discoveries of Mendel were announced in 1865, but their importance was not at first appreciated. In 1900 his laws of heredity were rediscovered, and this marked a new epoch in the study of heredity and the principles underlying hybridization. Since that time investigation has shown that Mendel's laws apply equally well to plants and animals. The explanation of these laws involves the behavior of the chromosomes in vegetative cell division and in fertilization. It is now possible to plan breeding experiments and attain the desired result in a fraction of the time formerly required, when breeding work was merely a process of crossing, planting, selecting, and trusting to luck for results. It is fair to say that the work of Mendel has revolutionized plant and animal breeding.

PROBLEMS

1. What kind of variations are brought about by cultivating and by adding fertilizers to the soil?
2. What kind of variations appear to have been most important in the production of the different kinds of plants that occur on the earth?
3. How would you attempt to secure a new mutant?
4. Why are there fewer variations when plants are propagated vegetatively than when reproduced by seed?
5. Corn is propagated by seeds. How would you plant and care for a corn seed plot in order to secure 100 per cent hybrid seed?
6. If a plant has three pairs of different chromosomes, how many different kinds of sperms and eggs can it produce?
7. If you were to cross a tall white-flowered pea with a dwarf purple-flowered pea, what would you obtain in the hybrid or F_1 generation? What per cent or fraction of the F_2 generation would be dwarf white-flowered peas? What fraction of the F_2 generation would be tall purple-flowered peas? What fraction of these tall purple-flowered peas would be homozygous for both characters? What per cent of the dwarf white-flowered peas would be homozygous for both characters?
8. If you were to cross a pure tall purple-flowered pea with smooth seeds with a pure dwarf white-flowered pea with wrinkled seeds, what chance would you have of obtaining a pure tall white-flowered pea with wrinkled seeds in the F_2 generation?

CHAPTER THIRTY-TWO

THE DISTRIBUTION OF PLANTS IN NATURE

THE distribution of plants in nature is determined by the hereditary qualities of the plant on the one hand and by the characteristics of the environment on the other. Some plants, like the dandelion and sheep sorrel, have such indefinite requirements that they can thrive in, or at least endure, the conditions in many different habitats. Most plants, however, have a far more definite set of requirements, and if any one of these requirements is not met by a habitat, the plant is excluded from the habitat. Orange trees, for example, cannot withstand freezing temperatures; the low creeping Arctic willows will not grow where the summer season is hot; cacti are excluded from soil that is wet and poorly aërated for even a few weeks each year; cat-tails die on land that is not submerged at least a part of each year; sycamore, poplar, and willow seedlings will not thrive in dense shade; alfalfa and certain species of clover thrive only in well-drained soils that are neutral or slightly alkaline (many limestone soils furnish both these conditions); rhododendrons, azaleas, blueberries, and cranberries grow well only on soils that are acid — they soon die on limestone soils.

In general, it may be said that plants requiring similar environmental conditions are restricted to certain regions of a continent and to certain habitats within these regions, because there only are environmental conditions suited to their hereditary structures and qualities and favorable to their complete development.

Vegetation. By vegetation is meant the plant covering of the earth or of its subdivisions. The plant covering of any region is a great organization of hundreds, perhaps millions, of individuals. Some of these are dependent only upon the conditions determined by climate and soil. Some are also dependent upon nitrate supply, water supply, or the presence of certain other

U. S. Forest Service

FIG. 189. A view in a mixed mesophytic forest in eastern Tennessee. The prominent trees in the picture are cucumber tree (*Magnolia acuminata*) and shagbark hickory (*Carya ovata*). The deciduous forest formation approximates its best development in this region.

plants from which they derive a food supply or from which they obtain shade.

One of the most familiar and important types of vegetation is a forest. A mature forest consists of several stories or *layers* of plants. The tallest trees form the canopy, and their leaves are exposed to full sunlight. Below these trees there are usually low trees and young trees that endure the shade, and are benefited at least during the seedling stage by the more even temperature and moisture conditions within the forest. Then there are tall and low shrubs, some of which thrive in the forest because of the accumulation of humus and the more constant water conditions that go with it. On the floor of the forest are many herbaceous plants and mushrooms of various kinds. Collectively, all the low trees, the tall and low shrubs, and the herbs make up

U. S. Forest Service

FIG. 190. Mature hemlock forest on a mountain slope in Pennsylvania. Note the sparsity of the undergrowth.

what is termed *undergrowth*. Besides these plants there are microscopic plants innumerable — some on the bark of trees, some on the surface of the ground, and others below the surface. All these plants are in one way or another affected by the other plants, and so we may very properly speak of the forest as an *organization.*

When we study forests still further, we find that there are many kinds in North America, and that when a particular kind of tree is *dominant* a definite group of low trees, shrubs, and herbs usually grow with these dominant trees, and that when another kind is dominant, a different set of plants make up the undergrowth.

Plant associations. Even a brief study of a mature forest, such as we have just described, brings out clearly the fact that plants in nature do not live as isolated individuals, but in communities, more or less definitely organized. The organization, to be sure, develops gradually through the carrying of seeds into the area, and through the elimination of those species of plants that cannot endure the environment. Each year some new plants are starting and others are dying; the population is continually changing, but in an orderly way which is determined by the conditions in the community and the kinds of plants whose seed is carried into it. Such communities of plants are called plant *associations.*

All the plants in one association have somewhat similar requirements, but usually each plant differs from the others in some one or more requirements. For example, some are shallow-rooted, others are deep-rooted; some complete their growth, flower, and produce seed in the spring; others require a longer growing season and flower in the summer or autumn; some require full sunlight, while others need partial shade. Plants that are associated, then, are species that have certain water, soil, light, or temperature requirements in common, and are enough different in other respects not to interfere materially with one another. In the naming of associations, we use the name of the plants that are most prominent and that dominate the community (*dominant species*). The other, less prominent plants are spoken of as *secondary species* of the association.

The water lily, bulrush, and cat-tail associations of ponds are small communities of plants dominated by these particular plants. The hemlock forest, the redwood forest, the yellow-pine forest, and the oak-hickory forest are all associations of large and small plants of numerous species that thrive under certain conditions.

Climatic plant formations. In any particular part of the country there are usually many different plant associations. Local differences of elevation, topography, soil, drainage, and slope exposure result in a diversity of local environments. Some plants fit into each of these environments better than others; consequently in each there arises a local community, or association, of plants. In general it has been found that throughout a region having similar climatic conditions, the group of plant associations is essentially similar. When we extend our study into regions with very different climates, we find a very different series of plant associations.

The plant associations of Indiana are very similar to those of Ohio because the climate in the two states is very similar. For the same reason the plant associations of Kansas and Nebraska are very similar. However, there is a vast difference between the plant associations of the Ohio-Indiana region and those of the Kansas-Nebraska region. This difference is primarily the result of difference in climate.

Different groups of plant associations, then, are characteristic of different climates. For this reason it is customary to group associations into larger units called *climatic plant formations*. The terms "evergreen forest," "deciduous forest," "prairie," "plains," and "desert" show the general recognition of these larger groupings of vegetation that are primarily determined by the light, moisture, and temperature conditions that make up climate.

Plant associations not permanent. Students of physiography are familiar with the fact that land forms are constantly changing.

Hills are eroded by wind and water, and their surface materials are being constantly carried to lower levels. Ponds and lakes are continually being filled by material that is carried into them by the wind and water, or by the material that accumulates through the death of the plants and animals living in the water. Streams enlarge their valleys, eroding here and depositing there, but constantly changing and wearing away the slopes and the valley bottoms. The plant associations that exist in these various physiographic situations are affected by all these changes; consequently the character of the association also changes.

In addition to the physiographic changes, the vegetation itself, through shading, brings about changes in light, temperature, and moisture. Humus accumulates in the soil, increasing the constancy of the water supply and affording better conditions for the growth of the bacteria and fungi which improve the available supply of soil salts. Animals, particularly earthworms and insects, also aid in these processes.

Habitats, then, are constantly changing; and in the course of years, decades, or centuries the conditions may be so altered that the kinds of plants now living in the habitat cannot survive and other kinds will have taken their places. This process of change in vegetation is called *succession*. Examples of succession may be seen in fields that have been abandoned and allowed to return to a wild condition.

In the forested regions of New England it is not uncommon to see areas embracing several hundred acres, once highly cultivated, but now, through abandonment, completely reverted to forest again. Other examples of succession may be seen along railroad cuts and fills, on sand dunes, and on sand bars and islands in streams. Here the newly exposed areas are occupied by associations of plants very different from the plants on areas that have been covered with vegetation for 10, 20, or more years. The youngest areas may contain mostly a great variety of annual weeds; the older areas are covered with perennial herbs and shrubs or with young trees. The order of succession is usually quite definite in a given region. By noting the seedling trees in a forest one can often predict what the composition of the forest will be 50 years from now, if it is left undisturbed.

The vegetation of a continent like North America, then, is made up of several great climatic plant formations, each of which is composed of many local plant associations. The plant associations are not permanent, but change as the habitats change, and are succeeded by other plant associations.

Plant realms. Taking the world as a whole, geographers distinguish three great realms that differ in their vegetation, mainly because of differences in temperature. These are (1) the *torrid realm*, where the temperatures are uniformly warm and frosts are unknown; (2) the *temperate realm*, where a warm growing season alternates with a cold period during which plant processes are slowed down, or stopped, each year; and (3) the *frigid realm*, where the cold is either continuous or alternates with a short, cool summer having almost uninterrupted light of low intensity. Vegetation is markedly different in these three realms. Under the most favorable conditions vegetation is densest and the number and variety of plants are greatest in the torrid realm and least in the frigid realm. Each of these realms, of course, is occupied by several or by many climatic plant formations, depending upon differences in climate. There are torrid forests, grasslands, and deserts, just as there are temperate forests, grasslands, and deserts.

Plant formations on mountains. High mountains occur in all parts of the world. The vegetation of the summits differs very materially from the vegetation at their bases. In polar regions the summits may be continually hidden in ice and snow, and the only plants that can grow there are microscopic ones that live on the surface and cause the so-called "red snow." Within the tropics the base of the mountain may be surrounded by tropical forest; higher up, temperate forests occur; and then comes a " timber line," beyond which only low-growing plants related to those of the frigid realm occur; the summits may be clothed in snow and ice. Increasing altitudes bring about the development of vegetation similar to, or corresponding to, the vegetation of higher latitudes.

Summary. A study of vegetation shows that the plants are naturally grouped into plant associations. The plant associations of any uniform climatic region are essentially similar and may be grouped into climatic plant formations. Climatic plant formations in turn are conveniently grouped into plant realms characterized by torrid, temperate, and frigid climates.

CHAPTER THIRTY-THREE

THE VEGETATION OF NORTH AMERICA

NORTH AMERICA extends from the North Polar Sea nearly to the equator, and consequently its vegetation includes climatic plant formations belonging to the frigid, temperate, and tropical realms. In this chapter the more important of these plant formations, and the factors which determine or limit their distribution, will be discussed. There are at least nine of these natural divisions of the vegetation of North America: (1) *Tundra*, (2) *Northern evergreen forest*, (3) *Deciduous forest*, (4) *Southeastern evergreen forest*, (5) *Prairie*, (6) *Plains grassland*, (7) *Western evergreen forest*, (8) *Desert*, and (9) *Tropical broadleafed evergreen forest.*

Climate — especially moisture, temperature, and light — determines the particular part of North America where each of these several types of plants may live. The habitats within the climatic formations determine the number and location of the plant associations. In the paragraphs that follow, the vegetation is described as it was before it was modified or destroyed by man. In the next chapter attention is called to the close correlation that exists between the climatic plant formations and the distribution of the industries directly dependent upon plant life.

The tundra formation. There is no more distinctive type of vegetation on the earth than the low-growing vegetation of the frigid realm, to which the name *tundra* has come to be generally applied. Originally used to designate the vast stretches of low, swampy, and rocky plains of northern Russia, this term is now applied also to the vegetation that covers the "barren grounds" from northwestern Alaska to Hudson Bay and eastern Labrador.

The tundra is a region of shallow, poorly drained soils, where the winters are long and the average temperature so low that the ground thaws only a few inches, or at most a few feet, during the

FIG. 191. The forest formations of North America. North of the northern evergreen forest is the tundra formation. On the unshaded areas south of it are the prairie, plains, and desert formations.

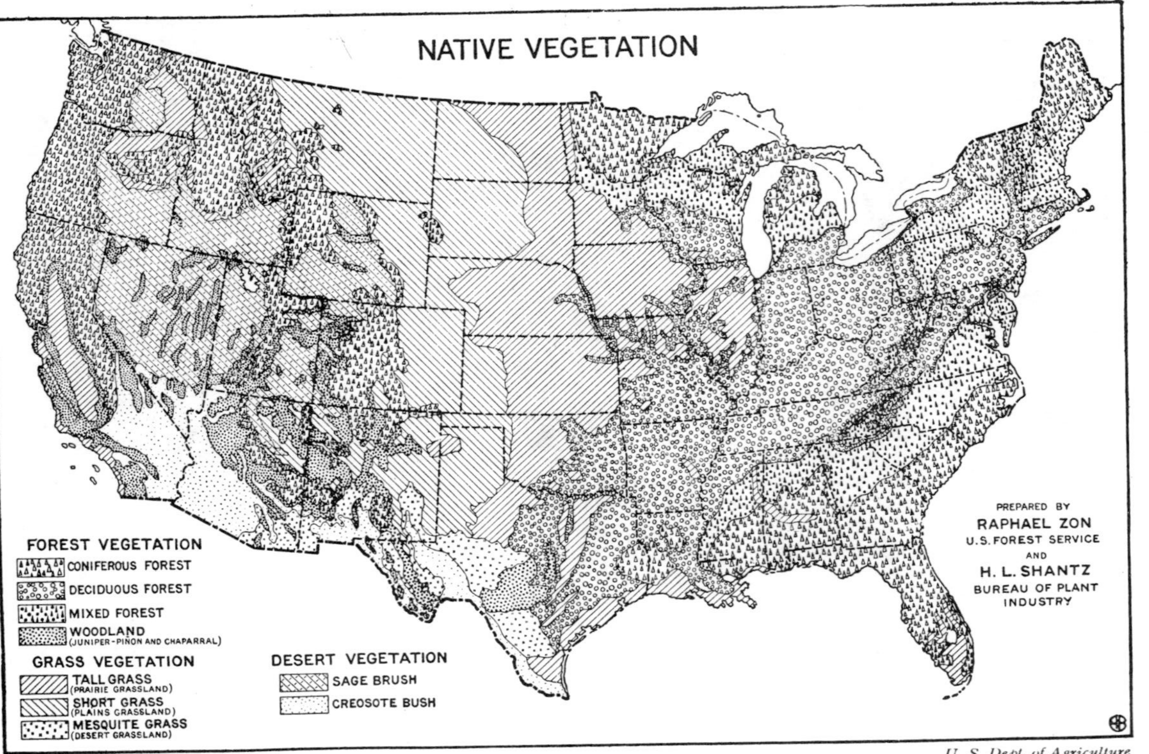

U. S. Dept. of Agriculture

FIG. 192. Map of the native vegetation of the United States, according to Zon and Shantz.

two or three months of summer. Consequently the only plants that thrive are the low plants that have shallow roots, like grasses and sedges, or that grow entirely on the soil surface, like the mosses and lichens. Northward, the tundra is limited by the polar seas and the areas of perpetual ice and snow; southward, by the northern evergreen forest. Even farther south, among the forests, are patches of tundra-like vegetation that remain in bogs and on bare rock outcrops. On the higher mountain summits everywhere are alpine areas covered with vegetation that is closely related to that of the tundra.

The winter season is characterized by intense cold, violent dry winds, and very light snows. The short growing period during the summer, the low soil temperature, poor drainage, and consequent scanty aëration of the soil are important factors in excluding most plants, particularly trees, from this region. The better-drained stony uplands are covered with grasses; the low places, with mosses, sedges, shrubs, and flowering herbs.

Most of the flowering plants are only a few inches in height. Many have leathery leaves and creeping or reclining stems, and are typical xerophytes. Cranberries, crowberries, and snowberries are examples of common low shrubs. Many of these plants are evergreen, and many of them are small compared with those of temperate regions. The willows are represented by several dwarf species that rise only a few inches above the soil.

The northern evergreen forest formation. Stretching from Newfoundland and Labrador to Alaska by way of the St. Lawrence Valley and the lower Hudson Bay region is the northern evergreen forest. This forest is composed of the white spruce, black spruce, paper birch, aspen, balsam poplar, tamarack, balsam fir, white pine, red pine, jack pine, arbor vitæ, and hemlock. The last five of these trees are found mostly east of Winnipeg. All the trees attain their greatest size in the region between northern Minnesota, Maine, and eastern Quebec. On the western plains of Canada, where the rainfall is reduced to

U. S. Forest Service

FIG. 193. Peat bog in northern Minnesota, with mature tamarack (*Larix laricina*) and black spruce (*Picea mariana*). In the foreground are alder (*Alnus*), sedges (*Carex*), and bulrushes (*Scirpus*). Throughout the northern evergreen forest region, thousands of square miles are covered by such bogs.

FIG. 194. Black spruce trees on the cliffs near Yarmouth, Nova Scotia, showing effects of winds from the Bay of Fundy.

15 inches, they are more or less confined to the stream margins, and in Alaska to the river valleys. On the best soils the white spruce, balsam fir, and paper birch grow in a dense mixture, forming the finest of northern forest types. The jack pine and white pine in some localities occupy sterile soils and form extensive forests. On mountain slopes and along the shores of the Great Lakes and the sea the black spruce is a common forest tree. In the numerous bogs and poorly drained areas the tamarack and black spruce dominate. On limestone outcrops and in the better-drained swamps the arbor vitæ is common. Where the original forest has been cut or burned over, there are extensive areas covered by birch and aspen poplar. These form only temporary forests that are later succeeded by pine and spruce.

From northern Minnesota to Nova Scotia the southern portion of the evergreen forest is mixed with the trees of the deciduous forest formation, especially on the best soils and at low elevations. There is an extension of the northern evergreen forest, made up of white pine, hemlock, yellow birch, and spruce, on the Appala-

U. S. Forest Service

FIG. 195. Red spruce forest on the high mountains of western Virginia, an extension southward of the northern evergreen forest.

chian Mountains southward through Pennsylvania to northern Alabama, where it is confined to the mountain summits.

The northern evergreen forest region is characterized by long winters with deep snows and a short, warm growing period of about 3 to 4 months' duration. The total rainfall varies from 40 inches in the east to 15 inches in the west, and the evaporation from a free water surface is equivalent to about one half to two thirds of the rainfall. The humidity is high, varying from 70 to 80 per cent of saturation. The snowfall begins before the ground is frozen, and where the snow is heaviest the ground remains unfrozen throughout the winter and the slow melting of the snow keeps the soil moist far into the summer. This is important, for it insures the trees an adequate water supply at all times. On the tundra to the north the soil is permanently frozen at a comparatively slight depth. South of the evergreen forest the ground freezes from 2 to 4 feet every winter, and this is possibly one of the factors which limits the southern extension of some of the evergreen trees. Another important factor is the competition of the deciduous hardwood trees. On good soil the hardwoods soon shade out the evergreens, with the exception of hemlock. On poor soils, sand plains, sand dunes, and sandstone cliffs the evergreens are more successful in maintaining a foothold.

In general, the soils of the evergreen forest region are shallow, and the drainage is poor except in the highlands. The soils are shallow because of the glaciers that once covered all of this part of North America to a thickness of 1 to 2 miles. The ice of these glaciers flowed toward the south, smoothing off the land surface at the north, and carrying away whatever soil there was in pre-glacial times. As the glaciers disappeared only 20 to 30 thousand years ago, there has been comparatively little time for soil to accumulate. From these facts we may infer that *during glacial times* the northern evergreen forest occurred farther south, from southern New Jersey to Kentucky and Nebraska, and has moved into its present region in geologically recent times (Fig. 311).

FIG. 196. White pine forest in northern Michigan, in which are scattered maples, birches, and aspens. A typical view in the northern evergreen forest.

The deciduous forest formation. Limited on the north by the northern evergreen forest and on the south by the Gulf coastal plain, a great forest of broadleafed deciduous trees extends from the Atlantic coast westward to the great plains of central Nebraska, Kansas, and Texas. This is the oldest forest on the continent and has occupied much of this region for several million years. Sometimes it was far more extensive, sometimes it was more restricted; but it has been practically continuous since the Cretaceous period of the earth's history.

This forest is dominated by oaks, hickories, elms, ashes, maples, chestnut, beech, sycamore, cottonwood, and tulip. It attains its best development on the mountain slopes in North Carolina and Tennessee and the lower Ohio River Valley. Under the most favorable conditions this forest attains a height of 150 feet, and some of its trees develop trunks 6 to 14 feet in diameter. In summer they spread an enormous area of green foliage; in winter the above-ground shoots consist only of cork-covered trunks and branches. On uplands and on the poorer soils the oak, chestnut, and oak-hickory forest types dominate. On the richer uplands sugar maple, beech, and tulip trees, with various other mesophytic species, occur. In the river valleys elm, ash, soft maple, birch, and sycamore make up the forest covering. Under the larger trees dogwood, redbud, sourwood, and numerous other shrubs decorate the second levels. On the ground are flowering plants that bloom before the trees have set their leaves. The autumn coloration is a notable feature each year at the close of the vegetative season.

The deciduous forest region is characterized by short, cold winters, with some snow, usually averaging less than 2 feet, and a frostless season of from 5 months at the north to 8 or 9 months at the south. The rainfall varies from 40 to 50 inches eastward and diminishes westward. Generally it exceeds the annual rate of evaporation; in the mountains it may be twice as great. The average relative humidity is less than in the northern evergreen

FIG. 197. Oak-hickory forest in central Illinois. In this region the deciduous forest formation and the prairies meet, the forests occupying the slopes and stream valleys, and the prairies the flat uplands.

forest, varying from 70 per cent in the east to 50 per cent at the western edge, where the forest extends along the rivers into the prairies and plains.

Toward the north and on mountain slopes, northern conifers like the white pine and hemlock occupy considerable areas, or they may be mixed with the broadleafed species. Southward, on cliffs, sandy plains, and shallow soils, many trees, like the shortleafed pine, pitch pine, and scrub pine, occupy pioneer habitats. On the Piedmont plateau region the shortleafed and longleafed pine are mixed with oak-hickory forests. The uplands immediately west of the Mississippi River were originally covered by oak and hickory forests, shading into walnut, elm, and beech of magnificent proportions on the more fertile soils. Here the shortleafed pine was mixed with the oaks and hickories. In many places the red cedar covered extensive areas of shallow, rocky upland.

Toward the west the deciduous forest occupies long, finger-like extensions covering the valleys in the prairie region, which finally become narrowed down to mere strips of elm, ash, poplar, and willows along the margins of the streams in the plains country. Toward the south the hardwoods compete successfully with the southern pines and occupy the better lands.

The southeastern evergreen forest formation. This forest centers on the Coastal Plain from eastern Virginia to eastern Texas. On the sandy uplands it is the home of the longleafed and shortleafed pine. In the swamps there are extensive areas of cypress; and along the streams and bayous near the coast, tupelos (sour gum), water oaks, pecans, sweet bay magnolias, and live oaks flourish.

The climate of this region is marked by average summer temperatures of 70 to 80 degrees and winter temperatures of 40 to 60

U. S. Forest Service

FIG. 198. Longleaf yellow pine encroaching on grassland in Florida. An example of succession.

U. S. Forest Service

FIG. 199. Bald cypress swamp near Memphis, Tennessee. Note cypress knees at the left. During the wet season the water covers the area nearly to the tops of the knees.

degrees. The relative humidity is high. The rainfall varies from 60 inches along the coast to 44 inches inland. At the coast this amounts to about 1.3 times the evaporation; at the inner edge of the Coastal Plain it is 1.1 times the evaporation. Snowfalls occur at rare intervals, but killing frosts occur over most of the area every year. For the most part, the soils are loose and arranged in belts parallel to the coast, except the alluvial deposits which extend up the Mississippi and other large rivers.

The climate is favorable for a broadleaf evergreen forest, but the poor soils of the Coastal Plain are better suited to the conifers and consequently the most extensive forest is dominated by the longleaf and other pines. The pine barrens are comparatively open woods on the dry or moist sandy plains; on better and on moist soils the forest is more dense and has an undergrowth of small oaks and other trees, some of which are evergreen. Among the undergrowth is the low-growing palmetto, which suggests an approach to subtropical conditions.

Along the eastern coasts white cedar swamps occur as far north as southern Maine. The cypress and tupelo swamps are common along the lower courses of the rivers from Chesapeake Bay southward, and extend as far inland as southern Illinois.

One of the remarkable plant associations of the South is that of the canebrakes, our only native representatives of the bamboos, which are so abundant in Asia. The canebrakes were formerly extensively developed on the low hills bordering both sides of the Mississippi flood plain, and in central Alabama as undergrowth on the oak-covered black-soil areas.

Just as the deciduous forest trees like the maple occur in the best habitats in many places as far north as Nova Scotia, so dense growths of oaks, beech, hickories, and magnolias occupy the most mesophytic habitats, called "hammocks," as far south as central Florida. As we go southward from the northern evergreen forest, the rate of evaporation gradually increases, and an increasing amount of rainfall becomes necessary to permit the growth of forests.

The tropical evergreen forest formation. The southern third of the peninsula of Florida, the West Indies, the lowlands of Mexico, and the eastern slopes of Central America are occupied by tropical forests. Where the rainfall is more than 50 inches, these forests attain magnificent proportions and great density. Where the rainfall is less, one finds tropical scrub and desert. As the trade winds of the tropics blow from the east, the greatest rainfall occurs on eastern slopes, and there the conditions for forest growth are at their best. On western slopes the rainfall

FIG. 200. Subtropical vegetation in southern Florida. Live oak covered with epiphytic bromelias. In the background, cabbage palmetto.

U. S. Forest Service

FIG. 201. Buttressed trunks of mahogany trees in southern Mexico.

is reduced slightly by low mountains and greatly by high mountains, and the vegetation changes accordingly.

In Florida the tropical forest is poorly developed, but its relationship to the tropical forest is shown by the presence of palmettos, palms, and other tropical trees. Along the coast are mangrove swamps, very similar to those found on all muddy coasts in the tropics. The Everglades constitute a vast area of shallow water largely occupied by saw grass, with narrow open-water channels forming a labyrinth of passages. Interspersed are many small islands covered with tropical trees, which support numerous epiphytic bromelias and orchids on their branches.

In the West Indies and in Central America the original tropical evergreen forest has been destroyed by centuries of *migratory agriculture*. This term is applied to a general practice in tropical countries, of clearing a piece of forest land and growing crops on it for a few years, while the returns are large and the weeds are easily controlled. When crop growing becomes more difficult,

U. S. Forest Service

FIG. 202. Tropical jungle in British Honduras. In the West Indies and Central America the original forest has been destroyed by migratory agriculture and the jungle has taken its place. The picture shows a large mahogany tree near the center.

the native moves on and clears another area. The sequel of migratory agriculture is the tropical jungle, with its dense, tangled, and almost impenetrable masses of vegetation.

In Dominica, Trinidad, Venezuela, and other northern states of South America, are remnants of the original broadleafed evergreen tropical forest. These forests are noted for their great variety of tree species and their freedom from dense undergrowth.

The prairie formation. Extending from North Dakota to Texas and eastward to Indiana is a roughly triangular region in which vast areas of level and rolling uplands formerly were covered with tall grasses from 3 to 10 feet in height, while deciduous forests dominated the river valleys. These are the true prairies. Toward the western margin the prairies were well drained, or even overdrained, but to the eastward they were interspersed with sloughs and temporary ponds which were also dominated by grasses.

During the summer the prairies were studded with the brightly colored flowers of scattered perennial herbs. In the fall the prairies were a vast sea of highly inflammable grasses, and often they were swept by fires that destroyed everything in their path. In winter they were bleak and exposed to the full sweep of the wind and drifting snow.

The dominance of the prairie grasses over this great area and the absence of forests was made possible by the climate. The most important climatic factors influencing plant growth are rainfall, temperature, humidity of the air, and wind velocity. The first — rainfall — represents the source of the water supply in the soil. The other three factors determine the rate of evaporation from a water surface. In the prairie region the rainfall is less than the amount of evaporation; it is about six tenths as great on the western side and eight tenths on the eastern border. The prairie region is characterized by high summer temperatures and summer droughts. Another characteristic feature of the prairie climate is the uneven distribution of the rainfall during the growing season. During one season the heavy rains occur in the spring, during others in midsummer or autumn. This leads to annual droughts either preceding or following the rains.

The soils are for the most part clays and sandy loams upon which there has accumulated since glacial times, through the comparatively slow decay of the prairie vegetation, several inches to several feet of black humus. But the nature of the soil was of less importance in the maintenance of the prairie than the climatic factors which controlled the moisture content of the soil. The prairies that were low and poorly drained were unfavorable to the growth of trees because of too much water in the spring and early summer. The more western prairies were subjected to too intense droughts in summer to favor the growth of trees. In the years that have passed since the prairies were first settled thousands of miles of tile drains and ditches have drained the ponds and sloughs, and the ground-water table is today several feet lower than it was originally. This has made possible the growth of trees in the wet prairies, where formerly they were absent. The absence of fires is also favorable to the extension of the forests.

Although there were several species of grasses common on the prairies, by far the most important is the "big bluestem." This grass formed an almost pure growth over the large areas, and in late summer was so tall and dense that cattle were lost to sight in it and their position could be told only by the swaying of the grass tops as they moved about.

On the sandy and more exposed dry prairies "bunch grass," or "little bluestem," 2 to 3 feet high, was most abundant.[1] As humus accumulated and the soil moisture was increased, these areas were invaded and often occupied by the "big bluestem."

In the wet areas of the prairie the "slough grass," 6 to 10 feet in height, was dominant. This grass was used frequently by the pioneers to thatch the roofs of their smaller farm buildings.

[1] Big bluestem is *Andropogon furcatus;* little bluestem is *Andropogon scoparius.*

Scattered throughout the prairies were large and small herbaceous plants, including milkweeds, sunflowers, rosin weeds, coneflowers, asters, and goldenrods. These plants gave color to the prairies at certain seasons. They never made up a large part of the original prairie covering, however, and they were most numerous on the borders between the prairie and the forest and on eroding slopes.

Among the explanations sometimes given for the treelessness of the prairies are the fires set by the Indians and by lightning. That these fires occurred in the autumn there is no doubt, and that they killed young trees on the forest edge and acted as a check to tree invasion there can be no doubt also. The prairies, however, preceded the prairie fires, and the fires could at best only delay forest invasion — not prevent it over such vast areas. Deciduous forests occurred throughout the prairies along the streams, on river bluffs, on valley slopes, and on the flood plains. The highly fertile character of the prairie soil caused them to be occupied by farms as rapidly as they could be broken and properly drained. Today patches of original prairie are far more difficult to find than patches of original forest.

In Illinois and Iowa the prairies occupy for the most part upland areas between the stream valleys. In Kansas and Nebraska, where the region that was dominated by the big bluestem reaches its western limit, the prairies were confined to the river valleys and lowlands.

The plains grassland formation. Between the Rocky Mountains and the prairies and from Saskatchewan to Texas is a vast rolling plain more or less dissected by streams and covered with grasses. West of the Rockies, from Montana to Washington and California, are similar areas of grassland bordering the forests.

The annual rainfall varies between 10 and 20 inches and is distributed irregularly in showers and occasional heavy downpours. As the depth of evaporation from a water surface is between 30 and 50 inches during this same period, the extent of the rainfall is only from two tenths to six tenths of the evaporation. Furthermore, in late summer this region is subject to prolonged hot dry winds from the southwest. At such times the temperature rises above 100° F., the humidity falls, soil moisture becomes very low, and all vegetation suffers through excessive transpiration. These " hot winds " were the source of great losses to the early settlers when they occurred before the crops were mature. On the high plains of western Kansas and eastern Colorado the soil is generally dry below a depth of 6 to 15 feet. The plains region is the home of occasional violent winds, tornadoes in summer and blizzards in winter. The snowfall is generally light but is subject to drifting and may become deep in the depressions.

The most characteristic grasses of the Great Plains are the buffalo grass (*Bulbilis* and *Boutelona*), the bunch grass (*Andropogon*), and the wire grass (*Aristida*). The buffalo grass is a turf-forming grass, a few inches in height, which affords highly nutritious forage. The bunch grasses received their name from the habit of growing in scattered dense tufts, especially in lands that have been disturbed by streams and wind erosion. The wire grass is a coarse grass, 2 feet in height, which also grows in tufts, usually mixed with other grasses.

Just as the deciduous forests stretch westward, occupying the valleys in the prairie region and the stream margins in the plains country, so the tall-grass prairies extend westward in the valleys, forming finger-like extensions between the tree-bordered rivers and the short-grass uplands.

At their western margin the plains are invaded on rocky slopes by the western yellow pine, and at the southwest by semi-desert scrub, consisting of mesquite, red cedar, and scrub oaks. The grasslands also grade into sagebrush, which occupies extensive areas from Colorado and Montana to the Great Basin and Sierra Nevada Mountains. Scattered over the plains are many small,

FIG. 203. Forests on a high mountain (Mt. Robson, British Columbia), showing timber on ridges and absent from valleys where the snow accumulates to great depths. On the talus cone are numerous avalanche tracks where the trees have been destroyed. The white line crossing the talus cone near the base is a trail.

W. S. Cooper

FIG. 204. Sitka spruce forest at Glacier Bay, Alaska. This forest has grown up since the retreat of the glacier about 100 years ago.

xerophytic, flowering herbs, as well as cacti, yuccas, legumes, and composites that relieve the gray-green monotony of the grasses by their vari-colored flowers.

We have seen that the treelessness of the *prairies* is due, for the most part, to an excessive transpiration rate in proportion to the soil-water supply; locally toward the east, to unfavorable soil drainage also. The treelessness of the *plains* is due to inadequate water supply; intense summer and winter droughts make it very difficult for tree seedlings to become established.

Western evergreen forest formation. The western Cordillera, extending from Alaska through the Rockies, the Sierra Nevada, and the Coast Ranges to Mexico, are clothed with conifer forests of pines, firs, spruces, hemlocks, and cedars. This forest roughly has three divisions: (1) the northern coastal forest, extending from Washington to southern Alaska, (2) the Coast Range and Sierra forest, extending southward to southern California, and (3) the Rocky Mountain forest, stretching from northern British Columbia to Arizona and Mexico. The region from Washington to western Montana is a meeting ground for species from all these divisions.

U. S. Forest Service

FIG. 205. Redwood (*Sequoia sempervirens*) forest in the mountains of northwestern California. The tree nearest the man is $45\frac{1}{2}$ feet in girth at the level of his hat. The redwood is found in the moist valleys of the Coast Range and is the tallest of all conifers.

U. S. Forest Service

FIG. 206. An alpine meadow in Cascade Forest Reserve. Fir and hemlock forests cover the rocky slopes. The alpine meadows occupy depressions at elevations above 9000 feet. The snow lies on them 8 to 10 months of the year; the soils are composed in large part of wet muck, and many of the plants are the same as those found on the tundra.

The Puget Sound region, the lower altitudes of southern British Columbia, and the coastal mountains of Oregon are the home of the most magnificent conifer forests in the world. Not only are the trees of great height (200 to 250 feet), but they have trunks 8 to 15 feet in diameter and they stand very close together. This great forest is the natural outcome of a moist climate with a rainfall of about 100 inches, together with mild winters due to the proximity of the Pacific Ocean. It is dominated by the Douglas fir, Western hemlock, and Western arbor vitæ. In spite of the thick growth of trees, there is a dense undergrowth of ferns, shrubs, and low-growing trees.

From Washington north to Alaska the forests on the western slopes are dominated by the Sitka spruce. Southward from Oregon, in the fog-laden valleys of the Coast Ranges to San Francisco Bay, the forest is composed of redwoods, the tallest of all conifers.

From this point southward the Coast Ranges are dominated by vegetation consisting of scrub oaks, hardleafed shrubs, and xerophytic grasses — collectively known as *chaparral*. The chaparral also forms a belt surrounding the central valley of California and the lower elevations of the southern California mountains. This is a region of winter rainfall and hot, dry summers.

Inland from southern Oregon and south along the Sierra Nevadas is an extensive forest of Western yellow pine, with Douglas fir, incense cedar, and sugar pine intermingled. In California this forest is restricted to the moist slopes above 1500 feet in the north and above 3000 feet in the south. Above the pine forest is a belt of firs and hemlock, and at the timber line the white-barked pine occurs. Between the pine belt and the desert is a belt of oak and digger pine, and at lower levels an extensive growth of chaparral. On rolling uplands between the large canyons on the western slope of the Sierras occur groves of the celebrated "Big Trees." At timber line here and elsewhere

W. S. Cooper

Fig. 207. Foothills and valley land in Arizona covered with grass and oak brush, affording grazing range for goats.

U. S. Forest Service

FIG. 208. Piñon forest in northern Arizona, with sagebrush and grassland in the foreground.

throughout the Western mountains, on flat areas are alpine meadows, where the snow accumulates to great depths in the winter. During July and August these meadows are covered with the most brilliantly colored flowers.

In general, alpine vegetation occurs at lower and lower levels as we go north, but much depends upon the local exposure to moisture-laden winds and whether the slopes face north or south. North-facing slopes are moister and cooler and the growing season is shorter than on slopes facing south.

The forest trees of the Rocky Mountains are closely related to those of the California-Puget Sound region. The upper limit of tree growth is about 9000 feet in Montana and 12,000 feet in southern Colorado. Since these mountains rise above an arid plateau region, there is also a lower limit to tree growth; this limit is between 4000 and 6000 feet. In the Canadian Rockies the forests are continuous in the broad valleys and mountain slopes. Southward, beginning at Montana, the broad inter-mountain valley is occupied by sagebrush semi-desert, and farther south by desert vegetation. The most characteristic tree of the entire region is the Western yellow pine. A close second is the widely distributed lodgepole pine. In Colorado the limber pine, and farther north in Montana the mountain pine, are locally abundant, as is also the Douglas fir.

In southern Colorado the yellow pine gives way to the nut pines and junipers in semi-arid places. The forests on the plateau of Arizona, and those above 5000 feet on the mountains of Arizona, New Mexico, and western Texas, are dominated by yellow pine bordered by belts of nut pine, juniper, and scrub oak at the semi-arid lower levels. Usually there are belts of grassland and sagebrush in the transition to the desert.

At higher altitudes and in the moist canyons of Colorado, Engelmann spruce and subalpine fir are abundant; farther north, firs, Western hemlock, and larch constitute important forest types. Grasses, composites, and legumes furnish the bulk of

U. S. Forest Service

FIG. 209. Desert scrub near the east end of the San Bernardino Mountains, California.

W. S. Cooper

FIG. 210. Small-leafed desert shrub vegetation on dunes, Monterey Bay, California.

the small flowering plants. Throughout the Rockies the streams are bordered by alders, willows, and poplars.

The southwestern desert formation. From the plateau of Mexico, extending northward into California, Arizona, Nevada, Utah, and Idaho, and eastward to New Mexico and western Texas, is the desert. This great region is, for the most part, a more or less broken plateau, with a rainfall of from 3 to 20 inches, and with an evaporation rate five to thirty times as great. Temperatures as high as 120° F. occur in the summer, and frosts are not unknown even in southern Arizona. Farther north the winters are severe, and, due to the great intensity of the sunlight, the summers are very hot. Toward the south there are two rainy periods, one in July and another in January. Following these rains the desert is green with a covering of summer or winter annuals that spring up quickly between the perennials and, within a few weeks, flower, fruit, and die.

At other seasons the vegetation is scattered, of a gray-green color, and consists of thorny and spiny shrubs, large and small cacti, fleshy-leafed agaves, yuccas, and other small succulent

W. S. Cooper

FIG. 211. Desert vegetation on Tonto Platform, Grand Cañon, Arizona, consisting of prickly-pear cactus, yucca, and low shrubs.

and woody perennials. Desert plants are either extreme xerophytes, or they are mesophytic short-lived annuals that complete their life cycle during a single moist period. The xerophytes include those with thick stems that accumulate enough water during the rains to carry them over dry periods, like the shallow-

W. S. Cooper

FIG. 212. A desert shrub, *Fouquieria splendens,* in leaf, near Tucson, Arizona. This plant is found over wide areas in the Southwestern desert.

W. S. Cooper

FIG. 213. Giant cactus and desert shrubs near Tucson, Arizona.

rooted cacti. The agaves accumulate water in their fleshy leaves, and the yuccas are deep-rooted. The shrubs for the most part are deep-rooted and live in soils where water flows or seeps from the better-watered mountains and elevations. Northward the succulent desert gives way to the sagebrush semi-desert; eastward and westward it passes into small-leafed desert scrub.

All the perennial plants show reduced leaf surfaces. Some have leaves only during the rainy season, and others, like the cacti, are quite devoid of foliage leaves. Heavy cuticles, bloom, and thickened epidermal cells are found on the agaves and yuccas.

In southern Mexico the desert gives way to semi-desert tropical scrub, which in turn merges into the tropical jungle that occupies more and more of the land through Central America to Panama. On the higher mountains subtropical oak and other hardwood forests pass into pine forests at still higher elevations. The highest peaks reach above timber line and have small areas of alpine vegetation.

CHAPTER THIRTY-FOUR

RELATION OF PLANT INDUSTRIES TO CLIMATIC PLANT FORMATIONS

THE climatic conditions that restrict each of the great plant formations to a definite region of North America also determine to a large extent the location of many industries dependent upon plants or plant products. It is self-evident that most industries that directly utilize wild plants are located near those plants. And for the same reason industries directly dependent upon crop plants are usually located in regions where the particular crop plants grow best.

Climate and the production of crops. Cultivated plants are affected in their development by climatic factors in much the same way as wild plants. Temperature, moisture, and light conditions must be favorable; and each of the important crop plants has its own requirements in this matter.

Crop plants must not only be able to grow in a particular climate, as wild plants do, but they must yield a profit to the grower. Peanuts, for example, can be grown in the Northern states, but the yield is so small that they are unprofitable; they produce the largest yields in the Southern states, where the temperature is high. A crop grown in the region whose climate is most favorable to that crop will produce a better quality and a greater return to the grower than the same crop grown in a less favorable locality. The distribution of crop plants is determined, then, by the same factors that limit wild plants, and in addition, by certain economic factors.

Soil factors and crop production. Within each climatic region the various crops, like wild plants, are further limited by the great variety of soil conditions. Some soils are poorly drained, others are over-drained, and still others show every gradation between. Soils may be lacking in some of the necessary mineral salts, or they may have some salts in excess. They may have a high humus content, or be nearly lacking in humus. Some

FIG. 214. View of rice fields in hills of the Philippine Islands. By the building of dams a mesophytic habitat has been changed to a hydrophytic one, in spite of steep slopes.

soils are acid, others neutral, and still others are alkaline. The slope of the soil may be so great, or the rocks so near the surface, that cultivation is impossible.

In other words, within each climatic region there are many plant habitats, some of which may be used for one crop, some for another, and some that are best left to produce pasturage or crops of trees. Consequently, any one crop usually occupies only a part of the climatic region in which it might be grown if the conditions in all localities were favorable. For example, tobacco is one of the most profitable crops of the deciduous forest region; but since its quality is greatly influenced by soil conditions, its cultivation on a large scale is limited to certain definite soil areas. Moreover, each of these soil areas is given over to the growing of some particular type of tobacco.

Crop centers. A study of the geography of crop plants will show that each crop has a region in which it is so profitable that a considerable proportion of the suitable land is given over to it. These regions are called *crop centers*. Moreover, the centers of production of most crops coincide in large measure with the climatic plant formations. Each plant formation has, therefore, become a center of production of a certain group of crop plants.

In colonial days all the possible crops were grown in every locality. As the West became settled and transportation facilities increased, the several crops were gradually moved into the most favorable regions, and farming in any one locality became more specialized. This movement is still going on and is of great importance for the future supplies of agricultural products. Disregarding market factors, those crops are most profitable which best fit both the climate and the soil. As a result of competition among farmers in different sections of the country, the production of a particular crop at times increases in certain localities and decreases in others.

Crop plants less restricted than wild plants. Crops may be so valuable that the grower can afford to make artificial habitats for the plants. He may irrigate the land with water from the mountains, or he may modify the climate by growing the crops under glass, or under shades and screens. Such devices lead to wide extension of the areas of crop production beyond their natural areas. The hundreds of tracts of irrigated lands in the Western states; the growing of tobacco under shade in Connecticut, Florida, and the West Indies; the growing of vegetables under palms in desert oases; and the growing of tropical plants and summer vegetables in greenhouses in winter are familiar examples of the extension of crop production into regions naturally unfavorable.

Extending areas of crop plants through plant breeding. There is still another reason why crop plants are less restricted than wild plants. This is the production of new plant forms through the activities of plant breeders. Ever since plants were first cultivated, men have tried to find better or more suitable varieties for cultivating in particular localities. As a result we now have hundreds of varieties of crop plants, some of which grow better in one climatic region and others in another. By producing or discovering new varieties, the areas of all the familiar crop plants have been greatly extended.

The factor of transportation. Some plant products — like the potato, for example — are bulky and the cost of transportation correspondingly great. Although potatoes grow best on sandy loam in the cool Northern states, they can be produced at a profit elsewhere, in spite of lower yields, when sold locally and transportation charges avoided, or when they can reach the market earlier. A map of potato production shows that potatoes are grown in quantity near all the large cities of the country.

Other plant products — like mahogany, oils, resins, rubber, and spices of tropical forests — are so valuable that they may be transported long distances before they are made into commercial products. Consequently, the industries dependent upon plants of this type may be far removed from the source of supply.

In spite of these exceptions to the rule, it is generally true that the great climatic plant formations are each characterized by certain groups of plant products and plant industries. In the following paragraphs the more important products and industries of each of the natural divisions of the vegetation of North America are discussed.

The tundra. This inhospitable region, lying far to the north, has few inhabitants except the Eskimos along the northern coasts. They derive most of their food from the seals and shore birds, though they do invade the tundra on hunting expeditions for caribou, musk oxen, and smaller animals.

Here more than anywhere else on the continent the vegetation of the sea is important to men. On this vegetation, especially the microscopic plants, the fish are dependent, and they in turn are fed on by seals, walrus, and shore birds — the primary food of the inhabitants. On land the lichens, grasses, and other plants furnish the food of the arctic hares, caribou, and musk oxen — the secondary food of the Eskimos. Direct use of plants is very limited, and plant industries are entirely wanting.

The northern evergreen forest. The excellent quality of the wood that is derived from the white pine, spruce, red pine, and arbor vitæ, and its value for building houses and ships, led to the early invasion of the forests of Canada, New England, and the Great Lakes region by lumbermen; and until 1900 the northern evergreen forest was the most important center of lumber production.

During the past 30 years the spruce has become a valuable wood for the production of paper pulp. Its freedom from resin, its white color, and its soft, smooth, and uniform grain make it the best source of white book and print paper. Hemlock is second in importance in the making of pulp for newspaper, wrapping paper, and other cheap grades of paper. Consequently, at the present time the paper-pulp industry centers in the northern forest.

U. S. Forest Service

FIG. 215. White pine (*Pinus strobus*) about 120 years old, with understory of balsam fir (*Abies balsamea*), in northern Minnesota.

The invention of new processes is gradually making it possible to use a variety of other woods, and also certain herbaceous plants, for the manufacture of paper; consequently the industry is now spreading to the Southern and the Northwestern coast states. All together, 600 million cubic feet of wood are used annually in pulp manufacture.

Another important product of the northern evergreen forest is tannin, which is used in the manufacture of leather. Formerly the bark of the northern hemlock furnished the bulk of this material and the nearness to hemlock forests determined the location of many of the large tanneries. This industry has been forced to move to other regions and find other sources of tanning materials. The bark of the western hemlock, the oak, and the chestnut are now used for this purpose.

The northern arbor vitæ has been one of the principal sources of telephone and telegraph poles because of its durability in the soil, its light weight, and its comparative strength.

Agriculture in the northern evergreen forest region is more or less limited to the production of hay and forage crops, and much of the remaining land is given over to permanent pasture. Rye and buckwheat are produced to some extent. This is the natural region for the production of spring wheat, but owing to the shallowness and poor quality of the soils it cannot be grown on a large scale profitably in competition with the northern prairie region. Potatoes thrive best in a cool, moist climate and are mostly produced in the states along the Canadian border and the east coast from Virginia northward. Cranberry production centers in Massachusetts, and the total crop amounts to upward of a half-million barrels. Large areas of pasturage lead to the production of dairy cattle and determine the location of great numbers of creameries and cheese factories in the Lake states and southern Canada.

The deciduous forest region. This great area is a region of plentiful coal, oil, and gas, and it has abundant water power for

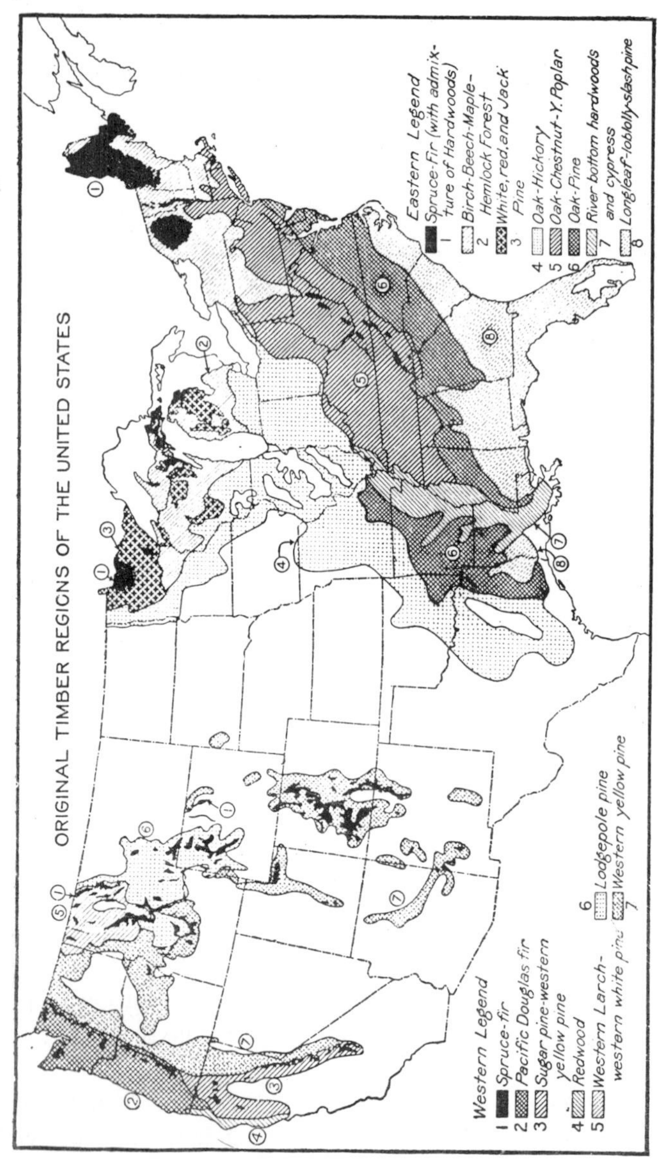

Fig. 216. Map showing the original timber regions of the United States.

manufacturing purposes. Furthermore, there are large areas of fertile soil that make agriculture profitable. Consequently it has become the region of the densest population of the United States.

The trees of the deciduous forest region are commonly known as "hardwoods." Oaks, maples, hickories, elms, ashes, chestnut, beech, sycamore, cherry, walnut, birch, basswood, and tulip constitute the important economic species. In consequence of the variety of products derivable from these hardwoods, these forests have been largely cut over, except on the more remote mountain slopes of the Appalachian system.

The oaks have furnished railroad ties and heavy beams for wooden structures. Oak is also used in large quantities, together with maple, birch, and walnut, in the manufacture of furniture. Nearness of supply led to the establishment of the center of the furniture industry in Michigan, New York, and Pennsylvania. Chestnut wood and bark and chestnut-oak bark have been most important sources of tannin in this region. Hickory, because of its great strength, is used for the handles of tools; and ash is important in the manufacture of vehicles.

Elm, beech, maple, chestnut, and birch furnish much of the material for staves in the manufacture of slack barrels for the shipment of cement, flour, sugar, apples, vegetables, and many other commodities. Elm also is the best wood for making the hoops of these barrels, because of its toughness and tensile strength. For tight cooperage — that is, barrels for the storage and shipment of liquids — white oak is the wood most desired. Cottonwood is one of the leading sources of excelsior and "wood wool" used for packing and for filling for mattresses and upholstery.

The making of syrup and sugar from the sugar maple was practiced by the Indians long before the advent of European settlers. The early settlers quickly took up the process and improved it. Today more than 2 million gallons of syrup and upward of 5 million pounds of sugar are produced. The sugar maple grows

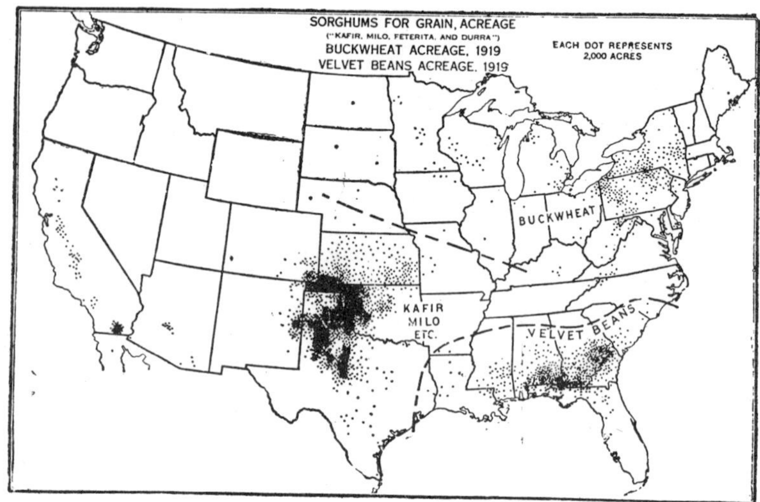

U. S. Dept. of Agriculture

FIG. 217. Map showing the acreage of sorghums, buckwheat, and velvet beans in the United States.

best in the states from Wisconsin to Maryland and Maine. The sap flows longest and the yield is greatest during a gradual northern spring, when there is freezing at night, thawing in the daytime, and a slow thawing of the ground. Such conditions are most perfectly attained in Vermont, New York, and northern Ohio, and the industry centers in these three states.

The distillation of hardwoods for the production of wood alcohol, acetate of lime, and charcoal is another industry that centers in Wisconsin, Michigan, New York, and Pennsylvania because of the large available supply of beech, birch, and maple, and the nearness to blast furnaces, which are the chief users of the charcoal. Much charcoal has been made in the past by simply driving out the volatile matter in the wood by slow combustion in pits. But in this way all the volatile matter was lost. Now the wood is heated in great retorts, and the by-products are far more valuable than the charcoal. Ash, oak, and hickory are also being used for distillation as the more desirable species become scarcer.

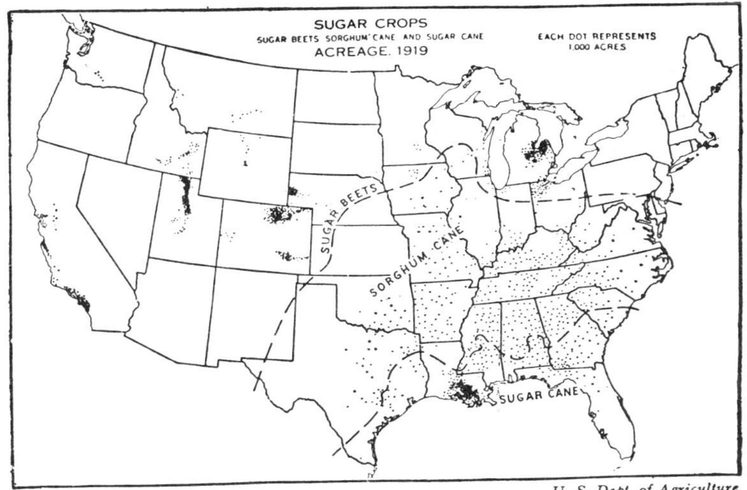

U. S. Dept. of Agriculture

FIG. 218. Map showing the acreage of sugar crops in the United States.

Agriculturally, the deciduous forest region is best suited for the production of winter wheat and corn; but due to the fact that the soils are far better in the prairies, the center of corn production lies in Illinois and Iowa.

Because of the numerous cities and industrial towns scattered from the Atlantic to the Mississippi River, market gardening and the production of cut flowers, ornamental plants, and nursery stock have been highly developed in this region.

The lands along the eastern and southern shores of Lakes Michigan and Erie are favorable localities for the growth of grapes, because killing frosts in the autumn are delayed by the warming effects of the lake. These areas are also favorable for the growth of peaches, because the lakes warm up more slowly than the land in the spring of the year; this retards the opening of the buds until danger of late spring frosts is past.

Apple, pear, peach, and cherry orchards are scattered over this region. These orchards produce a large part of the fruits of this type that are marketed in the eastern United States.

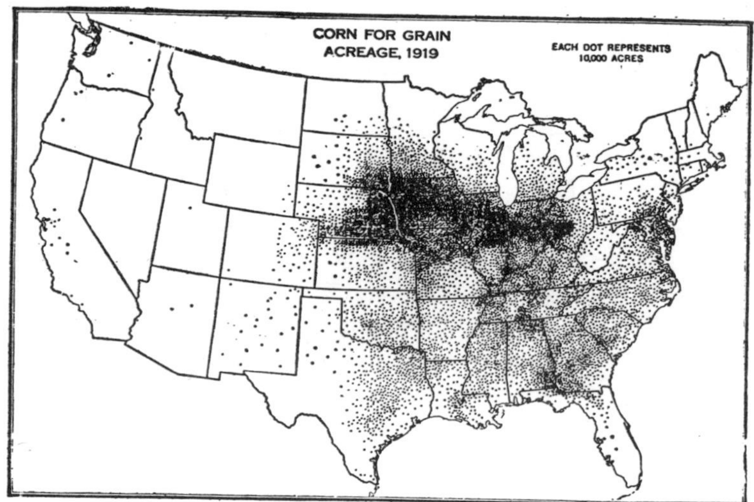

U. S. Dept. of Agriculture

FIG. 219. Acreage of corn grown for grain in 1919. This crop is raised over most of the deciduous forest region, but the greatest production is in the prairie region because of the fertility of the prairie soils.

The northern deciduous forest region was formerly the center of production of sugar from the sugar beet. These plants require the richest of agricultural land. Michigan was at one time the leading beet-sugar state; but there are now scattered factories from California and Washington to Ohio, and irrigation has recently made Colorado the leader in the production of beet sugar.

Certain soils of the deciduous forest region, from Kentucky and Pennsylvania to North Carolina and Virginia, are much utilized for the growing of tobacco. This is a highly profitable crop and gives a large return for each acre planted. In growing leaves for cigar wrappers it is important that they be large, thin, and have small bundles in the veins. Such leaves are secured, particularly in the Connecticut Valley, by growing the plants under canvas; this reduces the transpiration rate of the leaves, increases the size, and insures all the desired qualities. Burley tobacco

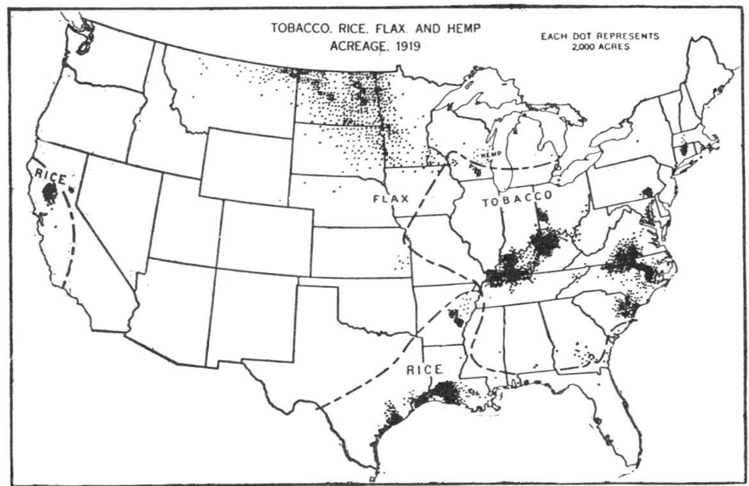

U. S. Dept. of Agriculture

FIG. 220. Tobacco, rice, flax, and hemp acreage, 1919. The regions in which tobacco is grown are determined by soil characters; each particular tobacco-producing area grows only certain kinds that develop on its soil the desired flavors. Flax is produced in the northern Great Plains region, because the drought and the prolonged illumination favor the production of seed. The necessity for hot growing seasons and inundation of the land for ice production explains why the growing of the crop centers in the areas shown on the map.

is mostly grown on the limestone soils of Kentucky and Ohio. Cigarette and light smoking tobaccos are grown on rather infertile sandy loams on the eastern Coastal Plain.

Sorghum, used for the manufacture of table syrup and sorghum molasses, is grown in the southern half of the deciduous forest region. It resembles corn in appearance, and the syrup is obtained by crushing it and evaporating the juice. In the North its sugar content is lower and its cultivation is not so profitable. Its cultivation is also excluded from soils rich in nitrates, because of the increased amount of bitter substances present under these conditions. As much as 50 million gallons has been produced in one year.

The southeastern conifer forest. The greatest plant industry of the Southeast has been the lumbering of the longleaf pine, the shortleaf pine, cypress, white cedar, and gum. Cypress is an especially valuable wood for use in building greenhouses, for it withstands without decay warm and moist conditions.

The southeastern conifer forest region has also been the center of production of turpentine and rosin, which are obtained from the longleaf yellow pine. A V-shaped cut is made through the sapwood of these big trees; the resin from the wood flows out slowly and collects in a large cup placed at the lower end of the V. The cups are allowed to remain for several months and then the accumulated resin is collected and distilled. The volatile oil, turpentine, passes over into a condenser and the rosin is left behind in the retort. The trees are tapped for three or four years and then cut for timber. The annual output amounts to more than 29 million gallons of turpentine and 3 million barrels of rosin.

Next to oak, the yellow pine is the biggest source of railroad ties. It is also used for poles and fence posts and for soft-wood distillation. The products of distillation in this case are charcoal, turpentine, pitch, and tar. Red gum is the principal wood used in the manufacture of cheap barrel staves, and pine is used for barrel heads. Arkansas is the leading state in the production of " slack " cooperage, and in the production of red gum, yellow pine, and cottonwood veneers, now so extensively used in packing crates, door panels, drawer bottoms, and chair seats. Sumac leaves are gathered in large quantities in the southern coastal plain, dried, ground up, and used as one of the sources of tannin. About a fifth of the paper pulp now comes from the southern yellow pine, gum, and cottonwood. Osage orange wood is a promising source of dyes for wool, leather, wood, and paper; the shades of these dyes varies from orange-yellow to olive and brown.

The greatest crop plant of the southeastern forest region is cotton. This plant belongs to the mallow family and is a native of the tropics. It requires high temperatures and can be grown only where the frostless season exceeds 6 months. Cotton lint is made up of the hairs that thickly surround the cotton seeds;

U. S. Forest Service

FIG. 221. Resins are formed by many plants and are produced in abundance by the pines. This illustration shows the collecting of resin in a forest of Southern longleaf pines.

these hairs vary in length from a half-inch to $1\frac{1}{2}$ inches. Formerly, only the fiber was marketed; but today the hulls of the seeds are used in the manufacture of hard paper board, the "meats" of the seeds are pressed for cottonseed oil, and the "oil cake" that remains after pressing is used for stock feed. The United States produces more than 11 million bales (500 pounds each) of cotton. Seven and a half million bales were consumed in this country in 1919; and at the same time 175 million gallons of oil and 2 million tons of oil cake were produced. Most of the oil is used in making soap, lard substitutes, and for salad oil.

Another coastal plain crop that has recently assumed great economic importance is the peanut. This legume has the peculiar habit of burying its developing ovularies in the soil by the

Bruce Fink

FIG. 222. A sugar-cane field in Porto Rico. The plant accumulates sucrose in its solid stems.

downward elongation of the flower stalk after pollination. The fruit does not develop normally unless buried. For this reason sandy soils are preferred for growing them. About one half of the peanut is composed of oil. The bulk of the crop is utilized in making confections and peanut butter.

Most of the rice produced in the United States is grown on low alluvial lands and on delta soils with heavy clay subsoils that can be readily flooded. Louisiana, Texas, Arkansas, and California produce more than 35 million bushels. The land is prepared and the seed planted as in growing other grain crops. After planting, however, the land is flooded; but before harvest time the land is again dried out to permit the use of machinery in gathering the crop.

The southeastern evergreen forest region produces most of the sweet potatoes. They are largely consumed locally and partly take the place of the Irish potatoes used in the Northern states.

About one third of the $2\frac{1}{2}$ billion pounds of sugar produced in the United States comes from sugar cane grown in Louisiana and Texas. Sugar cane is a large perennial grass, 8 to 15 feet in height, that accumulates sugar in its stems. After the leaves have been removed, the stems are crushed and the juice is evaporated. The refuse stalks are used as fuel and for making coarse paper.

Hay and forage crops of cowpeas and other legumes are important throughout this region, usually ranking third or fourth in importance, following cotton, corn, and sometimes oats. Because of its mild winters, this region is able to supply the Eastern city markets with the first berries, melons, and fruits of the season. Tea has been successfully grown in South Carolina, but the great amount of hand labor required in picking and preparing the leaves has prevented production on a large scale.

The prairie grass region. The northern prairies have become the leading region of spring wheat production. The unusually fertile soils and the bright sunshine in this region give ideal conditions for abundant growth, and the dryness of the climate in-

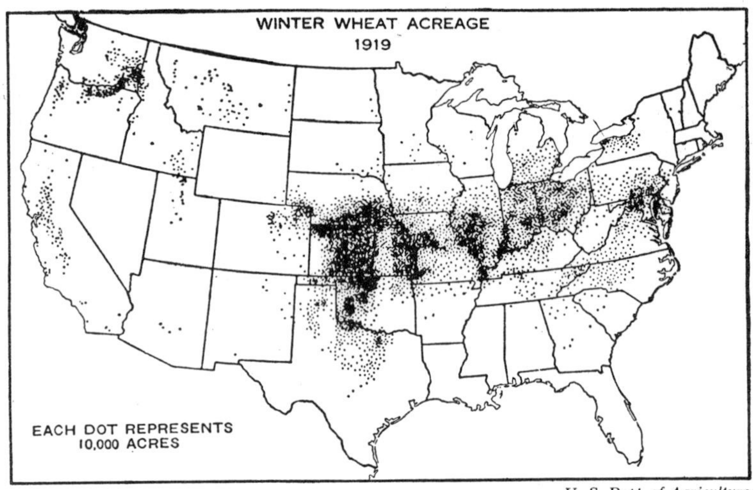

FIG. 223. Map showing the winter wheat acreage in the United States.

creases the hardness and amount of gluten in the grains, thus making the flour obtained from this wheat more valuable. Because of these facts and because there is easily available water power in this region, the greatest flour mills in the world are located at Minneapolis. Barley and flaxseed are also produced in large quantities on the northern prairies.

The bulk of the corn in the United States is grown on the rich black soil of the central prairies, from eastern Kansas and Nebraska to Ohio.

Because of the abundance of feed, the great cattle markets and the packing industries center in this same region. The various substances manufactured from corn and known as "corn products" — starch, oil, alcohol, glucose, and meal — are produced here on a large scale. The secondary crops of this region are winter wheat, oats, hay, and sweet corn. The areas in which sweet corn is grown abundantly determine the location of large canning establishments and also factories for the manufacture of cans.

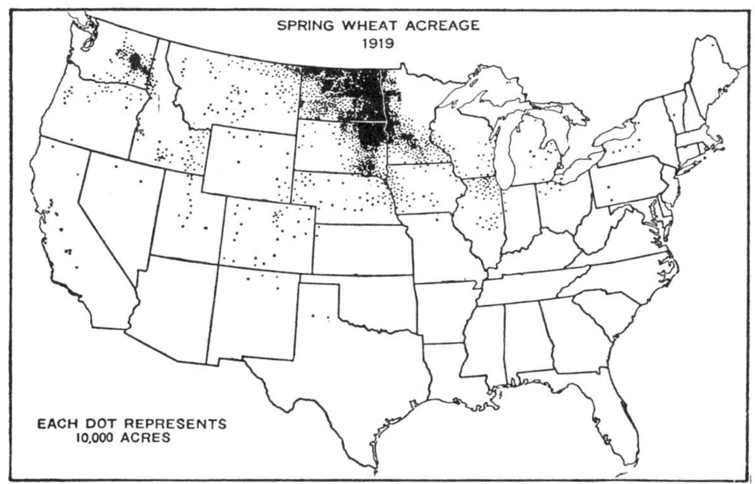

U. S. Dept. of Agriculture

FIG. 224. Map showing the spring wheat acreage in the United States.

The Great Plains. Originally the plains were the grazing lands of the continent, and as the buffaloes were killed off cattle took their place. At first the cattle were driven from one locality to another, whenever the grass gave out or water became scarce; and they were generally driven southward in the winter and northward in the summer. But as the land came into private ownership, ranches were established, and in addition to grazing attempts were made to grow crops. Durum or " hard " wheat was introduced from the steppes of Russia; thus wheat growing became possible 300 miles farther west than previously. From Africa came a grain-producing sorghum known as " kafir corn " and the millet called " milo," both of which thrive under plains conditions. The growing of broomcorn, another variety of sorghum, has become centralized in Texas, Oklahoma, and Kansas. The flowering branches of this plant furnish the straws for brooms.

The greatest forage crop of the central plains is alfalfa. This is a perennial, deep-rooted clover that thrives on well-drained

Cereal Investigations, U.S.D.A.

FIG. 225. A field of kafir corn in western Oklahoma; a good dry-land crop.

soils. On the plains uplands it may produce two or three crops of hay in a season; on the lowlands it may produce three or four, and in exceptional seasons, toward the south, five. Alfalfa, like all clovers, has bacterial nodules on its roots that accumulate organic nitrogen compounds. Thus by growing alfalfa in fields for several years, the fields not only yield a good return, but if the final crop is plowed under before planting other crops, the soil will be improved in its nitrogen and humus content. The growing of alfalfa has spread from the plains country into all the Western irrigated districts and to the prairie and deciduous forest regions. The drought-resistant millet grasses are of secondary importance as hay and forage crops on the plains and prairie border.

The western evergreen forest region. The plant industries associated with the western evergreen forest are, for the most part, lumbering and the manufacture of lumber products. As the timber of the Eastern states became scarcer and poorer in quality, the exploitation of these great Western forests began; and now products from them, such as rough lumber, shingles,

U. S. Forest Service

FIG. 226. Characteristic dense forest of Western hemlock and Douglas fir in Washington. The trees are from 5 to 8 feet in diameter.

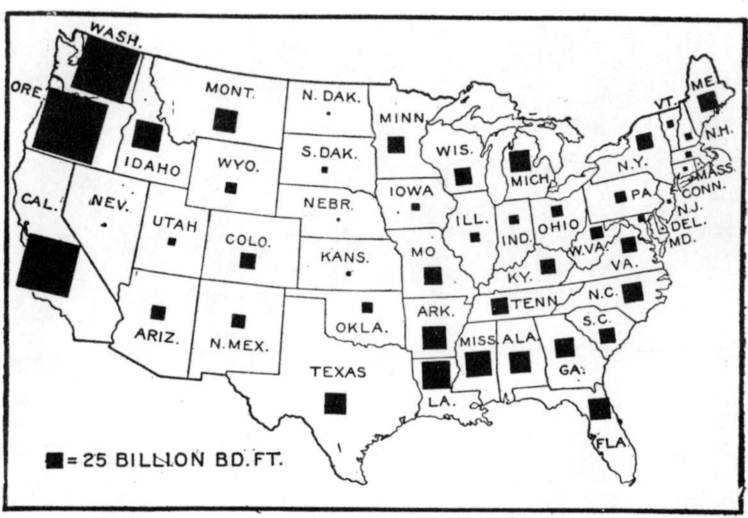

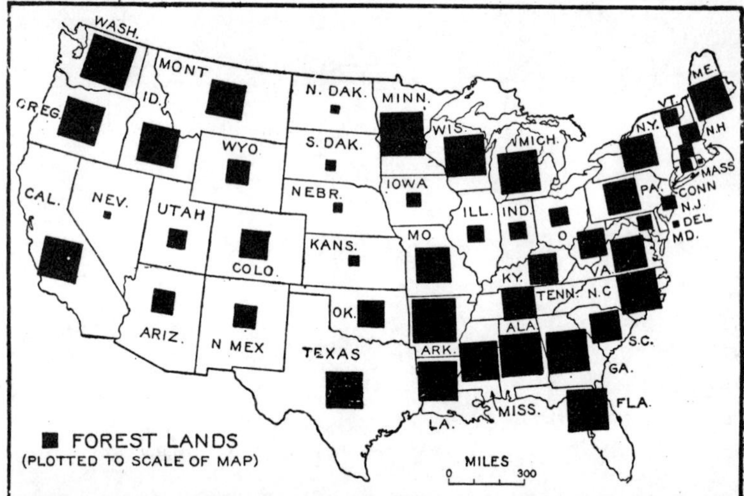

U. S. Dept. of Agriculture

FIGS. 227 and 228. The upper map shows the relative amounts of standing timber in the various states in 1919. The lower map indicates the areas of land in each state which should be in forest. Note that the largest areas, suitable only for forests, are located east of the Mississippi River near the greatest lumber markets and timber-consuming industries. Our tax laws should be revised in such a manner as to encourage the reforesting of these lands.

U. S. Forest Service

FIG. 229. Western yellow pine plantation, 11 years old; Lolo National Forest, Montana.

pulp woods, finishing woods, excelsior, and tannin, are rapidly taking a leading place, even in the Eastern markets. At the same time more and more of the wood-consuming industries are becoming established in the West coast states.

The numerous mines scattered throughout this region have consumed large quantities of wood for fuel and for mine props. The railroads, traversing great stretches of plains and desert as well as forest lands, have required a vast amount of timber for poles and crossties.

The Douglas fir, found in such abundance in Oregon and Washington, has proved to be adapted to a great variety of uses. The Western yellow pine, the lodgepole pine, and the sugar pine have come into the market for rough timber, fuel, and construction.

The giant cedar is the leading shingle wood of the United States, and the Sitka spruce is the leading source of paper pulp in the Western states.

Forest reserves and their uses. When the attention of lumbermen was turning from the Eastern forests to the timber on the Western mountains, the United States government still owned vast tracts of these forested lands. During the administration of President Roosevelt, a definite policy was decided upon by which many large areas of these forests were turned over to the management of the United States Forest Service. By this plan the forests are to be harvested as the trees come to maturity, and the methods of harvesting are those which will insure a constant supply of timber. Many of the forests in the drier regions have an undergrowth of grass. It is highly desirable that these grasses be utilized, and at the same time it is important that the development of tree seedlings shall not be permanently prevented.

U. S. Forest Service

FIG. 230. Typical section of a mountain slope in western North Carolina, after removal of forest. The binding effects of the roots have been removed, and the erosion of the soil is so rapid that it is difficult for seedlings to take hold. When the forest was cut, enough young trees should have been left to hold the soil and start a new lumber crop.

James B. Berry

FIG. 231. The piñon pine "protection" forest of the desert mountains of the Great Basin region. The stand is very open but is of great value in preventing wind and water erosion. In addition, it supplies the ranchers of the adjacent valleys with fuel wood.

Under proper supervision these grasslands are periodically grazed by sheep, and the Forest Service receives a fee for the grazing privilege. A third use made of forest reserves is the maintaining of water supplies for irrigating purposes. In many of the irrigated areas adjoining the mountains the rainfall on the mountains is insufficient to maintain the streams throughout the growing season. The snowfall, however, is very considerable in the mountains, and if the snow melts gradually it adds largely to the available water at lower levels. When the forests are removed, the snow melts rapidly and causes spring floods, which not only waste the water but also cause erosion on the steep slopes and destruction along the rivers at the bases of the mountains. Where forests remain, the run-off is slower and the water more evenly distributed.

Forest reserves, therefore, are areas set aside by the government to maintain a timber supply, to provide grazing lands, to control water supplies for irrigation, and to prevent destructive floods. They are not intended to prevent the cutting of mature timber or to keep suitable land from being settled and turned over to agriculture. These lands are being accurately surveyed, and whenever areas are found that are fitted to become farm-lands they are sold to settlers to use for farming.

The total amount of land in the national reserves is 160 million acres. In addition, many of the state governments have taken over smaller forest tracts within their borders for the same purpose. Out of the original 822 million acres in the United States only 137 millions of virgin timber remain. Only 6 billion cubic feet of timber are added by growth on all forested lands each year, while 25 billion cubic feet are being consumed. This is the same as saying that in the United States we are using the timber four times as fast as it is being replaced by growth.

Reforestation. Many of the lands owned by the United States and by the state governments have been partly or wholly denuded of forests, and are worthless for agriculture. After being lumbered, the tops and branches of the trees were not removed and destructive fires swept away all that the lumbermen at the time considered worthless. To make this land valuable, the foresters either replant the area with small trees grown in nurseries, or plant seeds which will develop into trees where they germinate. In areas where the forests have been only partly destroyed, the species that remain are usually worthless for timber purposes. In such places "improvement cuttings" are made; that is, the worthless trees are either cut for fuel purposes or cut and burned to provide room for the young trees that will make valuable timber in the course of time. Some trees, like the chestnut of the East and redwood of the West, sprout from the stumps when the trees are cut. Thus for several years these trees make use of the roots of the parent tree and the food materials stored in them. They possess a great advantage in reforestation over the trees that have to start from seed.

U. S. Forest Service

FIG. 232. Fighting a forest fire. During a recent 5-year period, fire destroyed 56 million acres of forest in the United States.

Fire patrol. Most of the Western forest reserves and of the smaller state forests are regularly patrolled during the summer and autumn to prevent the spread of fires started by careless campers or by lightning. Airplanes and permanent lookout stations on mountain peaks aid in this work.

In recent years almost as much forest has been destroyed by fire as has been cut into merchantable timber, and the fires have consumed not only the trees but also the forest humus or "duff," and made the areas unfit to establish valuable forests for generations to come.

Irrigation. Most of the large irrigation projects of the United States are located in the semi-arid regions adjoining the lower forest borders. Through the construction of dams and canals the water from the mountains is made available throughout the growing season. The soils of these areas are highly fertile, and

U. S. Dept. of Agriculture

FIG. 233. Irrigated and unirrigated sugar cane, showing the value of sufficient water in the growing of this crop.

because the supply of irrigation water is certain, these lands have become unusually valuable and produce a great variety of crops. The alfalfa, melons, and fruits of the Rocky Ford district of Colorado; the wheat of Idaho, Washington, and Oregon; the fruit orchards of Oregon and Washington; and the millions of acres of oranges, lemons, citrus fruits, grapes, English walnuts, almonds, figs, prunes, peaches, and apricots of southern and central California are made possible by irrigating systems that use the water from the adjoining mountains. The products of these irrigated lands are one of the principal sources of wealth of these Western states, as well as an important source of food for the country as a whole.

The southwestern desert region. The border lands of the desert, where they adjoin plateaus and mountain ranges, afford large areas of xerophytic grasses, sagebrush, and chaparral for grazing sheep, goats, and cattle.

The true desert regions also produce some vegetable products

W. S. Cooper

FIG. 234. Mixed forest of oak (*Quercus chrysolepis*) and fir (*Pseudotsuga mucronata*); Santa Cruz Mountains near Palo Alto, California.

of economic importance. The various species of cactus (*Opuntia*), especially those forms with but few spines, are valuable as forage for cattle. These plants contain at all times a considerable amount of water, and in a region of such little rainfall this is an important factor. When they are used as cattle food, it is customary to burn off the spines with torches before feeding.

Guayule is one of the minor sources of rubber. It is a small shrub which accumulates resinous material in the cortex of the stem. Most of the supply has heretofore been obtained from wild plants, but now its cultivation has been begun in Arizona and California. The agaves of northern Yucatan and the plateau of Mexico furnish the sisal and hennequin fibers used in making binder twine. Other agaves furnish fibers of less importance. The cultivation of agaves has spread to the West Indies and other semi-arid parts of the tropics. "Pulpue," the Mexican national alcoholic drink, is made by fermenting and distilling the juice of an agave.

U. S. Dept. of Agriculture

FIG. 235. Cutting leaves from the sisal, an agave, in the semi-desert of Yucatan, for making the fiber used in the manufacture of binder twine.

U. S. Dept. of Agriculture

FIG. 236. Drying sisal fiber in Yucatan.

FIG. 237. The coconut palm in fruit. A tree under the best conditions will yield a nut each day in the year.

The desert region produces abundantly when irrigated. The Salt River Valley, in Arizona, is a good example of what can be done under these conditions. As the valley is located in southern Arizona, however, it has a subtropical climate, and consequently many tropical crops can be grown. Alfalfa and cotton are the most important crops at present; the date palm has been successfully transplanted from northern Africa and is now being grown commercially. A great variety of other crops, such as olives, figs, avocados, and tropical pawpaws, are also cultivated, on a smaller scale.

U. S. Forest Service

FIG. 238. Semi-tropical aspect of the vegetation of southern Florida. Live oak festooned with Spanish moss (*Tillandsia*); the cabbage palmetto; and, in the foreground, young orange trees.

The tropical forest region. Southern Florida, Mexico, Central America, and the West Indies present a variety of forest and cultivated plant products unequaled by any temperate forest region. Yet it may be safely stated that only a beginning has been made in the utilization of tropical plant products; certainly the possible forest products are largely unknown.

Among woods of this region the most important is the so-called "mahogany" of commerce. Wood from not less than forty different species of trees are imported into the United States under this name. This gives some idea of the large number of hard, fine-grained woods, suitable for cabinet work and veneers, that occur in tropical forests. There are also many lighter woods in the tropical forests which are suited to general construction purposes, and eventually these will be exported to the United States.

Formerly much wild rubber came from the American tropics, but the yield is now so small as compared with the demand that most of the rubber used in the United States comes from the rubber plantations in the East Indies. Logwood and other dyewoods are of growing commercial importance. Tobacco is extensively cultivated. Sugar cane and cotton furnish the most valuable products of herbaceous plants. Cinnamon is derived from the inner bark of a tree now cultivated in the West Indies. The bark is carefully removed, piled in heaps, and allowed to ferment. When fermentation has reached a certain point, the bark is dried and prepared for the market. Coffee is obtained from the fruit of the coffee tree. The outer husk is removed and the two seeds or "beans" in each fruit dried and sent to market. Coffee is of some commercial importance in this region, but most of our supply comes from Brazil. Nutmegs are also seeds of a tree fruit. The fleshy part of the nutmeg fruit is discarded when fully ripe, and the seeds dried. The outer coat is then broken and removed and sold under the name of "mace." The inner part is the familiar nutmeg of commerce.

Tapioca is prepared from the starch of the cassava, which is related to our American milkweeds. Cocoa is derived from the seeds of the cocoa tree. Oranges, lemons, grapefruit, guavas, avocados, pineapples, and bananas are grown in large quantities for export. Along the coasts the coconut palm is extensively planted. It is valuable for its nutritious seeds and for the fibers in the husks that surround the seeds. Chicle, the coagulated sap of the naseberry tree, is a forest product used in the manufacture of chewing gum. Vanilla is obtained from the dried fruits of a climbing orchid native to America. Cloves are the unopened flower buds of a small tree cultivated in the West Indies.

At higher elevations on the mountains of the tropical forest region, wheat, corn, and beans are local sources of food for the natives. Deciduous forests furnish valuable timber, and the clearings afford rich pasture. The mountain slopes still higher up are covered with pine forests.

CHAPTER THIRTY-FIVE

WEEDS AND THEIR CONTROL

THE term " weed " is commonly applied to any undesirable plant, and to any plant growing out of place. Rye may become a weed in wheat fields. Red clover is very desirable in a field on the farm, but it becomes a weed when it springs up in a lawn. The most pernicious weeds, like the dandelion, cockle bur, Canada thistle, poison ivy, bindweed, plantain, and sand bur, are not desirable plants anywhere.

Weeds decrease the yield of crop plants, reduce the value of grain and seed crops, interfere with the growth and use of forage crops, and greatly increase the cost of agricultural production. Many weeds are conspicuous and unsightly on farms and lawns and thus depreciate the value of land. Some weeds are harmful or poisonous to stock, and others impart unpleasant tastes to farm and dairy products. Weeds may also harbor injurious insects and the bacteria and fungi that produce disease. One of the commonest sources of hay fever and asthma is the wind-borne pollen of ragweed, horseweed, and other weeds.

High reproductive capacity. Weeds are plants in which reproduction has reached the highest degree of efficiency. The sequoia may stand for the culmination of vegetative efficiency, the dandelion for efficiency in reproduction and dispersal. The dandelion produces good seed without pollination; if the stem is cut, the plant develops numerous new sprouts; if the root is cut into small pieces, each piece may sprout from either end or from both ends at the same time. The dandelion can thrive in a swamp, and it can withstand the droughts of a sand plain. The sequoia still occupies the comparatively small area to which it was restricted during the glacial period. The dandelion has in recent times spread to all parts of the world, and it occurs in most habitats, from the seashore to the alpine summits of mountains.

How rapidly a weed may spread is illustrated by the history of the Russian thistle. It was introduced into South Dakota in 1874 with imported flaxseed. By 1888 there were enough plants in the Dakotas to have it reported as a weed. In 1893 it was abundant around Chicago. In 1898 it was reported in all the states and provinces east of the Rockies, from the Gulf of Mexico to Saskatchewan.

The control of weeds. The measures taken to control weeds depend first of all upon whether the weed is (1) an annual, like crabgrass, smartweed, ragweed, or foxtail grass; (2) a biennial, like blueweed, bull thistle, or wild carrot; or (3) a perennial, like Johnson grass, Canada thistle, wild onion, or milkweed.

The first principle to be observed in controlling weeds is to avoid bringing weed seeds to the farm or lawn. All seeds planted should be inspected for weed seeds, and if they are present the seed should be either cleaned or discarded.

The second rule is that no weeds should be allowed to produce seed. Since annuals and biennials are propagated only by seeds, the strict observance of this rule will ultimately rid an area of these two classes of weeds.

The third principle of weed control is the prevention of the growth of shoots. Depriving a plant of its photosynthetic tissues leads to starvation of the underground parts. This principle is particularly applicable to perennial weeds with underground stems. The shoots may be destroyed by cutting, by spraying with poisons such as salt, copper sulfate, and petroleum, or by covering the area when small with roofing paper.

Preventing weeds from producing seed. A single plant of many common weeds will produce hundreds or thousands of seeds. Moreover, not all these seeds may germinate the first year, and seedlings may continue to appear for several years. Harrowing and cultivating farm lands not only improve soil conditions for the growing crop, but they also destroy countless numbers of weed seedlings, which in good soil is far more important. Mowing pastures and fencerows or pasturing off the weeds with sheep and cattle are efficient means of destroying weeds if practiced before they come into bloom.

Preventing the introduction of weed seeds. Weed seeds are introduced not only by the purchase and planting of uncleaned seeds but through other common practices. The use of fresh manure is a common source of weed introduction. Stock feeds made from screenings are likely to contain a large percentage of weed seeds. Finally, the seeds that may be spread by the wind from neighboring lawns and farms make the problem of weed control a community affair. One careless neighbor is a menace to the entire community, and he should be treated as we treat any one who maintains a nuisance. Coöperation through community organizations is essential to efficient weed control.

Poisonous weeds. Not only may weeds reduce crop yields, but some weeds are poisonous to human beings and to animals. Poison ivy is poisonous to many persons who come in contact with it. White snakeroot produces " trembles " in cattle and " milk sickness " among human beings who use milk from such cattle. Wild cherry leaves, especially when in a wilted condition, are poisonous to cattle. Larkspur and loco weed are poisonous to cattle and a great source of loss in Western pastures. Wild onions, garlic, and other less-well-known aromatic herbs produce unpleasant odors and tastes in dairy products. All these plants can be destroyed by persistent and intelligent effort, and the results are of such far-reaching importance that their eradication will in the end be profitable.

CHAPTER THIRTY-SIX

THE NON-GREEN PLANTS

THE green plants are called *autophytes* (Greek: *autos*, self, and *phyton*, plant), because they are independent or self-supporting. Given sunlight, they can make their own food from water, carbon dioxide, and mineral salts.

There are, however, great numbers of plants that lack chlorophyll and hence are not able to make their own food. Many of them, like the bacteria and yeasts, are microscopic in size; others, like the molds and mildews, are small but visible to the unaided eye; still others, like the puffballs, mushrooms, the bracket fungi that are seen on trees and logs, and the Indian pipe of the forest, reach a size comparable with that of many green plants.

Energy necessary for life. In the study of biology it is well to have in mind always that a perpetual-motion machine is no more possible in the living than in the non-living physical world. A living organism must have energy to carry on its activities and life processes. Green plants, non-green plants, and animals are alike in requiring an energy income for their life activities.

As we have seen, the green plant secures its energy from the sunlight. The energy of the light is not used directly in the operation of the vital mechanism, but it starts synthetic processes within the plant that end in the complex compounds we call "foods." These are then oxidized, and the energy required by the plant is released in the breaking-down process.

Lacking chlorophyll, non-green plants cannot use the sunlight in synthetic processes. They must, therefore, secure their energy from materials already built up so that it can be oxidized. This the great majority of them do either by living directly on other living organisms or by feeding on dead organic matter originally synthesized by green plants.

W. S. Cooper

FIG. 239. The dodder (*Cuscuta*), a yellow parasite belonging to the morning-glory family, grows on other plants not only in moist regions, but also in the arid coastal region of California. It is here shown growing on *Abronia*.

Some small non-green plants, however, secure their energy not from organic substances, but by oxidizing inorganic salts. The most important of these are the nitrifying bacteria in the soil that oxidize ammonia and nitrites to nitrates in their respiratory processes. With the energy thus secured they construct carbohydrates, fats, proteins, and organic compounds. These plants are autophytes as truly as the green plants. They live in the soil independent of other plants, and they can grow without organic compounds. Another group of bacteria common in sewage-polluted water is able to secure energy through the oxidation of inorganic sulfur compounds.

Vital syntheses. In this connection it may be noted that living organisms differ greatly in their synthetic powers. The green plant using the energy of the light in the first processes can build everything that it requires. Colorless plants, if given sugar or other carbohydrates that they can use, can construct fats and proteins, and, as we have seen, some of the bacteria can even synthesize their own carbohydrates. An animal can transform carbohydrates into fat, but it apparently lacks the power to make certain vitamins and several of the amino acids needed in protein synthesis.

Parasites. An organism that derives its food directly from another living organism is called a *parasite*. A parasitic plant may live inside the *host*, from which it secures food, as is the case with many bacteria and fungi. Or a parasitic plant may be merely attached to the host plant at one or more points. *Beech drops* are small, purple, flowering plants attached to the roots of beech trees. The *dodder*, or "gold thread," is a slender, yellow, climbing plant, related to the morning-glory, that becomes attached to the stems of a great variety of hosts by means of small, root-like structures (*haustoria;* singular, *haustorium*) that penetrate the cortex of the host and finally reach the conductive tissues (Fig. 239).

True parasites among the flowering plants are generally small; their leaves are mere scales, and the most prominent parts are the flowers and the reproductive structures. In color they vary from yellow to red and purple. Some apparently do not injure the host plant; others may injure and eventually kill the plant on which they grow.

Partial parasites. Some parasites contain chlorophyll and are able to manufacture at least a part of their foods. For example, the mistletoes occur on a great variety of trees from the Atlantic to the Pacific, and are very common particularly in the subtropics. The sticky seeds adhere to the bark of branches, and a root-like haustorium dissolves its way into the bark and forms a connection with the conductive tissues of the host trees. Mistletoes sometimes form much-branched masses of stems and foliage 2 or 3 feet in diameter. These plants have lost the power of growing on soils, and apparently are dependent on their hosts for their water supply and a part of their foods. They may be called *partial parasites*.

U. S. Forest Service

FIG. 240. A winter view in Texas, showing the mistletoe, an evergreen parasite, growing on the deciduous mesquite.

Saprophytes. A saprophyte is a plant that depends for its food on dead organic material. These plants live in the soil, on the dead bark and heartwood of trees, and on a great variety of plant products. They are exemplified among flowering plants by the Indian pipe, a common plant of moist woods. The body of the plant consists of a rather large root-like base from which colorless branches bearing flowers arise. From the humus in which it grows it secures water and organic compounds sufficient to furnish the materials and energy used in building its tissues. Like many of the flowering plants the root-like base of the Indian pipe is penetrated by fungi, which seem to be essential to its growth. Perhaps the fungi aid in transforming a part of

FIG. 241. Indian pipe (left) and pinesap (right), two saprophytes common in moist woods. The underground parts of the plants are penetrated throughout by fungous filaments, which enter from the humus in which the plants grow.

the humus into substances that are readily assimilated by the plant.

Among the bacteria and fungi there are thousands of saprophytes. They occur everywhere, and the amount of change that they bring about in the world is so great that it is impossible to overestimate their importance. Saprophytes are the direct causes of all decay and fermentation. They are present in the alimentary canals of the higher animals, and aid in the digestion of food. They are ever-present agents of destruction, and are the organisms that make cold-storage houses and refrigerators necessary. The canning, drying, and preserving industries are based on methods of eliminating saprophytes. The beer, wine, vinegar, and cheese industries depend upon the fermentations induced by carefully cultivated saprophytes. The tarring and creosoting of telegraph poles and railroad ties are made necessary by the universal presence in the soil of these destructive plants.

Among the one-celled plants there are some that can live either as green autophytes or as colorless saprophytes. There are many that may live either as saprophytes or as parasites. It is often very difficult, therefore, to classify plants among these three groups or to determine the exact sources of their food and energy.

The non-green plants, then, include a very large number and variety of plants. Autophytes are world-wide in their distribution; the occurrence of a species is limited only by climatic and habitat conditions. Parasites are widely distributed, but any species is limited by the occurrence of its particular plant or animal hosts. Saprophytes occur everywhere where organic matter exists. The non-green plants are not so conspicuous as the green plants, but they are of overwhelming importance to plants, to animals, and to man.

REFERENCE

MARSHALL, C. E. *Microbiology* (3d edition). P. Blakiston's Son & Co., Philadelphia; 1921.

CHAPTER THIRTY-SEVEN

BACTERIA AND THEIR RELATIONS TO LIFE

THE best-known and the most discussed of all the non-green plants are the bacteria. They are so intimately related to human welfare that most persons, even though they have never seen bacteria, know something about them. They are one-celled plants, at once the smallest in size, the simplest in structure, and the most abundant of all plants. They live in immense numbers in the water and in the upper layers of the soil, and they are blown about in dust in the air. Some are too small to be seen except with the highest powers of the microscope. Others may be seen with an ordinary laboratory microscope. They make up for the small size of the individual by their rapid multiplication and by the formation of colonies containing countless numbers of individuals. Bacteria are responsible for many of the diseases of men, animals, and plants, and bacteria affect our lives in almost countless other ways. All our modern methods of sanitation, quarantine, surgery, water supply, and sewage disposal, and much of our personal hygiene, are primarily based on our knowledge of the behavior of this group of plants.

Economic importance of bacteria. Economically the bacteria are of the greatest importance. Together with the fungi they are the principal cause of disease, decay, and the formation of humus. Bacteria bring about the ripening of milk in butter and cheese making, and they produce both the pleasant flavors in these products and the unpleasant flavors that develop in them with age. The bacteria are also the source of much of the available nitrogen in agricultural soils. The drying of hay, vegetables, and fruits, the canning and pickling of vegetables, fruits, and meats, and refrigeration and cold storage are methods of avoiding or making impossible the growth of bacteria. Thus a knowledge of these plants is fundamental to our understanding of thousands of details of our daily life.

Environmental conditions affecting bacteria. Like the higher and more complex plants the bacteria have certain rather definite

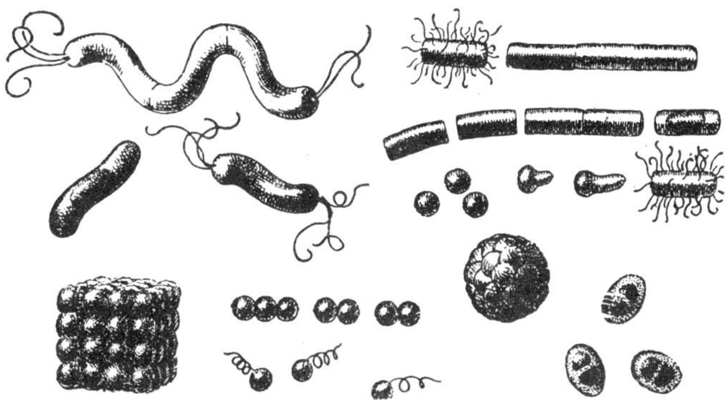

FIG. 242. Various forms of bacteria.

water, temperature, light, and nutritive requirements for growth and reproduction. The different species vary greatly in these requirements; consequently some kinds of bacteria are able tc live almost everywhere in nature.

Moisture. Since water makes up about 85 per cent of the bacterial cells, water is essential to their activities. Furthermore, since all of their nutrient materials are absorbed by diffusion, they must be surrounded by at least a film of water. The water or solution in which bacteria live is commonly called the *medium* (plural, *media*), and its properties are determined by the substances it contains.

For example, sugar and salts may be dissolved in the medium, thus determining its concentration. In dilute solutions the water and nutrient materials diffuse readily into the cells. In concentrated solutions (15 to 40 per cent) the concentration of water is less in the media than inside the cells, and water either does not pass in or diffuses out of the cells and the bacteria are unable to grow. They are affected in the same way as if they were dried. This explains why jellies keep more readily than preserves, preserves more readily than canned fruits, and canned fruits more readily than fruit juices to which no sugar has been added. The

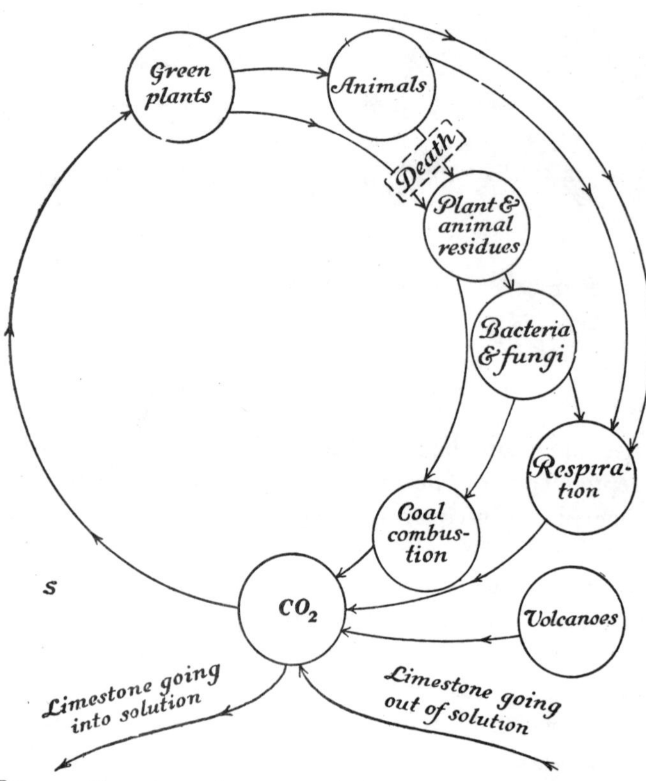

FIG. 243. The carbon cycle in nature. Bacteria and other saprophytes play a rôle opposite to that of the green plants.

first has a high concentration of sugar, the last a low concentration. Bacteria develop very slowly in the first medium and very rapidly in the last. In the laboratory, bacteria are cultivated on gelatine or on agar (seaweed jelly) plates. Many bacteria grow very slowly when the water content of the gelatine falls below 50 per cent. If the medium on which they live dries out, all the vegetative cells become inactive and death gradually follows. However, some bacteria, especially those found in soils, may be dried for days, months, and even years and remain alive. Most disease-producing bacteria cannot withstand desiccation, so that there is little danger of their being spread by dust.

Temperature. Bacteria have the temperature of the medium in which they live. Low temperatures retard the life processes, and high temperatures accelerate them. Likewise at low temperatures less amounts of food are consumed; hence they may live longer on a limited supply.

Between the highest and lowest temperatures at which an organism can live is a point at which it develops most rapidly, called the *optimum*, or best, temperature. Most bacteria grow best in temperatures between 70° and 100° F.

Very few bacteria grow well above 115° F. There are some, however, that live in rapidly decaying organic matter (e.g., in silos and self-heating hay) and in hot springs at temperatures as high as 175° F. — a most remarkable fact, when we consider that proteins which make up so much of the protoplasm commonly begin to coagulate at 145° F.

At the freezing point most bacteria grow very slowly. When freezing occurs and the medium becomes solid, diffusion of nutrients no longer takes place and all activities are checked. The bacteria may remain alive, however, for weeks and months in this condition.

Light. Bacteria living in nature in the soil, in decaying matter, in foods, and inside plants and animals are only temporarily exposed to the light. Most of them cannot withstand exposure to full sunlight for even a few hours. This action of sunlight is of great importance in the purification of rivers and in the destruction of bacteria on streets and sidewalks. Death is brought about either by chemical processes initiated by light within the cells or in the medium.

Oxygen. Bacteria are very sensitive to oxygen. Although it makes up 20 per cent of the atmosphere, bacteria are exposed only to the oxygen that dissolves in the water surrounding them. At room temperature this forms an extremely dilute solution (0.0009 per cent). If the oxygen content is increased artificially to thirty times this amount (0.027 per cent), practically all bacteria die. In other words, oxygen is about as poisonous to bacteria as formaldehyde and corrosive sublimate, two of the commonly used disinfectants.

Nevertheless, small amounts of oxygen favor the growth of most bacteria. On this account they are called *aërobes* (Greek: *aer*, air, and *bios*, life). Some bacteria, like the germ of lockjaw and the bacteria that produce the rancid taste of butter, can grow only when the oxygen content of the medium is extremely low and when there are organic substances available containing combined oxygen. These bacteria are called *anaërobes* (Greek: *an*, without). Anaërobic bacteria occur in poorly drained soils, in the bottoms of lakes, and in the deep waters of the ocean.

The effectiveness of hydrogen peroxide in dressing wounds and cleaning teeth depends upon the fact that it releases oxygen readily.[1]

Food supply. Almost all bacteria require organic foods, and live usually as saprophytes or parasites. They all depend upon the oxidation of a part of these foods for their energy. They differ widely in their food requirements and in their effects upon the medium in which they live.

The most important exceptions to this general rule of requiring organic foods are found in the *nitrifying* bacteria of soils. These resemble green plants in the fact that they can synthesize organic compounds from CO_2, H_2O, and mineral salts, but differ in that they cannot utilize sunlight.

Structure and reproduction. Bacteria consist of one-celled individuals, that occur usually in masses on or in the food-

[1] Hydrogen peroxide changes to water and oxygen on exposure to the air:

$$2\,H_2O_2 \longrightarrow 2\,H_2O + O_2$$

Hydrogen peroxide ⟶ water + oxygen

containing medium. They are so small that the details of cell structure are not well known. The protoplasm is surrounded by a cell wall probably composed of cellulose and chitin. In some forms protoplasmic threads, called *flagella* (singular, *flagellum*) extend through the cell wall and provide organs of locomotion. The flagellate forms are active individuals, that become stationary later and lose the flagella.

Many bacteria have each cell further surrounded by a gelatinous sheath. Sometimes the sheaths of many individuals coalesce, forming slimy scums on stagnant water and on objects in the water.

When all the conditions are favorable, bacteria may multiply very rapidly. This is accomplished by cell division, the cell simply pinching in at the middle and separating, forming two new individuals. As the daughter cells quickly grow to the size of the original, this process may be repeated in 20 minutes to an hour. A little calculating will show that if this process continued for 24 hours there would be hundreds of million-millions of individuals.[1] Of course, long before any such number can accumulate, the water and food supplies are consumed and the products of their activities accumulate and cell division is stopped. If this were not true, the whole organic world would be turned to bacteria over night.

Spores are formed by many bacteria, by the contraction of the

[1] Starting with one bacterium and counting a generation every half hour, the number at the end of a day would be 281 million-millions, or about one pint of bacteria. Starting the second day with one pint of individuals all multiplying at the same rate, at the end of 48 hours there would be 281 million-million pints of bacteria, or about 32 cubic miles. At the end of the third day there would be enough to fill the ocean basins 3 million times, or sufficient to make 33,000 bodies the size of the earth.

Why do not bacteria capture the earth? First, because they produce acids and other harmful substances in the medium, that stop their development; second, because they can obtain only the food that diffuses to them from infinitely small distances beyond their own cell walls; third, because they soon meet unfavorable temperature, moisture, or light conditions; and fourth, because they are eaten by microscopic animals in large numbers. A short life is the rule among bacteria.

protoplasm into a rounded mass at one end, or near the middle of the cell, and by the secretion of a secondary spore wall. In this condition the protoplasm contains less water and is highly resistant to drying, to high and low temperatures, and to poisons which readily kill the ordinary bacterial cells. It is because the spores of certain forms withstand the temperature of boiling water that steam pressure is used in sterilizing cans of corn, beans, peas, and other vegetables. Most of the common disease-producing bacteria, however, do not produce spores.

Forms of bacteria. Some of the largest bacteria form long rows, or filaments of cells. These may be found commonly in stagnant water or in streams that carry sewage. Among the small forms it is customary to call the rod-shaped cells *Bacillus* (plural, *bacilli*) the round ones, *Coccus* (plural, *cocci*), and the spiral forms *Spirillum*. Some of these type-forms are shown in Figure 242.

Bacteria and sanitation. The bacteria of decay help to keep the surface of the earth clean. They change the highly complex organic substances that form the bodies of plants and animals into simple substances that may be used again by other plants in building foods. When plants and animals die, their bodies are gradually transformed by the bacteria into carbon dioxide, water, and mineral salts. The sewage that is turned into our rivers is chemically changed and disposed of in the same way by these minute plants. The great increase in the number and size of our cities has made it necessary to build large sewage-disposal plants where the bacteria can act rapidly and efficiently. This prevents the pollution of streams and keeps the water suitable for city water supplies.

The modern processes of filtering and sterilizing the water supplies of cities are carried on partly to remove sediment and partly to remove disease-producing bacteria. Adding minute quantities of alum and chloride of lime to the water and then filtering it through sand not only renders the water clear but removes from it disease-producing bacteria. The most dreaded of all the water-borne diseases is typhoid fever, and the cities are now much freer from this disease than are the country districts where people depend upon well water. Surveys in some of the Middle Western states showed that from one fifth to one third of the wells examined contained large numbers of bacteria derived from surface drainage. In such wells there is always danger that the surface waters may bring in disease-producing bacteria, especially typhoid germs derived from human sources.

Other sanitary practices, such as quarantine, disinfection, admitting plenty of sunshine into living rooms, cleaning walls and floors, removing dust, cooking food, washing and scalding dishes, pasteurizing milk, and keeping food supplies in refrigerators, are all related to the control or elimination of bacteria.

Bacteria and disease. When certain bacteria grow in the body, they produce poisonous substances called *toxins*. These interfere with the normal working of the bodily processes and cause illness. The body under these circumstances produces substances called *antitoxins*. These are substances formed by the cells of the body, which neutralize the effects of the toxins, either by combining with them chemically, or by rendering the cells insensitive to the toxins. In this way they protect the tissues until the bacteria are destroyed by leucocytes (colorless blood corpuscles) or in other ways. Not all persons are equally susceptible to infectious diseases. A person may be immune to a disease because his blood contains the corresponding antitoxin or is able to produce it, or because his body is insensitive to the bacterial toxins. Some of the commoner bacterial diseases are tuberculosis, pneumonia, grippe, diphtheria, typhoid fever, colds, lockjaw, and "blood poisoning."

A fundamental fact that should be learned in this connection is that no one can contract a bacterial disease unless he comes in contact with the particular bacterium which causes that disease.

Furthermore, persons rarely contract bacterial diseases unless they come in contact with another person carrying the disease. With the exception of lockjaw and wound infections, diseases are rarely spread by clothing, dust, or other objects. Apparent exceptions to this statement are typhoid and diphtheria, carried by water, milk, and other foods when handled and contaminated by a diseased person. Typhoid may also be carried by flies that have visited infected matter.

Natural barriers to disease. The natural means of defense against disease are somewhat similar in the higher plants and in animals. The plant, in addition to protective chemical substances within its cells, has an epidermis which renders the entrance of bacteria difficult. Bacteria are able to enter, however, if the epidermis is bruised or broken. Plants probably suffer from bacterial diseases as much as do animals. Most of the well-known plant diseases, however, are produced by fungi. Of the bacterial diseases of plants, the twig blight of pear and apple, the cucumber wilt, and the crown gall of various plants are perhaps best known. Some of these diseases are transported from one plant to another by insects.

Bacteria in the dairy. Milk is an ideal medium for the growth of bacteria. This makes necessary the most careful handling of milk, especially when it is used directly as food. The bacteria get into the milk from the cow, from the stable, from the vessels into which the milk is put, and from the persons who handle it. Evidently the cows should be kept clean, and the stable should be as clean and free from dust as possible. The vessels with which the milk comes in contact should be sterile. The *dairymen* should have clean hands and clothes, and above all they *should be free from infectious diseases*. Because bacteria multiply very rapidly at high temperatures, the milk should be chilled at once and kept on ice. To make butter and cheese of fine flavor, pure cultures of the proper bacteria are added to the milk and allowed to develop for a time.

In order to avoid the danger that lies in the use of milk contaminated with disease germs, milk that is shipped into the large cities is usually pasteurized before being sold. This treatment kills most of the bacteria, destroying all the kinds that produce disease in human beings. By "pasteurization" is meant the heating of the liquid to 150° or 160° F. for from 10 to 30 minutes. This does not kill the spores, but they are to a large extent prevented from developing by the subsequent cooling that the milk receives.

The preservation of foods. The greatest losses that occur in the utilization of crops are connected with the distribution of the products to the consumer. Much of the food produced never reaches the consumer, because bacteria and molds render it unfit for use before it can be distributed through the markets. There are four methods of preventing this loss: (1) cold-storage warehouses and refrigerator cars are used to keep foods below the temperature at which bacteria grow appreciably; (2) fruits, vegetables, or other foods are packed in cans, and the cans are then sterilized by heat and are sealed so that they are bacteria-tight; (3) food products are dried to make it impossible for bacteria to grow in them; and (4) foods like meat and fish are treated with salt or with some other chemical that will prevent the growth of bacteria. Refrigeration enables us to preserve foods for weeks and months. Canning and drying make foods available after months and years of storage.

Soil bacteria and humus. In the process by which the bacteria of decay destroy animal and vegetable bodies, the humus represents the products of partial decomposition, particularly of cellulose. Carbohydrates, fats, proteins, and related compounds are all subjected to bacterial action. Some are oxidized, and some are split into less complex substances. Among the many products of decay are hydrogen, marsh gas, organic acids (e.g., acetic, butyric), ammonia, hydrogen sulfide, carbon dioxide, and water. Usually the production of the final products CO_2,

H_2O, and nitrogen are delayed by the formation of rather stable intermediate products. These form the humus of soils.

Some of the bacteria of decay are of importance in industrial processes, as in the retting of flax and hemp fibers and in the preparation of hides for the making of leather.

Spontaneous generation and bacteriology. Not many years ago it was thought, even by the most learned persons, that the minute plants and animals that occur in stagnant water and that cause decay and fermentation arose spontaneously in the water. It was the experiments of Pasteur (1862) and Tyndall (1869) that finally proved that the organisms get into liquid media from the air. It was these studies that led to the discovery of the relation between bacteria and disease. The experiments of Lister (1860) led to the use of antiseptics (Latin: *anti*, against, and *septicus*, putrid) in surgery. Modern methods of sanitation, the control of diseases, and antiseptic surgery have all been developed since 1860. It is quite impossible for us to realize to what extent the dangers to life have been removed through the development of the science of bacteriology. This science has also made it possible to make use of bacteria in many important industries.

Methods of killing and controlling bacteria. Long before the discovery of the importance of bacteria, many methods of preserving foods, of caring for wounds, and of avoiding disease had been tried. They were very crude when compared with those that have been perfected since bacteria have been carefully studied. In the following table some of the methods of control are listed, and opposite them are a few domestic and industrial applications. Can you add to the list?

1. Cleanliness	Washing, keeping down dust, certified milk, disposal of garbage, sewage disposal
2. Ventilation	Sleeping porches, open-air schools
3. Sunlight	Purification of water supplies
4. Drying	Hay, fruits, vegetables, milk, eggs, pemmican

5. Refrigeration	Meats, fruits, vegetables, dairy products
6. Antiseptics:	
Common salt	Meat packing, cleansing mucous membranes, surgery
Acetic acid	Packing and pickling
Hydrogen peroxide	Cleansing wounds, preservation of milk
Chloride of lime	Purifying water supplies
Formaldehyde	Fumigation, seed treatment
Corrosive sublimate	Sterilizing wounds, surgery
Iodine	Sterilizing wounds, surgery
7. High osmotic pressure by salt and sugar	Curing meats, preserves, jellies
8. Sterilization by heating to boiling point	Canning and cooking, seed treatment, sterilizing surgeons' instruments
9. Pasteurization	Milk, beer, wine
10. Sealing	Canning, sterile bandages, and dressings
11. Precipitation by alum	Water supplies
12. Vaccination	Typhoid, bubonic plague, "colds"
13. Antitoxins	Diphtheria, tetanus
14. Avoiding contact with infected persons	"Colds," influenza, and other diseases

REFERENCES

DUCLAUX, E. *Pasteur: The History of a Mind.* W. B. Saunders Company, Philadelphia; 1920.

SMITH, E. F. *Introduction to Bacterial Diseases of Plants.* W. B. Saunders Company, Philadelphia; 1920.

VALLERY-RADOT. *The Life of Pasteur.* Doubleday, Page & Co., Garden City, New York.

CHAPTER THIRTY-EIGHT

SOIL BACTERIA AND THE NITROGEN CYCLE

NEXT to carbon, hydrogen, and oxygen the most important element used by plants is nitrogen. Agricultural crops on mineral soils are very frequently limited by an insufficiency of this element. As we shall see, the occurrence of nitrates in soils is due almost entirely to the action of bacteria and fungi. Owing to differences in their modes of life, several groups of nitrogen bacteria are distinguished, all of which play an important rôle in the nitrogen cycle in nature.

Nitrifying bacteria. In order to manufacture proteins, seed plants must have a supply of nitrogen, usually in the form of nitrates. There may be other nitrogen compounds in the soil, but they are for the most part unavailable until certain *nitrifying bacteria* change them to nitrates. Ammonia is one of the nitrogen compounds produced in the process of humus formation. If the soil is moist, the temperatures high, and the drainage sufficient to provide an adequate air supply, ammonia will be acted upon by certain bacteria and changed to nitrites, which in turn are changed by other bacteria into nitrates. These are oxidizing processes, and the energy liberated is used by the nitrifying bacteria in the various chemical syntheses necessary to transform CO_2 and H_2O and the soil salts into food and into tissue substances. These plants are as truly autophytic as the complex green plants. They require nothing but inorganic substances to maintain themselves. Wherever ammonia occurs, the nitrifying bacteria soon make it available for green plants.

Saprophytic nitrogen-fixing bacteria. Still other bacteria bring about a process known as *nitrogen fixation*, by which nitrogen is actually taken from the air and built into compounds which are added to the soil. The nitrogen-fixing bacteria are, with a few exceptions, the only plants that can take nitrogen from the air and combine it to form nitrogen compounds. They flourish

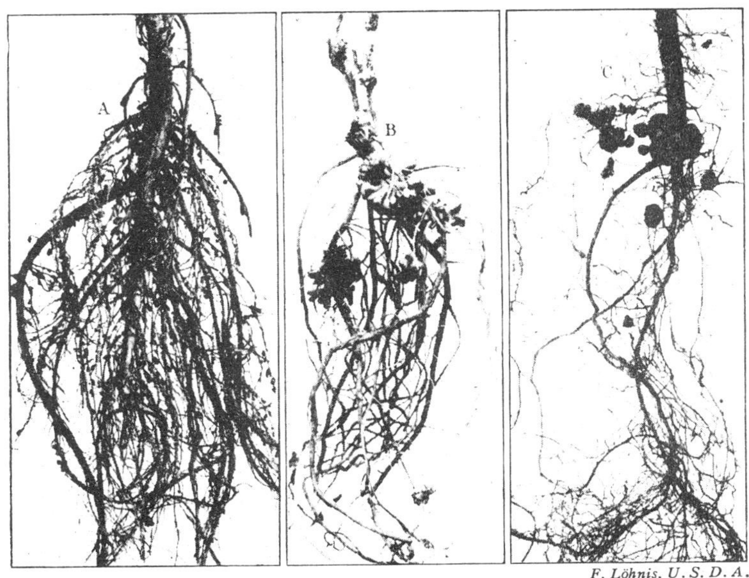

FIG. 244. Nodules containing nitrogen-fixing bacteria on the roots of legumes: *A*, red clover; *B*, sweet pea; *C*, soy bean.

only in rich, well-drained soil. They are of great importance in agriculture because nitrogen is the most expensive of all the elements that are bought for fertilizers. Their relation to the humus is very different from that of the nitrifying bacteria just described. To fix nitrogen (N_2) — that is, break up the molecules — requires much energy. The nitrogen-fixing bacteria secure this energy by oxidizing the carbon compounds (especially carbohydrates) found in the humus. It is estimated that 100 pounds of humus must be oxidized for every pound of nitrate formed in the soil. The bacteria that carry on this process are true saprophytes.

Bacteria and legumes. Clover, alfalfa, beans, soy beans, and peas belong to a family of plants called *legumes*. They increase the nitrogen in soils on which they are grown, and for many years they have been used in crop rotations, following wheat or

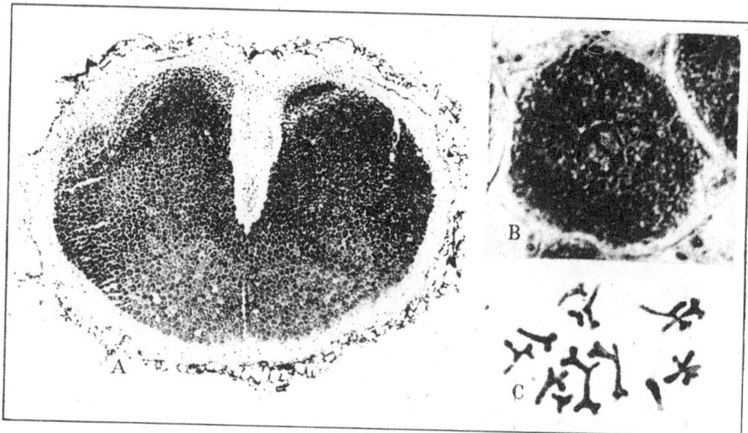

F. Löhnis, U. S. D. A.

Fig. 245. *A*, cross-section of root nodule of a legume; *B*, a single root cell showing nitrogen-fixing bacteria within it; *C*, branched bacteria from a nodule.

corn. The practice of using legumes in crop rotations was followed long before the real cause of the increase in soil nitrogen was discovered, and even before it was understood how the different elements in the soil contribute to its fertility. By experience it was learned that other plants flourish on land after leguminous plants have been grown on it, and for this reason the farmer included legumes in his scheme of crop rotation.

It is now clearly understood that nitrogen compounds accumulate in leguminous plants only because of the presence of certain *nitrogen-fixing* bacteria. These bacteria occur in many soils, and when the legume is planted and develops roots, they invade the cells of the root. This causes the infected parts of the root to enlarge, forming nodules. If a nodule from a clover or alfalfa root is crushed and examined under a microscope, it will be found to be filled with bacteria. These bacteria are parasites and take their food from the legume. A part of it they use in building their tissues; the remainder is oxidized and the energy used in changing nitrogen from the soil air into nitrogen compounds, just as the other nitrogen-fixing bacteria mentioned in the preceding section do. The nitrogen compounds thus formed are used by the host plant, and when the latter is plowed under and decays, the nitrogen compounds are made available for a succeeding crop of wheat or corn.

The nitrogen cycle. It may be well at this point to call to mind all the facts we have learned concerning the uses of nitrogen and the transformations that it and its compounds undergo in nature, involving the nitrogen of the air and the nitrogen compounds of soils, of plants, and of animals (Fig. 246).

If we start with green plants, carbohydrates (see diagram) combine with (1) nitrates and form (2) amino acids. These are built up into (3) proteins of plants. Plant proteins are the sources of amino acids used by animals in building (4) animal proteins.

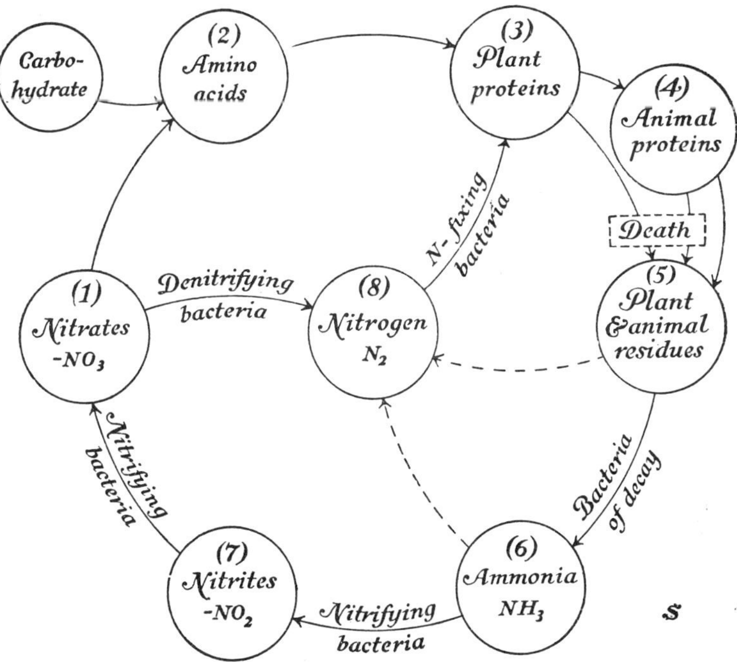

FIG. 246. Diagram of the nitrogen cycle in nature.

Animal and plant proteins are a part of the residues (5) left by death of the plants and animals.

Bacteria now become active agents of disintegration. By hydrolysis, reduction, and oxidation the complex substances of the cells are broken up into simpler compounds. Among these is (6) ammonia. This compound, if liberated in soil in the presence of water and carbon dioxide, forms ammonium carbonate ($(NH_4)_2 CO_3$).

A second group of bacteria, the *nitrifying* forms, use the ammonia and oxidize it to a nitrite ($-NO_2$). The nitrites are in turn oxidized by other nitrifying bacteria to nitrates ($-NO_3$). The cycle is complete and we are back where the process started.

Notice, however, that we have only used nitrogen that occurred in organic matter, and that some has been lost by the way, by going into the atmosphere as nitrogen gas. Furthermore, the above cycle is not the only possibility. The soil may not be well drained and aërated, and the (1) nitrates are then attacked by *denitrifying* bacteria and broken down to (8) nitrogen gas. Thus it is released and becomes unavailable for the higher green plants. So if there were not some means of securing an additional nitrogen supply, the land would become poorer and poorer as time went on.

The nitrogen-fixing bacteria provide this additional supply. In well-drained neutral soil the saprophytic varieties that obtain energy by oxidizing carbon compounds in the humus fix enough atmospheric nitrogen to form the compounds used in building their own cells. At their death these compounds become available to other bacteria and the first cycle is repeated with the addition of nitrates from the air to the soil. The bacteria that live in the nodules of legumes also build the free nitrogen of the air into organic compounds, and when the bacteria die the nitrogen compounds become directly available to the legume plant. If legumes are plowed under, the cycle starts over again with plant residues (Fig. 247) and in a few months it has come around to the nitrates.

Bureau of Agriculture, P. I.

FIG. 247. Cowpeas, which, like other legumes, accumulate nitrogen from the air and build it into organic compounds. The advantage of growing a legume in any system of crop rotation is that these compounds may be added to the soil by plowing the plants under.

So the ever changing nitrogen compounds pass from the soil to higher plants, to animals, to a succession of bacteria, and by the changes that they undergo they increase or decrease the fertility of the soil. If we understand each of these stages in the cycle and the conditions that favor the changes from one stage to another, and the injection of atmospheric nitrogen into the cycle, we can improve the nitrate content of soils and greatly increase crop yields.

REFERENCE

MARSHALL, C. E. *Microbiology* (3d edition). P. Blakiston's Son & Co., Philadelphia; 1921.

CHAPTER THIRTY-NINE

FUNGI

Of the plants without chlorophyll the most conspicuous are the fungi. They form an exceedingly large and diversified group, ranging in size from microscopic one-celled forms almost as small as the larger bacteria to the large, fleshy mushrooms and to the massive bracket fungi found on tree trunks and logs, which may weigh 20 or 30 pounds. Among the most important fungi are the yeasts, molds, mildews, smuts, rusts, and mushrooms.

All the fungi derive their food either from living plants, from animals, or from dead plant and animal tissues and their products. The yeasts, molds, and most of the mushrooms are saprophytes. These and their bacterial associates are the chief agents of fermentation and decay. The smuts, rusts, and some of the mildews are *parasites* on the seed plants. They produce injurious effects (diseases) on the host plant, which result in serious losses to the farmer and gardener and in reducing the supply of plant products for every one.

Some of the mushrooms are edible and furnish small quantities of pleasantly flavored food for man and animals; some of the fungi found on roots undoubtedly aid in the nutrition of the plant on which they grow; and others produce diseases among annoying insects and help to destroy or keep them in check.

The vegetative body of a fungus. The vegetative bodies of most fungi are composed of branching filaments called *hyphæ* (singular, *hypha*). In the molds and mildews these fine threads are readily visible with a magnifier, and under favorable conditions for growth form a soft, cottony layer in or on the substrate where the fungus is growing. The whole mass of hyphæ which make up the vegetative body of a fungus is called a *mycelium.* Sometimes the hyphæ, as in the yeasts, are very short; they are composed of cells that separate readily, and the filaments are rarely composed of more than a few cells. In the fleshy fungi

FIG. 248. Stages in the development of the common edible pink-gilled mushroom. Note the underground vegetative body of the plant.

the hyphæ are massed together, although each grows more or less independently of the others.

In decaying logs or in masses of fallen leaves one often finds cord-like strands of a white, brown, or black color. Not infrequently these may be traced for considerable distances and found to be connected with a puffball or other mushroom. They are parts of the mycelium, and absorb and conduct food to the fruiting bodies and growing parts. Sometimes these underground strands of the mycelia accumulate food and become greatly enlarged and act as storage organs. In parasitic forms that grow in contact with roots (*mycorhiza*) they probably absorb and transfer food into the host plant.

Food supply. When the fungus lives on the soil, its food is derived from the soluble organic materials like sugars, soluble proteins, and amino acids from the plant and animal matter occurring there. Moreover, the fungus may give off enzymes, that act on organic matter and bring about processes that change much of it to soluble substances. This is nothing less than a kind of external digestion, which makes the organic substances capable of diffusing into the cells of the fungus and provides material for

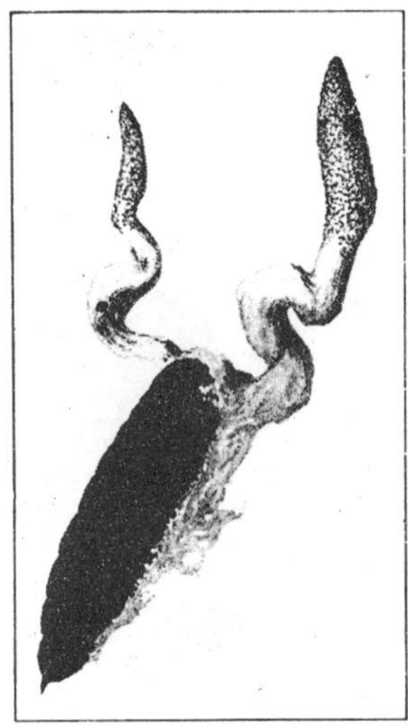

FIG. 248 *a*. A fungus (*Cordyceps*) parasitic on the pupa of the tobacco worm.

building tissue and supplying energy. The large surface exposed by the innumerable hyphæ is obviously advantageous in making contact with the food substances.

When the fungus lives within or on the tissues of another plant, a part of its hyphæ extends among or into the living cells of the host, and the food accumulated there becomes a source of food to the fungus. In this case also, by the secretion of enzymes, external digestion may occur before the food passes into the fungus hyphæ. The growth of the hyphæ through the cell walls of the host plant comes about in some cases by the liberation of enzymes that dissolve the walls ahead of the growing tip of the hypha, in others by the mechanical breaking of the tissues. Fungi, like most plants without chlorophyll, must have access to complex carbon compounds for food and energy, and in some cases they must have nitrogen compounds also.

Conditions for growth. Many fungi grow best in partial shade. Since they require a certain amount of moisture, fungi are usually most abundant in damp places. The most rapid destruction of organic matter also occurs in such situations. The decay of wooden beams under porches and in mines, the rotting of fruits and vegetables in cellars, and the disintegration of partially buried railroad ties are familiar examples of the results of favorable conditions for the growth of fungi. Bacteria also are present in such situations, and their growth and activities may go along with those of the fungi and hasten the final destruction. In the desert, where drought and intense light hinder the growth of these organisms, timbers may withstand exposure for centuries.

Reproduction. The development of the vegetative body culminates in the production of numerous fruiting bodies. Among the simpler forms of fungi the reproductive structures, or fruiting bodies, may consist merely of specialized reproductive cells cut off from the ends of hyphæ, or of ends of hyphæ that become enlarged and form several or many reproductive cells inside them. The reproductive cells are called *spores*, and the cells in which they are formed are called *sporangia* (singular, *sporangium*). Probably no other group of plants compares with the fungi in the variety of its reproductive bodies. Certainly no other plants produce such enormous numbers of spores in comparison with the size of the plants. As a consequence of the small size and great number of these spores, they are carried long distances by the wind and scattered everywhere. Furthermore, when the spores germinate and develop a new mycelium, a second crop of spores may be produced within a few days. In this way, under favorable conditions, as many as 120 crops of bread mold may be grown in a year.

Among the more complex and fleshy fungi the mycelium may grow for weeks and months before reproductive structures begin to develop. There may be large structures formed by the consolidation of large numbers of hyphæ, part of which later produce spores. The fruiting bodies may take the form of small disks, saucers, cups, hollow capsules, solid balls, toadstools, woody brackets, or irregular coralline masses.

Germination of spores. The spores of many of the parasitic fungi germinate readily when placed in water. However, unless the sporeling is in contact with its proper host plant, it fails to develop further and dies. Of course, many of them have been grown on nutrient media.

It is difficult to germinate the spores of many of the saprophytic fungi, like the puffballs and toadstools, because of their very exacting requirements. Such spores may be germinated only in a nutrient solution containing certain sugars, proteins, and organic acids. The germination of all fungus spores takes place only in the presence of moisture and oxygen and at suitable temperatures.

Distribution of fungi. Many fungi are world-wide in their distribution. Others, however, have rather definite temperature requirements which limit their development to certain regions and to certain seasons of the year. If the spores cannot withstand freezing temperatures, the fungus will not thrive in cold temperate and arctic regions. If the vegetative development requires warm temperatures, the fungus may be suppressed by low temperature even when the spores are present in abundance. The efficiency of cold-storage houses and refrigerators in conserving foods depends in large part upon the temperature requirements of the fungi and bacteria.

Groups of fungi. Of the nearly 50,000 described species of fungi, there are three conspicuous groups, commonly known as the Tube fungi, the Sac fungi, and the Basidium fungi.

The Tube fungi (*Phycomycetes*) receive their name from the fact that the hyphæ are for the most part tubular and without cross-walls. They include the molds, water molds, and downy mildews.

The name of the Sac fungi (*Ascomycetes*) is derived from the peculiar way in which the spores are produced. There are usually two, four, or eight spores formed inside a sac-like body. To this group belong the yeasts, the powdery mildews, most of the lichens, and certain of the fleshy fungi.

In the Basidium fungi (*Basidiomycetes*) spores are produced usually four or two in number on the end of a club-shaped hypha, which is called a *basidium* from the Latin word for club. In this division of the fungi are the smuts, rusts, and most of the fleshy fungi, such as puffballs and toadstools.

THE TUBE FUNGI OR PHYCOMYCETES

Among the commonest examples of the tube fungi are the bread molds and water molds. These molds are usually white, filamentous plants that are of great economic importance because of the damage that they do to foods during storage or shipment. Like

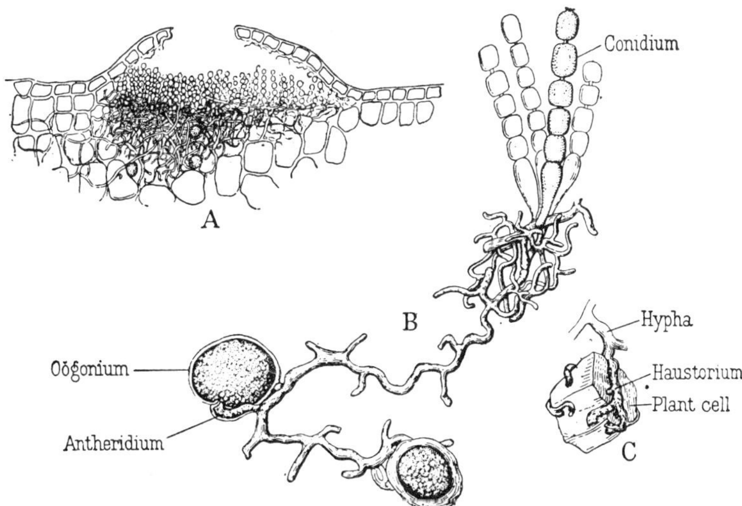

FIG. 249. One of the tube fungi, *Cystopus*, a parasite on leaves of green plants: *A*, section of pustule developed on a leaf showing fungus producing conidia (under the broken epidermis) and zygotes (below, in the mesophyll of the leaf); *B*, part of fungus dissected out of the mycelium shown in *A*; *C*, mesophyll cell surrounded and penetrated by absorbing hyphæ. *Cystopus* produces swimming spores when the conidia and zygotes germinate.

bacteria, the spores of molds are in the air and in the dust everywhere, and foods of all kinds are thus continually exposed to them. If the temperature is warm and the food is moist, they germinate and, together with bacteria, soon destroy the food. The same measures that will prevent the growth of bacteria in foods will prevent the growth of the molds, which are usually associated with them.

Bread mold. When a spore of bread mold germinates, a tube-like hypha develops. This hypha soon branches profusely. Some of the branches penetrate the bread and become absorbing organs, others spread over the surface of the bread like the runners of a strawberry plant and at intervals develop clusters of upright hyphæ which terminate above in sporangia. Just beneath each cluster of sporangial hyphæ a much-branched

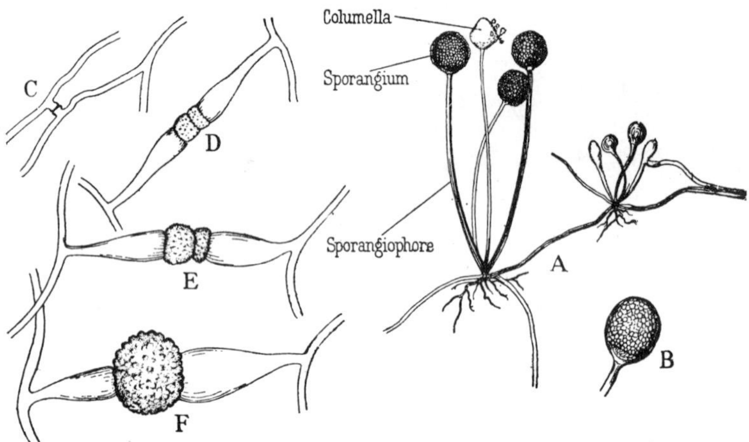

FIG. 250. Bread mold (*Rhizopus*): *A*, showing general habit of growth; *B*, sporangium enlarged; *C*, *D*, *E*, and *F*, stages in the formation of a zygospore.

rhizoid grows downward into the bread, affording anchorage and a food- and water-absorbing surface.

Under favorable moisture and temperature conditions the mycelium developed from a single spore may entirely surround a slice of bread with a white, fluffy, cottony growth, dotted with countless sporangia filled with black spores.

The bread mold may also reproduce in another way. When filaments from different strains of the fungus grow in contact, they may each form small, lateral, club-shaped branches. Then each of the branches is cut off by a wall near the inner end, forming a cell, the contents of which act as a gamete. Finally, the cell wall between the two gametes unite, forming a *zygote* or *sexual spore*. A heavy, rough black wall forms around it and it becomes a resting spore. Under favorable conditions, after a period of time, it germinates and produces a new mycelium with sporangia and asexual spores. The bread mold derives its water and food from the moist bread through the hyphæ and rhizoids. The rhizoids secrete enzymes in the bread, which change the starch, fat, and protein into various soluble substances that diffuse into the fungus hyphæ and pass to all parts of the mycelium. Try an experiment with a small piece of moist bread in a covered tumbler and see how long it takes for the bread to be consumed. These molds are for the most part saprophytes.

There are molds, however, that are both parasites and saprophytes. Perhaps you have seen goldfish growing in an aquarium, that were given to turning sidewise somersaults in the water and rubbing their sides on the gravel in the bottom of the jar. You may also have noticed cobwebby filaments attached to their scales; perhaps there were enough of them to make the sides appear white. This growth is one of the water molds (*Saprolegnia*) common in ponds and streams and often a cause of great losses at fish hatcheries. There are other related molds that attack flies and other insects and cause their death.

THE SAC FUNGI OR ASCOMYCETES

The sac fungi differ from the tube fungi in being composed of hyphæ made up of short cells. They exhibit a great variety of forms, from those like the green molds to fleshy forms like the morels. This is the largest division of the fungi and includes more than 30,000 species.

The outstanding feature of the sac fungi is the method of producing spores inside sac-like sporangia called *asci* (singular, *ascus*), which are the terminal cells of upright hyphæ. Usually the sacs are grouped in clusters; sometimes they stand upright side by side and form a layer over a part of the plant body. Many species also produce spores that are pinched off at the ends of short hyphæ. These spores are called *conidia,* and are important in spreading the fungus during the growing season. They may be produced in such abundance that an infected leaf has the appearance of being covered with white powder.

The best known of the sac fungi are the yeasts, the green molds (*Penicillium* and *Aspergillus*), the powdery mildews, the cup fungi (*Peziza*) and morels (*Morchella*), and the lichens.

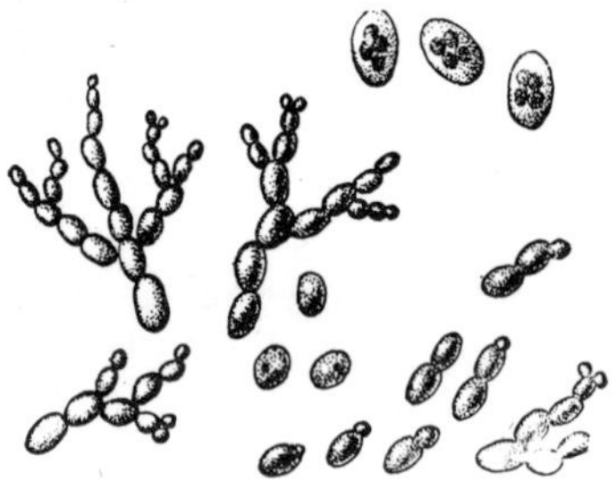
FIG. 251. Yeast (*Saccharomyces*): cells and branching filaments. Above are three cells, each containing four resting spores.

The yeasts. The yeasts are small, one-celled plants that multiply very rapidly. In the making of bread, they are of primary importance. When properly mixed with flour and water they develop in all parts of the dough. The yeasts have within them enzymes which oxidize part of the sugar that is present into carbon dioxide and alcohol. In this way the yeast obtains its energy. The carbon dioxide accumulates in bubbles and causes the dough to rise and become " light." When the dough is put into a hot oven, the alcohol is vaporized, and together with the carbon dioxide it is driven off into the air. The high temperature kills the yeast, bakes the dough, and changes some of the starch into its soluble form, dextrin, which makes it more readily digestible. Sour bread is produced when the yeast that is added contains acid-forming bacteria which change part of the alcohol into acetic acid.

Yeasts and bacteria are the organisms that change fruit juice into " hard " cider and vinegar. Yeast first changes the sugar in apple juice to carbon dioxide and alcohol, and bacteria further oxidize the alcohol to acetic acid, thus forming vinegar. Yeasts are also used in the manufacture of beer and wines.

Yeast fungi may readily be grown by adding a bit of yeast to a 5 per cent sugar solution in a test tube. The branching groups of cells may then be examined under the microscope. The manner of forming new cells among yeasts is unique, in that the new cells start as small protuberances (buds) from the older cells. These buds gradually enlarge until they attain their complete growth and separate, forming new individuals. The

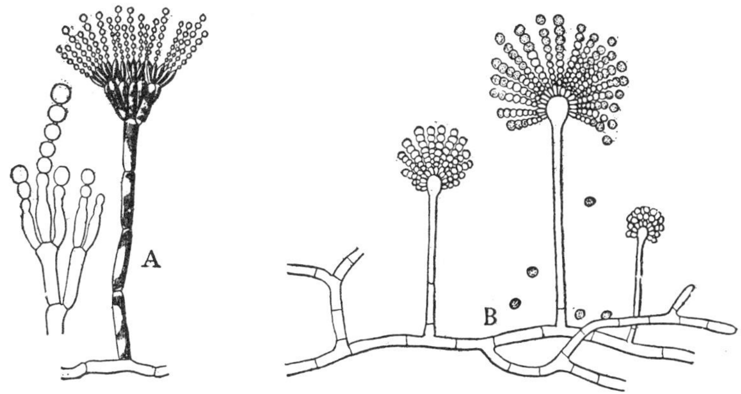

FIG. 252. *A*, blue mold (*Penicillium*); *B*, green mold (*Aspergillus*). Both show the hyphæ, upright branches, and conidia. (*After Frank.*)

alcohol formed in the test tube by the yeast may easily be detected by its odor.

The green and blue molds. Among the most widely known of the fungi are the green and blue molds. They are conspicuous destroyers of food. Fruits, vegetables, bread, and other starchy materials; canned fruits, preserves, and jellies; and even smoked meats, are all subject to the attack of the fungi under favorable moisture and temperature conditions. Some of the blue molds, on the other hand, are used in the manufacture of cheese. The flavors of Roquefort and Camembert cheese, for example, are largely due to the blue mold present.

The blue and green molds are so common and so widely distributed that their spores are present in the dust and air everywhere. When they fall on moist food they germinate, forming a hypha which soon branches profusely and forms a disk-like mycelium. After two or three days the upright hyphæ toward the center begin to produce spores, and the green color is due to the colored spores.

The method of spore formation is quite simple. The upright hyphæ of the blue mold branch several times near the end. In this way a broom-like tuft is formed, the branches of which terminate in rows of spores (*conidia*). In the green mold numerous branches arise at the enlarged ends of the upright hyphæ, forming a globular mass of spores.

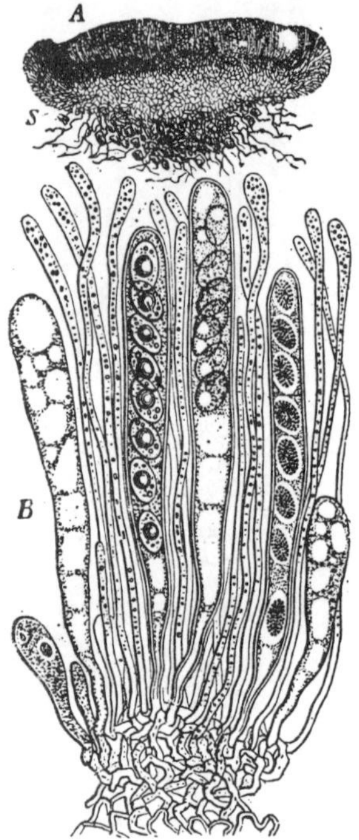

FIG. 253. *A*, vertical section of an ascomycete, *Peziza;* *B*, enlarged view of fruiting layer, showing asci- and ascospores. (*After Frank.*)

Under certain conditions these fungi may form round, capsule-like fruiting bodies in the material on which they are growing. Within a capsule is a group of sacs (sporangia), each of which contains eight spores. It is the presence of these sacs and sac spores that show the relationship of the blue and green molds to the sac fungi rather than to the bread molds.

The powdery mildews. Closely related to the green mold are the powdery mildews which form conspicuous white cobwebby patches on the leaves of roses, lilacs, willows, dandelions, and many other plants in late summer. They too produce innumerable spores from simple upright hyphæ. These form the dust-like powder which suggested their common name. Late in the season, after the union of two hyphæ, larger fruiting bodies may be formed, and within them groups of sacs and sac spores. These appear on the leaves as black dots and usually have a number of appendages. The powdery mildews are external parasites and derive their food from the epidermal cells of the leaves or stems of the host plant.

Cup fungi and morels. These are fleshy forms found in open forests on the soil and on decaying wood. In the cup fungi

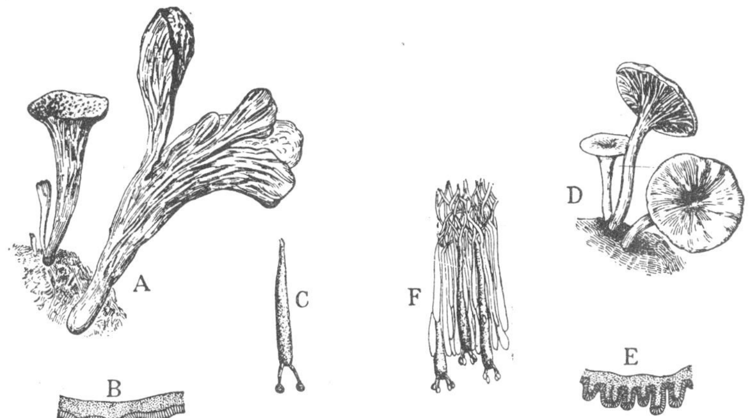

FIG. 254. *A, B, C, Craterellus,* showing habit of growth, a cross-section of the wall, and a basidium with two spores; *D, E, F, Cantharellus,* showing stipe and pileus, a cross-section of the pileus showing gills, and a part of the gill showing basidia with four spores.

the ascospores are produced in a layer covering the inside of the cup. In the morels the spore-bearing layer covers the elongated and much-wrinkled top of the hollow fruiting body. Morels are edible when fresh and by many persons are much esteemed for their peculiar flavor. Truffles also belong to this group. They resemble puffballs in form, but grow underground. They are much sought after in France and are collected by the use of pigs, and of dogs trained to locate them.

Lichens. Among the parasitic fungi are some that live on such one-celled green algæ as Protococcus. The fungus forms the plant body and completely envelops the algal cells. These forms constitute the lichens, which are gray-green, irregular-shaped plants that are common on the bark of trees, on rock surfaces, and occasionally on the soil (Fig. 255). Like other fungi they produce fruiting bodies, small disk-like or cup-shaped elevations, in which sac spores are produced in great numbers.

Fig. 255. A group of lichens: *Parmelia* (on tree); in the middle foreground, *Peltigera;* the remaining forms are species of *Cladonia.*

Lichens also multiply vegetatively by the breaking away of small bits of the thalli.

BASIDIUM FUNGI OR BASIDIOMYCETES

The second largest division of the fungi includes the smuts, rusts, and toadstools, numbering more than 20,000 species. The distinctive feature of the Basidiomycetes is the formation of one, two, or four spores at the end of a short, club-shaped hypha called a *basidium.* In the smuts and rusts the basidium and basidiospores make up a separate plant; in the toadstools the basidium and its spores are formed on or in the fleshy fruiting body of the fungus.

The smuts. The smut fungi of the small cereals have a mycelium extending throughout the tissues of the host plant, which was infected during its seedling stage. When the plant "heads out," the smut causes the grains to enlarge, and the smut hyphæ consume the food usually stored in the grain and then produce black spores in such abundance as completely to occupy or replace the grain. These spores have heavy walls and are capable of living over to the next season. In some cases this involves

W. S. Cooper

FIG. 256. Lichens covering the branches of a Monterey cypress, Point Lobos, California.

passing the winter in the granary and in the field. When the grain is harvested and threshed, the smut is spread to the good

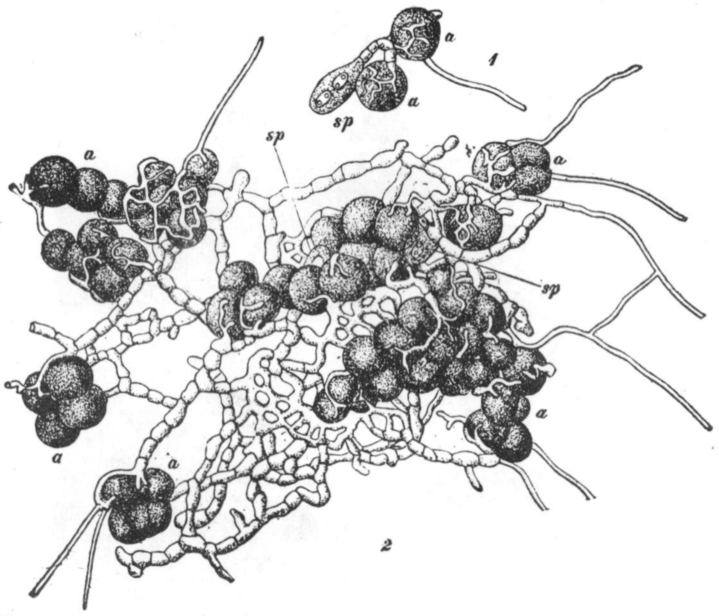

FIG. 257. Early stages in the development of a lichen, showing relation of the hyphæ to the algal cells. (*After Bonnier.*)

grain. Unless the grain that is used for seed the following year is suitably treated with disinfectants, the smut spores will be planted with the grain the following year. The spores of the commoner smuts germinate readily in water and produce a filament of four cells, each of which then produces a small, thin-walled spring spore (basidiospore) which may infect the young seedling. The smuts of the small cereals usually are carried over from one season to the next in the form of spores on the grain or of hyphæ inside the grain.

Corn smut behaves somewhat differently, in that the entire plant is not usually traversed or invaded by the mycelium. The spores last over from one season to the next in the soil and germinate there, producing the basidia and basidiospores which are blown about and infect the new crop of corn. The infected region of the corn plant usually swells, forming a large, glistening white ball. This later turns black and disintegrates, liberating myriads of spores to be further scattered by the wind. When the soil in a field becomes greatly infected with corn-smut spores, the best way to avoid further trouble is to plant another kind of crop for a year or two. Seed treatment is of no value.

Office of Cereal Investigations. U.S.D.A.

FIG. 258. Corn smut. The large white masses of tissue protruding from the stem contain the black spores. The mycelium of this smut does not extend far from the point of infection.

The rusts. Among the most serious diseases affecting wheat, rye, barley, and oats are those produced by the fungi known as the *rusts*. These fungi are called rusts because plants that are infected with them develop yellow and brown spots that have the appearance of iron rust. The rusts occur wherever grains are grown, and they cause millions of dollars' worth of damage to crops every year.

The rusts are parasites that live inside the host plants and injure or destroy the tissues that are concerned in food manufacture. Their life history is peculiar in that the fungus usually produces disease on two different kinds of host plants. The stem rust of wheat, for example, produces patches of red spores (ure-

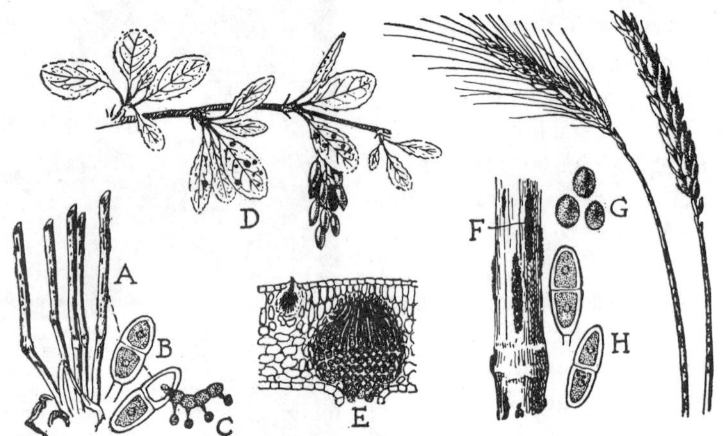

FIG. 259. Life history of the stem rust of wheat. In fields where wheat has been grown, the stubble (*A*) carries over the winter black spores (*B*), that germinate in early spring, producing smaller spores (*C*). These infect the leaves of the common barberry (*D*). In the leaves of the barberry the fungus grows and produces cup-shaped cavities filled with spores (*E*) that are carried by the wind to wheat fields and infect the wheat plants. After growing in the wheat a short time, the fungus produces first the red spores (*G*) that spread the disease to other wheat plants, and later the two-celled black spores that carry the disease over the winter again.

diniospores) which will infect other wheat plants. It produces also black spores (teliospores) which live over winter on the stubble and straw, and which germinate the following spring and produce spring spores (basidiospores) that infect the barberry. On the barberry leaves the fungus produces a cup-like depression within which a fourth kind of spores (æciospores) are formed. These spores will not germinate on the barberry, but they will infect wheat. Thus the stem rust of wheat spreads from one wheat plant to another by means of red spores. The following season it may spread from wheat stubble to the barberry by basidiospores produced by the teliospores, and from the barberry back to the wheat by still another kind of spore (Fig. 259).

In the Northern states, from the Dakotas to New England, the barberry stage is of special importance in the life history of the rust. In the southwestern United States, where winter wheat

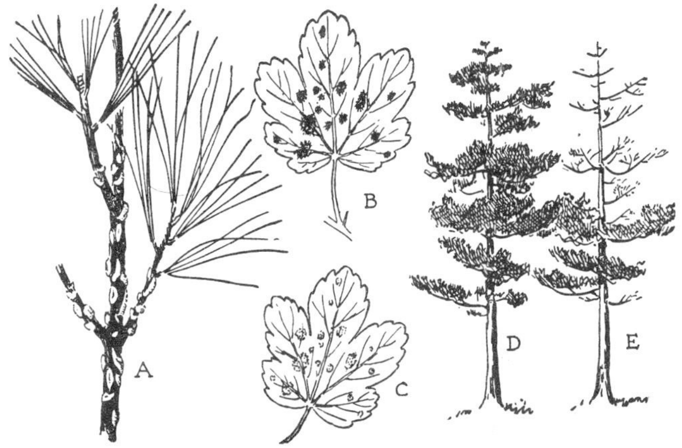

FIG. 260. The white-pine blister rust. The fruiting bodies on the white pine (*A*) produce spores that infect the leaves of the gooseberry (*B* and *C*). On the gooseberry leaves the fungus produces at first yellow spores that will infect other gooseberry plants, and later brown spores that carry the disease back to the pine. When a pine (*D*) is infected by the disease, the younger parts soon die (*E*).

is grown, the red spores produced during the summer may be carried from one field to another by the winds and infect the wheat over wide areas. In the Northern states the destruction of all barberry plants has been undertaken, and this work has already reduced the amount of infection. The hope of entirely controlling wheat rust, however, probably lies in breeding new varieties that are immune to the disease. This has been accomplished in Kansas producing a variety known as "Kanred wheat."

Apple rust. Other rusts also live on two host plants, and because of this double life and the fact that the fungus grows on the inside of its host, they are very difficult to control. The rust on the red cedar produces the so-called "cedar apples." In the spring these swell and protrude masses of teliospores that germinate at once and form basidiospores, which infect the leaves of the apple tree and may do great damage to them.

Fig. 261. Group of fleshy fungi. Beginning at the top, are species of ***Hydnum***, puffball, bracket fungus (*Fomes*), coral fungus (*Clavaria*), poisonous ***Amanita***, cornucopia fungus (*Craterellus*), ***Russula***, and earthstars (*Geaster*).

Pine blister rust. Recently the blister rust of the white pine has been brought to America, and it threatens to destroy what remains of our white-pine forests. The alternate host plants of this fungus are the wild and cultivated gooseberries and currants. Attempts are being made to prevent its spread westward, both by cutting the diseased white pine and by systematic destruction of the wild and cultivated currants and gooseberry bushes in newly infected regions.

Another common rust is frequently seen on raspberry and blackberry bushes along roadsides; it colors the under sides of the leaves with its bright, orange-red spores. In this instance no alternate host is known.

Mushrooms and toadstools. The largest and most complex of the fungi are the mushrooms and toadstools. They are common in fields and woods and for the most part live on decaying wood and organic matter in the soil. There is no real distinction between mushrooms and toadstools. Some of them are edible, others are indigestible, and some are deadly poisonous. Edible forms are cultivated on a large scale in caves and abandoned mines, and on a smaller scale in cellars. Wild forms should not be eaten unless they are gathered by persons competent to distinguish the different species, many of which are similar in appearance but very different in their effects when eaten.

The mushrooms as they are gathered are only the fruiting bodies of the fungi. The vegetative part of the plant consists of bundles of hyphæ extending in all directions throughout a large mass of soil on which the fruiting bodies appear. It may take several years for the underground vegetative part of the fungus to develop, while the fruiting bodies may develop in a few days. It is the enlargement of the fruiting bodies that persons have in mind when they speak of "mushroom growth." This expression leaves out of account the months or years of growth during which the materials were accumulated that led to the sudden production of the fruiting body.

The spores of mushrooms are produced in unthinkable numbers either inside the fruiting body (puffballs) or on the under side of the umbrella-shaped cap (mushrooms). A large puffball has been estimated to contain 7000 billion spores; the shaggy-mane mushroom (*Coprinus*) about 5 billion spores on each pileus; and one of the bracket fungi produces about 100 billion spores each year for several years.

REFERENCES

DUGGAR, B. M. *Fungous Diseases of Plants.* Ginn & Co., Boston; 1909.

HARSHBERGER, J. W. *Mycology and Plant Pathology.* P. Blakiston's Son & Co., Philadelphia.

HESLER, L. R., and WHETZEL, H. H. *Manual of Fruit Diseases.* The Macmillan Company, New York; 1917.

McILVANE, C., and MACADAM, R. K. *One Thousand American Fungi.* Bobbs-Merrill Company, Indianapolis; 1912.

STEVENS, F. L. *Fungi Which Cause Plant Disease.* The Macmillan Company, New York; 1913.

STEVENS, F. L., and HALL, J. G. *Diseases of Economic Plants.* The Macmillan Company, New York; 1921.

CHAPTER FORTY

PLANT DISEASES

It is difficult to define a plant disease, although it is usually not difficult to distinguish in a particular example between a healthy and a diseased plant. The difficulty of definition arises from the fact that diseases or abnormal conditions are produced by so great a variety of causes, and the effects or symptoms are so diverse, that it is impossible to include all examples of diseases and at the same time exclude others that are merely effects of unfavorable environmental conditions.

Plant diseases are usually defined as derangements of the normal structures and physiological processes of plants or parts of plants. The formation of galls and tumors, the development of gray, brown, and black spots on leaves, the sudden wilting and drying of shoots, and the rotting of seedlings are familiar external signs of disease.

Prevalence. Plant diseases have been recognized since the earliest historic times. They have been extensively studied, however, with a view to their control only during the last half century. They could not be well understood until the life histories of the bacteria and fungi were discovered.

The increasing prevalence of destructive diseases in America is the natural result of the more extensive cultivation of crop plants. The present world-wide exchange of plants and plant products has resulted in the accidental widespread transference of the bacteria, fungi, insects, and other organisms that injure plants.

Losses from plant diseases. The present losses from plant diseases are enormous when we view them for a state or for the United States as a whole. Sometimes on a single farm a disease may reduce the yield so greatly that the crop was grown at a loss instead of a profit. Some idea of the extent of the injuries inflicted may be gained from the following table of estimates by the United States Department of Agriculture for 1917:

Crop	Yield	Loss by Plant Diseases	Loss in %	Loss in Dollars
Wheat . . .	650,828,000 bu.	64,227,000 bu.	9.	128,000,000
Oats . . .	1,587,286,000 bu.	154,120,000 bu.	8.8	103,000,000
Corn . . .	3,159,494,000 bu.	175,368,000 bu.	5.2	224,000,000
Potatoes . .	442,536,000 bu.	117,167,000 bu.	20.9	143,000,000
Sweet potatoes	87,141,000 bu.	41,706,000 bu.	32.4	46,000,000
Cotton . . .	10,949,000 bales	1,866,000 bales	14.5	256,000,000
Peaches . .	45,066,000 bu.	14,459,000 bu.	24.3	21,000,000

These figures are still further increased by the losses caused by bacteria and fungi during storage and shipment. When any of these products reach the market, their price is increased in proportion to these losses.

Control of plant diseases. The control of plant diseases may be said to have begun with the discovery of the efficiency of the Bordeaux spray mixture in 1862. In 1896 the formaldehyde treatment of oats seed for smut was introduced, and more

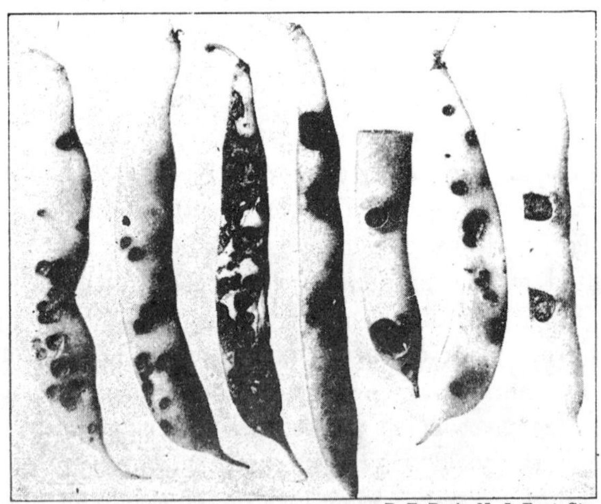

R. F. Poole, N. J. Expt. Sta.

FIG. 262. Bean anthracnose. The disease is due to a fungus which attacks all parts of the plant. The fungus is carried over winter in the seeds.

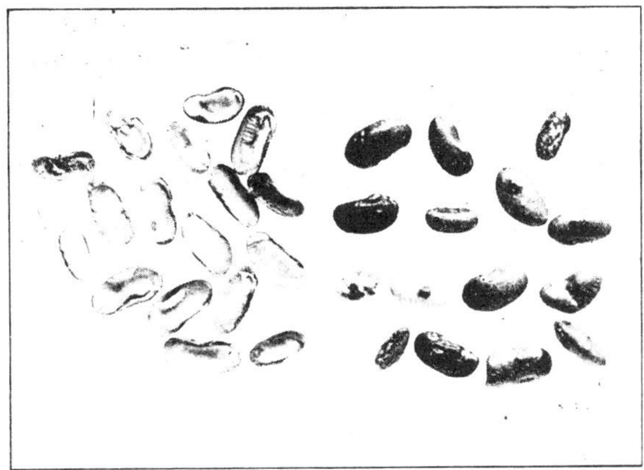

R. F. Poole, N. J. Expt. Sta.

FIG. 263. Bean seeds affected by anthracnose. The principal means of controlling the disease is to reject all seeds for planting that show the characteristic reddish spots.

recently the lime-sulfur sprays have been found an efficient preventive of certain diseases. All sprays are directed toward killing of the parasites before they invade the host tissue.

Another method of disease control is exemplified by the removal of all common barberry bushes, now being extensively carried on in the Northern wheat-growing states to decrease the losses from stem rust of wheat. In the Northern states also attempts are being made to check the ravages of the white-pine blister rust by removing the wild currants and gooseberries which form the alternate host of the pine-rust fungus. The removal of red-cedar trees from the vicinity of an orchard will prevent apple rust.

The development of disease-resistant varieties of crop plants is a third method of control that has been successfully used. The Kanred wheat is highly resistant to rust. A strain of cabbage has been selected that is resistant to cabbage yellows. New immune varieties of watermelon, tomato, and cotton have in certain sections replaced the older varieties which were susceptible to " wilt " diseases.

Rotating crops on farms so that there is small chance of the disease-producing organisms being carried in the soil from one season to the next has been found an efficient deterrent of certain diseases. The root rot of tobacco, wilt of potatoes, and smut of corn are some examples of soil-borne diseases.

Sterilizing soils and treating seed with fungicides before planting are used to control diseases that attack seedlings. Finally, diseases of one state or country may be excluded from another by the enforcing of a strict quarantine. At the present time nursery stock grown in Europe is excluded from the United States. Nursery stock that is to be shipped is also carefully inspected by government experts for symptoms of certain diseases, and only healthy stock may be shipped from one state to another.

Symptoms of disease. The external signs of diseases are of many kinds, among which the following classes may be noted:

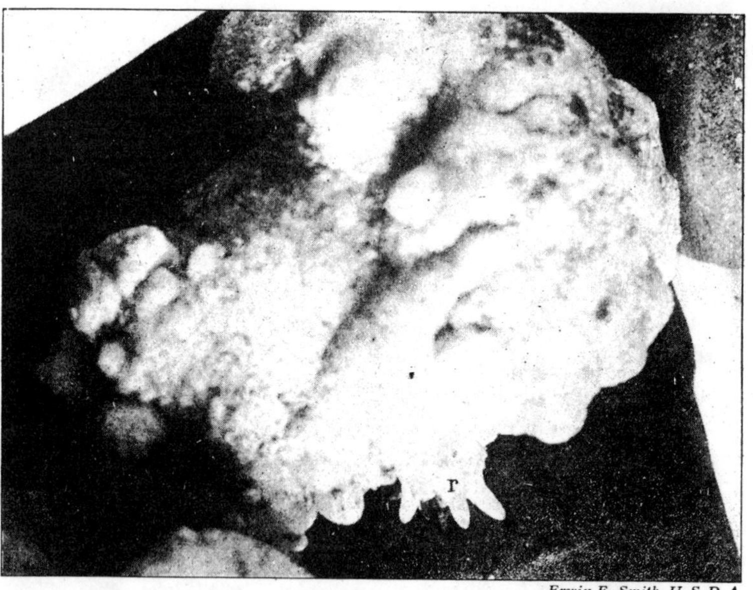

Erwin F. Smith, U.S.D.A.

FIG. 264. Tumor on a leaf of *Bryophyllum* inoculated with *Bacterium tumifaciens*. Roots are developing from the tumor at *r*. This disease is commonly known as "crown gall."

Erwin F. Smith, U.S.D.A.

FIG. 265. Tumors developed on leaf of tobacco by inoculating with *Bacterium tumifaciens*. From the tumors, leafy shoots are growing.

(1) *Malformations* of leaves, stems, roots, and fruits. Galls, knots, "witches' brooms," and curled and wrinkled leaves are common examples.

(2) *Cankers*, or rough sunken spots on stems and on branches of trees. They may be caused by frost injuries. Sometimes they are produced by a fungus, as in the brown rot of the peach tree and the black rot of apple.

(3) *Blight* is the term applied to the sudden dying of leaves, shoots, and blossoms. The fire blight of apple and pear are known wherever these trees are grown in America.

(4) *Leaf and fruit spots* are the result of local injuries to leaves and fruit through the growth of a parasite in or under the surface.

(5) *Wilts* include various diseases that are first noticeable by a sudden wilting of the leaves or of the complete plant. Cucumbers, cotton, cowpeas, and watermelons are particularly subject to diseases of this type.

(6) *Yellowing* of foliage or development of yellow spots and blotches characterizes a disease of peaches and the mosaic disease of tobacco.

Causes of diseases. Bacteria and fungi are the most common causes of plant diseases. Certain insects are almost equally injurious, and not infrequently act as carriers of disease-producing bacteria and fungi. Insects also are the commonest cause of galls. Nematode worms are sources of injury to the roots of many plants, particularly in greenhouses and in tropical and subtropical countries.

Just how these various parasites affect the plants is not thoroughly known. Some of them certainly withdraw sufficient food from the host plant to cause injury. Others seem to produce poisonous substances that injure the tissues or kill them. Still others in some way stimulate the tissues and cause abnormal growths.

Weather influences. The temperature and moisture conditions frequently influence the prevalence of diseases. Water is necessary for the germination of spores. Severe epidemics of potato blight, brown rot of stone fruits, and rots of grapes and

A. B. Stout

FIG. 266. Cucumber plants attacked by bacterial wilt. The plants wilt because the vessels are plugged by the bacteria that cause the disease.

Office of Forest Pathology, U.S.D.A.

FIG. 267. Effects of chestnut bark disease. A familiar sight everywhere within 200 miles of New York City, where the disease was first seen in 1905.

apples occur usually only in wet seasons. Dry weather is favorable to the spread of the spores of the loose smuts of cereals from an infected plant to the blossoms of another. When the seed is planted, untreated, the following year an abundance of smut results.

Low temperatures at the time of planting favor certain diseases, and high temperatures favor others. For this reason the prevalent diseases of one season may be very different from those of the following season.

Soil influences. The most usual effect of soil is in harboring bacteria and fungi from one season to another and in this way transferring the infective agent from one crop to the next. This is particularly important in causing infection with the fungi that cause the damping off of seedlings and the various forms of root rot.

The character of the soil, whether acid or alkaline, may also influence the growth of the fungi or the host in such a way as to increase or decrease the amount of the infection. The scab of potatoes is increased by liming.

Plant pathology. The foregoing paragraphs are sufficient to indicate the complexity of the problems of plant pathology, which is *the science that treats of plant diseases and their control*. They may also serve to show the importance of studying and understanding the life histories of fungi and bacteria, of their effects on their host plants, and of discovering new and better methods of eradicating them. One of the most important functions of the federal and state agricultural experiment stations is the promotion of research for the control of plant diseases. In view of the enormous annual losses to growers of plants, and the improved yields already obtained through the discovery of control measures, the expenditure of large sums of money for the agricultural experiment stations is more than justified. Every one profits by these investigations through more abundant and cheaper food supplies.

In the following paragraphs a few common diseases are described. More specific information concerning them may always be obtained by writing to your State Experiment Station.

Fire blight is one of the commonest diseases of pear, apple, and quince trees. It becomes noticeable in early summer through the turning brown of the leaves of twigs on these trees as though they had been scorched. The infective agent is a bacterium which lives in the interior of the affected branches, and unless it is checked as soon as it appears, it may extend into other tissues of the plant. The bacteria are apparently spread by insects. After a rain the infected branches bear numerous drops of a gummy nature containing countless numbers of bacteria, and insect visitors carry the bacteria from infected branches to blossoms and other twigs. The bacteria pass the winter in the living tissues at the edge of the cankers on the larger branches, and in the following spring these may become a source of further infection to near-by trees. The only known method of control is the cutting and immediate burning of all infected branches. These branches should be cut well below the darkened portions, because the bacteria usually extend some distance down the water-conducting tissues. Care must also be exercised in keeping the knife or saw from coming in contact with the diseased portion of the branches. Cankers on the older branches should be cut out during the fall or winter to prevent infection the following season.

Damping off of seedlings. Gardeners and nurserymen are much troubled by the damping off of seedlings in seedbeds. The fungi concerned in this process occur in the soil, and some are related to the common bread mold. The seedlings when first attacked become transparent, fall to the ground, die, and are finally consumed by the fungus. These fungi reproduce by spores, some of which are thick-walled and carry the plant over winter in the soil. Damping off may be controlled by sterilizing the soil of seedbeds. This may be accomplished by treating the soil with a 10 per cent solution of formalin and covering the bed with a piece of oilcloth for 24 hours.

A. B. Stout

FIG. 268. Damping off of seedling lettuce. This disease is due to molds and other fungi that attack the stems at the surface of the soil and cause the seedlings to rot.

In greenhouses and small seedbeds the sterilization is frequently accomplished by forcing live steam from a boiler through the soil by means of an inverted galvanized iron pan. The sterilization of soil not only kills the damping-off fungi but many other disease-producing organisms, both plant and animal. Moreover, it kills weeds and has a beneficial effect upon the soil.

Clubroot of cabbage. Many plants belonging to the mustard family, particularly cabbage, turnip, and cauliflower, are subject to a disease that causes swellings on the roots and impairs the efficiency of the roots. When young seedlings are infected many die, and those that continue to live never produce normal plants that can be marketed. The fungus belongs to one of the lowest groups of fungi, the slime molds, and produces enormous numbers of spores that winter over in the soil. In Europe the disease is a constant menace to the cabbage industry. In America it is not so prevalent. The disease may be partly controlled by liming the soil and by changing the location of the

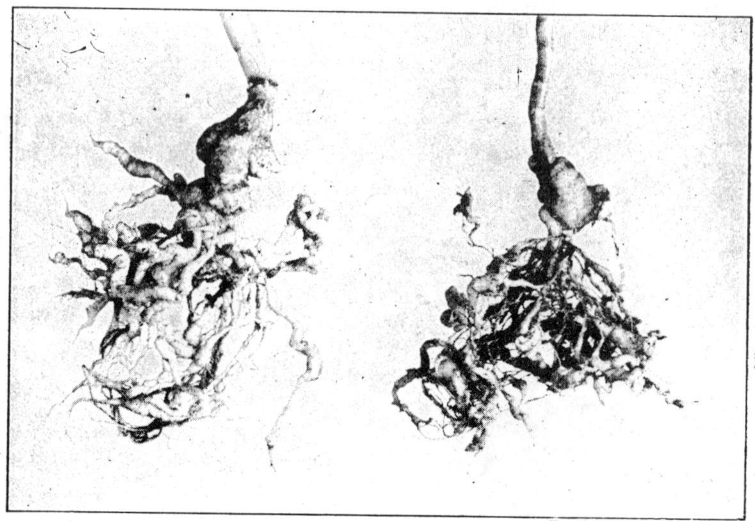

Erwin F. Smith, U.S.D.A.

FIG. 269. Clubroot of cabbage, caused by infection with a slime mold, *Plasmodiophora.*

cabbage field from year to year, alternating with crops that are not affected by this fungus.

Black knot. This disease is frequently seen in plum and cherry orchards and is made conspicuous by the presence of greatly thickened portions of twigs and branches. The spores of the fungus causing it seem to be carried by the wind, and infection takes place through the bark. As soon as the hyphæ of the fungus penetrate the inner tissues, the twigs begin to enlarge and the outer bark is broken. On the swollen surface in the spring and early summer one may see with a magnifying glass hyphæ producing spores.

During late summer another type of small, rounded fruiting body develops on the knots, which last over the winter. In the spring these bodies contain numerous sacs, each with eight spores inside it. These constitute a second means of spreading the disease. Where the disease is very prevalent, as in the Atlantic coastal states, it is difficult to control because it affects the wild as well as the cultivated cherries. The removal and destruction of all knots by burning has been found effective, and may best be done during the winter when the trees are leafless and before the winter spores are ripe.

The smuts. The smuts, described in the last chapter, are very destructive to cereals, but effective methods of control have been found. Corn smut differs from the other smuts in that its spores are shed from the lesions on the plant and remain over the winter on the soil. Hence the only methods of control available are the rotation of crops and the removal and burning of infected plants from the field as soon as they are noticeable.

Among the small cereals there are two types of smut, one of which results in the total breaking down of the grain and glumes into fine powder and spores. These are the so-called "loose-smuts." Other species of smut do not produce as complete a destruction of the seed coats and glumes. These are known as the "covered smuts." Experiments have shown that the covered

H. B. Humphrey, U.S.D.A.

FIG. 270. Smuts of small grains: *A*, loose smut of wheat; *B*, loose smut of barley; *C*, covered smut of barley. In each figure one of the heads is free from disease.

smut of oats, wheat, and barley and the loose smut of oats may be successfully prevented by treating the seed with formaldehyde before planting. This is possible because the plant is carried over the winter by spores on the grain or by masses of spores (smut balls) among the grain.

The *loose smuts of wheat and barley* cannot be successfully prevented by formaldehyde treatment, because the fungus is carried over from one year to the next by means of living hyphæ inside the grain. These fungus hyphæ, however, cannot withstand a temperature of 125° to 130° F., while the seeds of wheat and barley are uninjured by a 10-minute exposure to this temperature. Consequently a method of control has been devised by which the grain is dipped for 10 minutes into water carefully maintained at 129° F. Detailed directions for both the formaldehyde treatment and the hot-water treatment may be obtained from your State Agricultural Experiment Station. The hot-water treatment and the subsequent drying of the grain are so difficult to perform that this method of prevention is rarely used for the general crop. It is used, however, for treating grain that is to be planted for the production of seed for the following year.

Galls. Among the most striking examples of abnormal development of tissues and organs are the galls produced on a great variety of plants by insects, and more rarely by fungi and bacteria. Almost every one has seen the large papery galls of oak leaves, the velvety gall of the rose, the cone-like shoots of the pussy willow, and the swellings in the stems of goldenrod. These are all brought about in some unknown way by insects living in the plant tissues.

Downy mildew. A downy mildew commonly occurs on leaves and stems of grape leaves and may cause a reduction of the grape crop by injuring the leaves and causing them to drop. In some cases the fungus may attack the green fruit, causing it to wither and drop to the ground.

Brown rot of stone fruits. One of the most destructive diseases of cherries, plums, and peaches is the brown rot. The fungus causing this disease is carried over the winter on the mummied fruits hanging on the trees or lying on the ground beneath. Beginning in June, spores are carried to the developing fruits where they germinate, and decay follows the growth of hyphæ, resulting in brown spots and finally the withering of the entire fruit. When infection occurs late, at the time of gathering, this disease may cause serious losses during the marketing of the fruit.

Brown rot may be controlled by carefully removing all mummied fruits at the end of the year, and by spraying with a mixture of lime and sulfur at proper intervals during the season.

Mosaic disease. This disease has been especially injurious to tobacco, but it also affects tomatoes and to a less extent a great variety of wild and cultivated herbs. The external signs are a light-green mottling of the leaves or distorted and stringy leaf development. The cause of the disease is not known. It may be transferred to healthy plants by injecting juice from a diseased plant. In the field it is apparently spread by insects. Until the parasite is known and the exact manner of its transfer from one plant to another has been discovered, control will be impossible.

E. S. Schultz

FIG. 271. Mottled and crinkled leaf of potato affected by mosaic disease. (See also Figure 128, page 223.)

CHAPTER FORTY-ONE

THE CLASSIFICATION OF PLANTS

WHENEVER an attempt is made to describe the plants that occur on the earth, it becomes necessary at once to adopt some scheme of classification. About 250,000 plants have been distinguished, and they vary so greatly in size, structure, physiological requirements, and life histories that it is obviously impossible to describe them as a whole.

Since the earliest times students of plants have proposed schemes of classification which would group together plants having somewhat similar structures and life histories. At first these attempts were very artificial and unsatisfactory because so little was known about the plants themselves. During the past century and a half, great progress has been made in studying the plants that are now living and the plant forms of past geological ages now found as fossils in the rocks. The large amount of data thus accumulated has made it possible to build classifications that more nearly approach actual or natural relationships. Back of all modern classifications is the idea that the plants of the present have been derived through modification from the plants of the past.

Terminology. Since the time of Linnæus it has been agreed among botanists that all the *individual* plants which are essentially identical in structure and life history shall be grouped together as a *species* and given a two-word name. Thus all the millions of corn plants are grouped together as one species, *Zea mays*. Species have long been recognized and many of them have been given common names, such as Kentucky bluegrass, black mustard, cottonwood, black walnut, and white pine.

Groups of closely related plants having many characters in common have also been recognized, such as oak, willow, hickory, and pine. These larger groups are called *genera* (singular, *genus*). In each of these groups several or many species are distinguished. For example, the oaks are commonly separated into white oak,

FIG. 272. Four species of oak: *A*, white oak (*Quercus alba*); *B*, bur oak (*Quercus macrocarpa*); *C*, red oak (*Quercus rubra*); and *D*, pin oak (*Quercus palustris*). Note that the species differ in shape of leaves and in size and form of acorns, but that they have many characters in common. (*After E. L. Moseley.*)

black oak, blue oak, red oak, live oak, all differing from each other and from other oaks.

Common names are quite satisfactory for ordinary purposes

and when used to name plants of a given locality, but they are very unsatisfactory when used to name plants in widely separated localities. For example, the white and black oaks of California are not the same trees as the white and black oaks of Pennsylvania. The white pine of New England is not the same as the white pine of Colorado, and both differ from the white pine of Idaho. Consequently *taxonomists*, or students of classification, have been forced to give each kind of plant a scientific name. This name consists of two words having a Latin or Greek form: (1) a genus name and (2) a species name. The name for the oak genus is the old Latin word for oak, *Quercus;* for the species called "white oak" in the eastern United States, *Quercus alba* (Latin for "white"); for the California white oak, *Quercus lobata.*

Larger groups. Similar individuals, then, are grouped into species, and species having certain fundamental characters in common are placed together in a genus. In a similar way genera are grouped into *families*, and families are grouped into *orders.* Several *orders* taken together form a *class*, and a group of *classes* forms a *phylum* (or *division*) of the plant kingdom. In some of the largest phyla it may be convenient to recognize secondary divisions in each of these groups, such as subclass, suborder, subfamilies, and subgenera. Highly variable species are sometimes subdivided into *varieties.*

The following diagram showing the higher groups to which the oaks belong will help to make these groupings clear:

THE PLANT KINGDOM

A. Phylum — Angiospermæ
  1. Class — Dicotyledoneæ (Includes more than 25 orders of plants whose embryos have 2 cotyledons)
    a. Order — Fagales (Includes Birch and Beech families)
      (1) Family — Fagaceæ (Beech family) (Includes genera of Beech Oak, Chestnut, etc.)
        (*a*) Genus — Quercus (Oak) (Includes more than 200 species mostly in North America and Asia)

The great plant groups. We have already described two of the major plant groups or phyla: the Bacteria (*Schizomycetes*) and the Fungi (*Eumycetes*). In succeeding chapters the more important characteristics of other major plant groups will be discussed. These groups are the *Algæ* (including several distinct phyla), the *Bryophytes* (mosses and liverworts), the *Pteridophytes* (ferns, horsetails, and club-mosses), the *Gymnosperms* (conifers and cycads), and the *Angiosperms* (flowering plants).

Because of the great diversity of the plant kingdom, the number of groups in it is very large. For this reason only brief generalized descriptions of the major groups can be given in a text like this, and all consideration of some phyla must be omitted. Furthermore, it may be necessary to group together other phyla under series names that do not necessarily imply relationships. For example, *Thallophyta* is often used to designate all plants below the mosses and liverworts — all plants with a vegetative body undifferentiated into leaf-like, stem-like, or root-like organs. The algæ constitute the series of chlorophyll-bearing phyla placed among the Thallophyta, and the fungi include the non-green phyla. Yet in grouping algæ and fungi together we do not mean to imply relationships between them.

CHAPTER FORTY-TWO

THE ALGÆ

THERE is a large assemblage of chlorophyll-bearing plants usually small and comparatively simple in structure, known as *algæ* (singular, *alga*). Some species are unicellular and microscopic, with cells so simple in structure that they are comparable to those of bacteria; other species are multicellular, filamentous colonies, often branched and attaining lengths of several inches; a few species have thick, leathery, vegetative bodies, composed of several distinct tissues and varying from a foot to many feet in length. The algæ are of peculiar interest in showing various methods by which complex plants may be derived from simpler forms.

The algæ include many diverse types of plants which are grouped together because their vegetative and reproductive structures are simple when compared with other groups of chlorophyll-bearing plants. All algæ reproduce by cell division, and usually also by spores.

Classification of algæ. For convenience of description the more important algæ may be divided into five groups, four of whose common names are suggested by the characteristic color of the plants in each group. The classification, however, is based upon more fundamental characters than color; namely, (1) the structure of cells, or vegetative body; (2) the reproductive structures; and (3) the life history, or the series of events that occur in the life of the plant, beginning with the germination of a spore and ending with the formation of similar spores.

THE BLUE-GREEN ALGÆ OR MYXOPHYCEÆ

These are the simplest known autophytes. A majority of the plants contain a water-soluble blue pigment in addition to the chlorophyll. Both this pigment and the chlorophyll are dispersed in the protoplasm and are not in definitely organized bodies like the chloroplasts of higher plants. The cells lack true nuclei. One of the striking characteristics of the group is the abundant formation of mucilage by the cells, leading in many instances to the production of gelatinous colonies of cells and to simple and branched filaments. About 1200 species have been described.

Occurrence. The blue-green algæ occur in abundance in all parts of the earth. They are the prominent algæ of the tropics and the polar regions, and they may impart their color to the landscape. Most of these algæ are in fresh water; a few are found in salt and brackish water along coasts. Many of the forms are violet, red, gray, or brown in color. The Red Sea owes its name and color to a red species of "blue-green" algæ. The so-called "water bloom" is frequently a sudden development and accumulation of certain blue-greens near the surfaces of lakes and ponds. Pond waters at such times have a distinct greenish color and have been known to poison cattle and horses.

Blue-greens occur on all moist soils, and in some parts of the tropics as epiphytes also. Some of these algæ form papery layers on the soil surface which are very hygroscopic and aid in retaining the soil water. The soil algæ are commonly associated with the nitrogen-fixing bacteria, and as they die they set free carbon compounds useful to the bacteria. The bacteria, in turn, at their death liberate nitrogen compounds useful to the algæ and the higher plants.

Blue-greens often become troublesome weeds on soils in greenhouses — our artificial tropics.

Resistance to unfavorable conditions. Some blue-greens thrive in conditions where no other autophytes can live. This resistance is partly due to the high water-retaining capacity of their gelatinous walls, and perhaps partly to the nature of their proteins and their simplicity of organization.

Blue-green algæ have been known to remain alive in dry soil samples for a period of 50 to 70 years. They withstand being frozen for several months.

Certain blue-greens constitute the principal vegetation of hot springs, and are known to live in temperatures between 150°

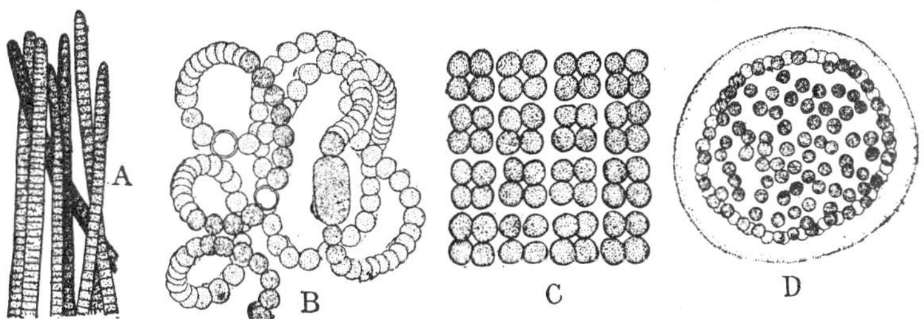

FIG. 273. Blue-green algæ: *A, Oscillatoria; B, Nostoc; C, Merismopedia; D, Cœlosphærium.* (*B, D, after G. M. Smith; A, C, after J. E. Tilden.*)

and 185° F. These algæ become encrusted with minerals from the water, and in this way the brightly colored rock basins are formed about such springs — as, for example, the Hot Springs terraces in the Yellowstone National Park. When fresh-water streams are polluted by sewage, by poisonous wastes from manufacturing processes, and by drainage from coal mines and oil wells, the blue-green algæ survive long after all other plants except certain bacteria are killed. Many blue-greens grow more luxuriantly when they have access to organic matter; that is, they are partial saprophytes. These forms aid in destroying sewage in streams and are of great economic importance; for they, with the bacteria, constitute a first step in transforming such waste materials into food for fishes and other aquatic animals.

Blue-green forms in lichens. Many of the genera of lichens (page 413) have blue-green algæ as the food-manufacturing part of the fungus-alga complex. When this host happens to be one of the highly gelatinous algæ like *Nostoc*, the lichen also forms a jelly-like mass.

Common genera. Some of the commonest genera among the blue-greens are: *Oscillatoria*, filamentous forms with short cylindrical or disk-shaped cells; *Glœocapsa*, gelatinous unicellular forms; *Nostoc*, gelatinous forms in which the cells are formed in chains held together in large, irregular masses by the gelatinous sheaths; *Anabœna*, with short, curved chains of cells, very frequent in "water blooms"; and *Tolypothrix*, filamentous forms that are often highly branched.

THE DIATOMS OR BACILLARIACEÆ

The diatoms include about 12,000 species of one-celled algæ, remarkable for their abundance in moist places everywhere, their small size, and their beautifully sculptured cell walls. These algæ differ from the other algæ described in this book in having siliceous cell walls. The cell wall consists of two overlapping valves, like the two halves of a pill box. In cell division the two valves are moved outward, and following the division of the protoplasm, two new valves are formed between the two halves of the original cell. Diatoms have distinct nuclei and chloroplasts. Many of them are motile, but some have a stage in which they are fixed by chitinous stalks and gelatinous sheaths to under-water objects.

Attention is directed to the diatoms because of their great economic importance. They are one of the most important sources of food, directly or indirectly, of marine and fresh-water fishes. The gizzard shad, for example, consumes directly enormous numbers of diatoms, while the hake feeds upon a series of aquatic animals all of which directly or indirectly have diatoms as their ultimate food supply. No matter how long the chain of animals, up to the fish, the fundamental organism is the diatom which changes inorganic compounds into food.

Diatoms occur in vast numbers in the upper layers of the ocean and in all kinds of streams and ponds. They sometimes multiply rapidly in reservoirs, and when they subsequently die become a source of annoyance by producing bad odors and tastes in drinking water. Extensive deposits of diatom shells occur in many parts of the world, and these shells are the basis of powders and soaps for polishing metals. Diatomaceous earth is also used as an absorbent of nitro-glycerine in making dynamite, and in the manufacture of fireproof linings and walls.

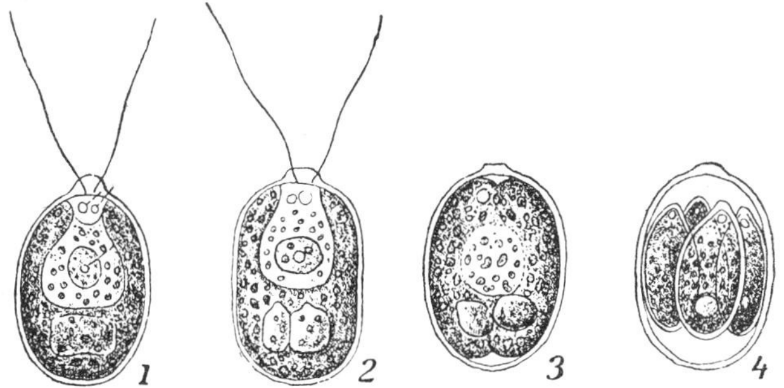

FIG. 274. *Chlamydomonas*, a simple free-swimming form of green alga. Four stages in the reproduction of the plant are shown above. (*After Dill.*)

THE GREEN ALGÆ OR CHLOROPHYCEÆ

Probably no other group of plants exhibits as wide a range of forms, structures, and life histories as the green algæ. So great is this diversity that it is exceedingly difficult to describe the group as a whole. More than 5000 species have been described, and these are scattered among many families, the interrelationships of which are far from clear.

The green algæ include unicellular forms, simple and branched filaments, plates or extensive sheets, and small gelatinous masses. All possess chlorophyll inclosed in definite chloroplasts, which also have an almost endless variety of form. The cell walls are variously formed of cellulose and pectic compounds. Some of the forms have an outer layer of chitin on the wall. Most of the green algæ produce motile cells at some stage in their development. In the simplest forms the motile stage is the most prominent one in the life history.

The green algæ are of peculiar interest, because among them are not only the simplest plants containing a definite nucleus and chlorophyll-bearing structure in the protoplasm, but also the types of plants from which the more complex land plants are thought to have been derived. The product of photosynthesis which accumulates most frequently in the *Chlorophyceæ* is starch, supplemented in many instances with oil. Both the starch and the oil occur in greatest amounts in cells and parts of plants that become reproductive structures.

Multiplication occurs by cell division in the unicellular forms and by the fragmentation of filaments in more complex forms. Thick-walled cells are frequently formed from vegetative cells, and these are highly resistant to drought and cold. Rounded spores may also be formed by the contraction of the contents of vegetative cells and the subsequent secretion of a new cell wall. These spores usually remain dormant for weeks or months before germinating and starting growth anew. In this way the plants live through drought and winter conditions.

Reproduction may also take place by swimming spores. These are naked protoplasts, with two or more *cilia*, or *flagella*, that vibrate and propel the cell through the water. A swimming spore may consist of the entire content of a vegetative cell, or several or many spores may be formed by several successive internal divisions of a vegetative cell. In any event they pass out of the original cell wall and swim about for a few minutes or a half hour, and then become attached to some object. They then begin to divide and form new cells.

If certain green algæ are kept in the dark 12 to 24 hours, and are then brought into the light, swimming spores will appear in profusion in a half hour or an hour. Transferring algæ from cold water to warm water will frequently produce the same effect. Swimming spores reproduce the plants rapidly and help to spread them during favorable conditions.

Sexual reproduction. The green algæ present a remarkable series of forms, with every gradation of sexuality. In sexual reproduction the essential fact is the union of two gametes to form a *zygote*. The gametes in the simplest forms are similar free-swimming protoplasts, resembling swimming spores but smaller in size. In the more specialized forms the male gamete is a small, free-swimming cell (sperm), and the female gamete (egg) is a large, non-motile cell containing a large amount of food materials. The cell in which the sperms are formed is called an *antheridium* (plural, *antheridia*), while the cell that forms an egg is called an *oögonium* (plural, *oögonia*). The spore formed by the union of a sperm and an egg is called an *oöspore*.

In the peculiar group to which Spirogyra and Zygnema belong, the gametes unite through a tube that forms between the two gamete-producing cells. There are neither swimming spores nor free-swimming gametes.

Distribution. Green algæ are comparatively abundant in tropical seas and decrease rapidly in number along the northern coasts. On land they are most numerous in fresh-water ponds, pools, and streams of the temperate zone. The small forms occur in every conceivable habitat. Some of them cause the red patches on snow banks in the arctic regions and on high mountain tops.

Protococcus. On the partly shaded, moist sides of trees, rocks, buildings, and fences everywhere, there occur patches that look as if they had been stained green. If a little of this stain is scraped off and examined under a microscope, it is seen to be made up of little, rounded green cells. Each cell consists of a cell wall, cytoplasm, and nucleus. In the cytoplasm is a large green plastid which almost fills the cell.

When the cells are examined, certain of them will be found to be elongated; some of these may be dividing into two. Sometimes there are two or more cells still clinging together, showing clearly that they have just been formed by division. These groups separate readily when the cover glass is tapped, and each single cell may go on living quite independently of the others. The plant, therefore, consists of a single cell which carries on all the essential processes of life and is able to reproduce itself. Moreover, it is a highly successful plant, for Protococcus occurs in moist places in all parts of the world, from the tropics to the polar regions, in habitats of many different kinds.

Chlamydomonas. This alga is one of the simplest unicellular plants and differs from Protococcus in being free-swimming during its vegetative phase. Chlamydomonas occurs in fresh-water ponds, ditches, and roadside pools. Reproduction occurs both by forming resting cells, by swimming spores, and by gametes. This genus is of interest because it seems to present better than any other the characteristics of the primitive plants from which all the green algæ and possibly the higher plants may have been derived.

The pond scums. If examined in the spring or fall, almost every pond and little stream will be found to contain many kinds of algæ. Some of these are merely masses of rounded cells like the cells of Protococcus. Others have the cells arranged in rows, forming simple filaments. In still others the filaments are highly branched and the plant body may be several feet in length. Some of the forms are embedded in a gelatinous matrix. All these various kinds of algæ taken together are popularly called the "pond scums." They are forms of algæ most commonly observed. As pond scums they are most unattractive, but seen through a microscope they present varied and beautiful examples of cell architecture.

Many of the pond scums are at first attached to under-water objects, but during warm weather they break loose and come to the surface, where they form a green or yellowish-green surface layer. All cells carrying on photosynthesis give off oxygen, and the bubbles of oxygen that come from the filaments cling to them and help to buoy them up. Furthermore, bubbles of air from the water when it becomes warm also collect in the masses of algæ and help to support them at the top of the water. The pond scums are generally considered unsightly, and not a few persons think them poisonous. In reality, they are quite as harmless as lettuce. The danger in drinking from ponds lies, not in the green scums, but in the presence of certain disease-producing bacteria that may have been carried into the ponds by surface water. Several thousand different species of algæ are concerned in the formation of pond scums. Microspora may be studied as an example of the more simple filamentous forms.

Microspora. The Microspora plant is a filament made up of cylindrical or barrel-shaped cells placed end to end. Each cell carries on all its own food-producing and energy-producing processes. During early spring, as food is manufactured, the cells enlarge and divide. The division is always in the same direction, however, and the cells remain attached to each other, so that the growth and division of the cells cause the filament to increase in length. This long, slender line of cells is easily broken, and the plant may be multiplied by the breaking of the filaments into parts.

Spores in Microspora. Microspora produces swimming spores and resting spores. These are special cells that reproduce the plant. A swimming spore is formed by the contents of a cell in the filament contracting into an ovoid body. At one end of this body two cilia, which are small, hair-like propellers, are developed. The wall of the original cell then breaks and the swimming spore is set free. After swimming about in the water for a short time, it becomes attached to some object under water, loses its cilia, and grows into a cylindrical vegetative cell. This cell then continues to grow and divide until a new filament is formed. The advantages of swimming spores are that they multiply the plant, and by their ability to swim they enable the plant to spread to new locations that it might not reach without these motile cells.

The resting spores are usually formed in the spring after the active period of vegetative growth has passed. At this season the cells in the filament stop dividing and food accumulates in the form of starch and protein granules. The protoplasm in each cell then contracts into a spherical form and secretes a heavy cell wall about itself inside the original cell wall. In this way the cells of a filament form a row of ovoid or spherical heavy-walled resting spores. Usually the walls of these spores become yellow or brownish. The resting spore remains dormant until the late fall or early spring. Then it germinates; the outer wall that incloses the spore breaks and the protoplasm and delicate inner wall push out and form a cylindrical vegetative cell, which continues to grow and divide, producing a new filament.

Microspora, then, in addition to the vegetative multiplication of the cells shown by Protococcus, has swimming spores that multiply and spread the plant, a stage that recalls Chlamydomonas. It also has resting spores that undergo a dormant period, after which, when favorable conditions for growth appear, they produce a new plant. Its life cycle and that of other similar algæ include (1) an active chlorophyll-working period, during which the plant grows and enlarges its body and accumulates food; (2) a reproductive phase, which closes with the production of resting spores; and (3) a period of dormancy, during which only the resting spores are alive. The length of the dormant period for a particular alga is practically the same, whether it lives in a permanent pond or in a pool that dries up in summer.

Ulothrix. Another green alga occurring on the margins of lakes, in running streams, and in clear springs is Ulothrix. It has a filamentous body similar in many respects to Microspora, and like that form it is attached to rocks and other objects. Its methods of reproduction, however, are more numerous and more complex than those of Microspora, and they will serve to exemplify the reproductive processes of many other forms of algæ. When the filaments are mature, the protoplasm within some of the cells divides into two, four, or eight parts, each of which contains nucleus, cytoplasm, chloroplast, and vacuole. Each of these parts becomes oval in shape and develops into a swimming spore with four cilia. An opening appears at one

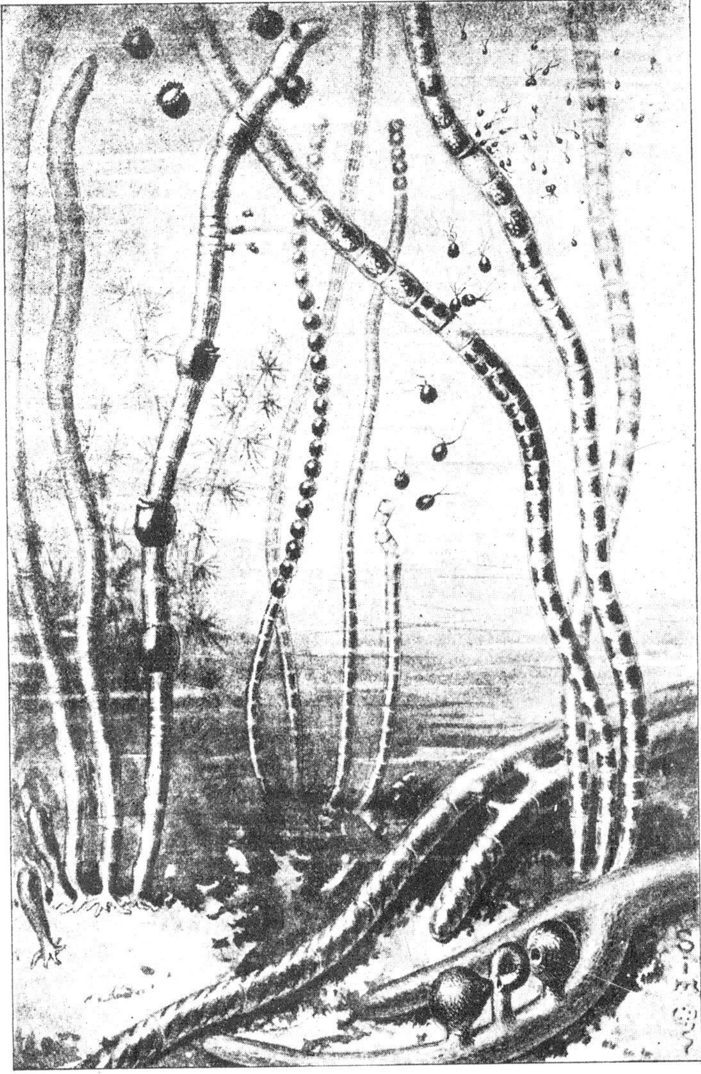

FIG. 275. Fresh-water algæ. The upright filaments are, from left to right: *Œdogonium*, producing swimming spores, eggs, and sperms; *Microspora*, forming resting spores and swimming spores; and *Ulothrix*, forming swimming spores and gametes. The horizontal filaments are *Spirogyra* (left) and *Vaucheria* (right). Highly magnified.

side of the original cell wall, and a few minutes later the swimming spores pass out from the cell cavity and swim away. Sometimes all the cells in a filament produce swimming spores at about the same time, and hundreds of these small green bodies may be found moving about in the water. At the end of from 15 to 30 minutes the swimming spores settle down on some object and become attached. By the end of a day the cell formed from each spore has divided and produced the first two cells of a new filament.

The protoplasm of other cells, of the same or other filaments, continues to divide until 16, 32, 64, or more bodies have been formed. These are called *gametes.* They are similar to the swimming spores but much smaller, and each possesses two cilia for swimming. Like a swimming spore, each of them leaves the old cell through an opening in the wall. The gametes swim about for some minutes and then unite in pairs. They are attached at first only by the ciliated ends, but later the two gametes fuse. The body thus formed may grow directly into a new filament, or it may produce swimming spores from each of which a new filament is formed.

Œdogonium. Œdogonium is another filamentous alga that flourishes in ponds and streams. In early life the filaments are attached, but large masses of them will often be found free in ponds and stagnant pools. From the cylindrical vegetative cells, large swimming spores are formed. Gametes also are produced. These are of two distinct forms, male and female. Plants belonging to the Œdogonium group may be used to exemplify reproduction in many other algæ, whose gametes are essentially like those of more complex plants.

At the time of production of the gametes, some of the cells in the filament enlarge, become rounded, and accumulate starch and other food material; also, a small opening is formed in the cell wall. The content of this cell is the female gamete or egg, which like other eggs has in it a store of food.

Other cells of the filament are cut up into very short cells by the formation of transverse walls. In each of these short cells there are formed two small gametes, which escape from the filament and swim out into the water. These are the male gametes, or sperms. Fertilization takes place when one of the sperms enters through the opening in the oögonium and unites with the egg. The egg and the sperm may be of the same filament or of different filaments. The product is an oöspore (egg spore). After a dormant period this produces four swimming spores that start new filaments.

In Œdogonium, therefore, the sex cells are of two kinds quite distinct in structure and function. The egg is a large, stationary cell filled with food. The sperm is a small, swimming cell that moves to the egg and accomplishes fertilization by uniting with it. The product is an oöspore which germinates and produces four swimming spores. These start a new generation of the filamentous plants.

Reproduction among the algæ. The methods of reproduction among the algæ that we have studied are representative of those found in the entire group. The three general types are:

(1) *Vegetative multiplication.* By means of cell division all masses, filaments, or highly branched plant bodies are produced. If the individual cells separate from each other after division, as in Protococcus, many new individual plants are produced; and when filaments and branched forms are broken, as in Microspora, a new individual plant is produced by each part.

(2) *Reproduction by spores.* Vegetative cells form thick-walled resting spores which carry the plant over to the next season. Another kind of spore is the swimming spore, by means of which the plant secures immediate reproduction and spreads to other parts of the pond or stream. These spores are formed directly from vegetative cells, or by the division of vegetative cells. There is no union of cells as there is when sexual spores are formed.

(3) *Sexual reproduction.* A sexual spore, or oöspore, is formed by the union of two gametes. The gametes may be similar in size and appearance, as in Ulothrix, or they may be unlike, as in Œdogonium, where one gamete accumulates a large food supply and the other is small and motile. In either case, the one gamete corresponds to the sperm and the other to the egg that is found in higher plants. The union is the process of fertilization. The oöspore may germinate immediately, but more often it remains dormant for a period of weeks or months.

Other genera of green algæ. The most beautiful of the larger green algæ are the *Draparnaldias*, having a main filament with little plumose tufts of lateral branches. Closely related are the *Stigeocloniums*. Both are frequently found in springs and small temporary streams.

Among the most readily recognized forms are the species of *Spirogyra*, with their spiral chloroplasts — sometimes as many as sixteen bands in each cell; and *Zygnema*, with cells marked by two large, radially branched chloroplasts.

Vaucheria is a common group found in pools, ditches, and streams and on moist soil. These algæ are remarkable in having no cross-walls in the long and much-branched vegetative filaments. In the warmer seas are a number of genera related to Vaucheria, that attain considerable size.

Species of *Cladophora* are highly branched. They are coarse forms, found attached to rocks in lakes and swift-flowing streams everywhere.

Plankton algæ. The microscopic plants and animals that float or swim in all bodies of water make up what is known as the *plankton* (Greek : *planktos*, wandering). It includes hundreds of species of algæ, that multiply rapidly and go through their life cycles in a few days. These algæ are so minute that they can be collected only by passing the water through silk bolting cloth, or filter paper. Nevertheless, they are quite as important as the filamentous algæ as a source of food for small fish and minute water animals.

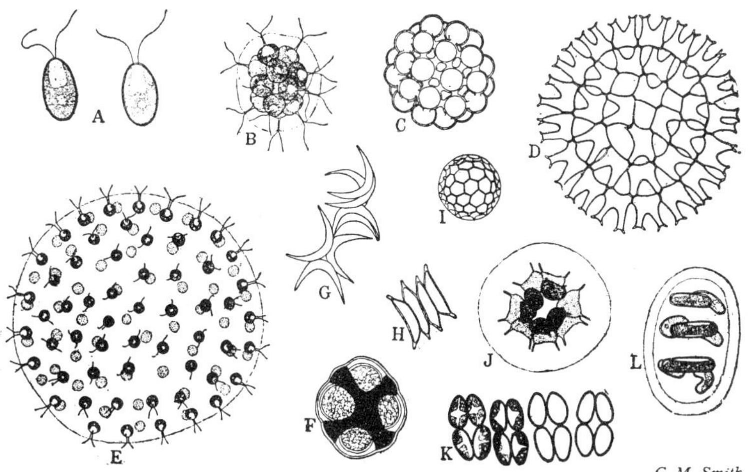

FIG. 276. Plankton algæ: *A, Chlamydomonas; B, Pandorina; C, Cœlastrum; D, Pediastrum; E, Pleodorina; F, Glœotænium; G, Selenastrum; H, Scenedesmus; I, Trochiscia; J, Sorastrum; K, Crucigenia;* and *L, Nephrocytium.*

The importance of the algæ. Both green and blue-green algæ are generally considered a nuisance in ponds and streams, and they are commonly thought to have no economic importance; but the fact is that these pond scums are the primary food supply of all the water animals. They bear the same relation to aquatic animal life that the herbaceous plants bear to animal life on the land. Nearly all the water animals, from minute insects and crustaceans to the largest fishes, ultimately depend upon them for their supply of food. For, like the land plants, these small water plants manufacture food, and the animals that live in the water must feed either on them or on other animals that get their living from the plants. Without the algæ the fish would soon disappear from our waters, because their primary food supply would be cut off. A decrease in the number of fish in a lake frequently follows the draining of its swampy margins, for the

FIG. 277. Food relations of aquatic life. No matter how long the chain of animals is up to the fish, the fundamental food organisms are the algæ that transform inorganic materials into foods.

algæ thrive best in shallow water, and it is from the algæ that the small animals on which the fish feed secure their food.

But while the algæ are a source of food for water animals, they are also a source of annoyance in reservoirs in which drinking water is stored. When they accumulate in large quantities and die, they cause the so-called "fishy taste" of water. This trouble has been to some extent controlled during recent years by the exclusion of light from small reservoirs, and by the addition of small amounts of copper sulfate to the water in large reservoirs. Copper sulfate is very poisonous to algæ, even in quantities of one part to a million parts of water. Since animals are not injured by such small amounts, the water may be used without harm for drinking purposes.

Periodicity of algæ. The fresh-water algæ show somewhat the same periodicity of development, reproduction, and dormancy that is shown by the more familiar land plants. There are six general seasonal classes that may be distinguished. There are *winter annuals*, whose spores germinate in the autumn and which increase during winter thaws by cell division and swimming spores, and whose life cycles culminate in the production of oöspores and resting spores in March and April. During the summer the vegetative plants disappear and only the spores live over in the mud.

The *spring annuals* constitute by far the largest wave of algæ. The spores germinate in autumn, winter, and early spring, and reproduction reaches its maximum in May and early June. Most of these plants disappear by July.

The *summer* and *autumn annuals* germinate in spring and have longer vegetative periods before fruiting.

There are also *perennials*, like Cladophora, that live over from one year to the next and produce spores at various times of the year. These form the long, green streamers that one sees in swift streams, on dams and waterfalls, and attached to objects in lakes.

Finally, there are the *ephemerals* — short-lived unicellular or colonial forms of the plankton and wet soils. Here a new generation may arise every few days. They reach their greatest abundance in late summer.

Algæ are more numerous in seasons when the water levels are high. They also fruit most abundantly under these conditions. The periodicity determines what species will be found associated at any time of the year. Ponds that dry up in early spring obviously can have only winter annuals, while ponds that last until June will have both the winter and spring annuals.

THE BROWN ALGÆ OR PHÆOPHYCEÆ

The brown algæ are with few exceptions marine plants. They possess in addition to chlorophyll a brown pigment which masks the green color. They attain their greatest dimensions along rocky coasts where the temperatures are low. The vegetative body, or thallus, is in many species larger and is far more complex in structure than in the green algæ. Some of the plants attain lengths of several or many feet, and internally the plants show distinct tissue systems. There are three distinct lines of develop-

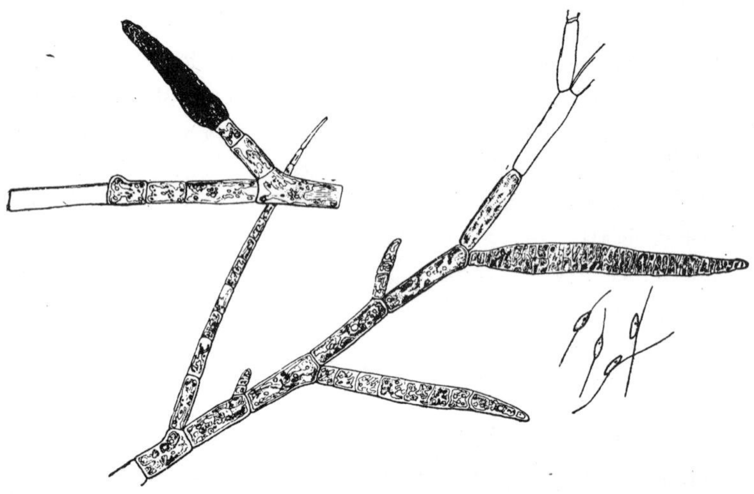

FIG. 278. A filamentous brown alga, *Ectocarpus*. On the tips of three of the branches are many-celled sporangia, which develop zoöspores. This alga is common along the Atlantic coast, growing as an epiphyte on the coarser rockweeds.

ment in this group: the filamentous forms (*Ectocarpus*), the highly branched rockweeds (*Fucus*), and the large stalked forms with flat blades (*Laminaria*). About 1000 species are known.

The filamentous forms. There are a large number of branching, filamentous forms that are not very different from some of the green algæ. These reproduce by swimming spores, and by zygotes formed from swimming gametes. The swimming spores and the gametes of the brown algæ differ from those of all other groups in having two cilia laterally placed.

The rockweeds. A second group are the rockweeds or bladder wracks (*Fucus*), which cover the rocks between tide levels. These are thick, leathery, highly branched plants with internal air sacs at intervals and with reproductive structures in the

W. S. Cooper

Fig. 279. *Postelsia*, a brown alga, on the rocky coast of Santa Cruz County, California. The plants are covered by water at high tide. Below, near the water, are other brown algæ.

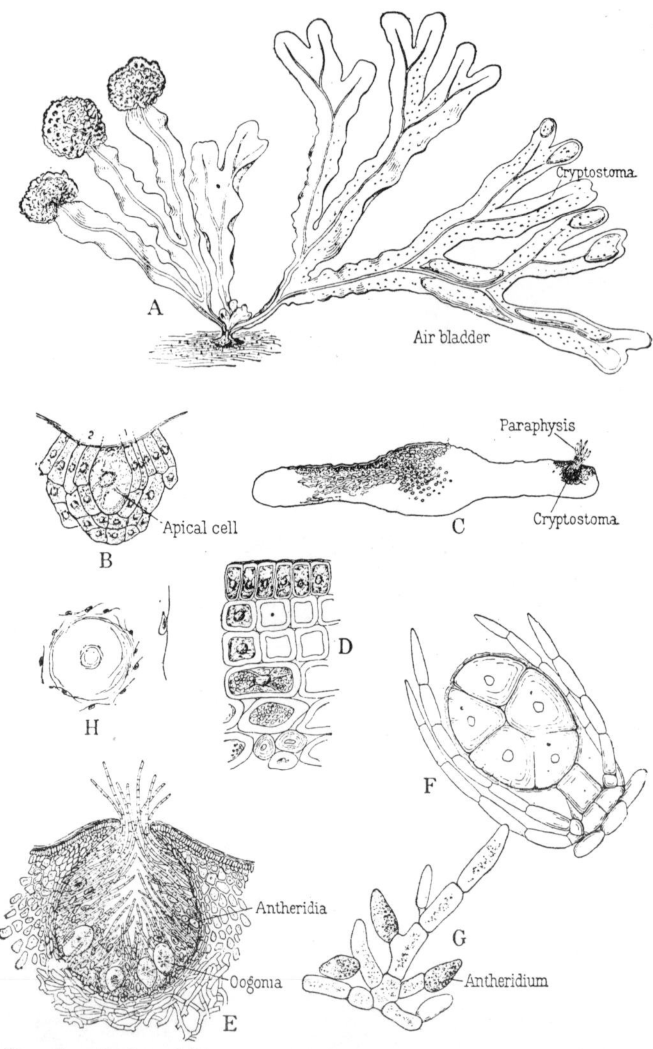

FIG. 280. Rockweed (*Fucus evanescens*): *A*, sketch of plant showing dichotomous branching, air bladders, and fruiting bodies; *B*, growing point; *C*, cross-section of thallus; *D*, enlarged view of a part of the cross-section, showing tissues; *E*, section of a conceptacle from a fruiting body, in which antheridia and oögonia are forming; *F*, oögonium with eight eggs; *G*, branch with three antheridia; *H*, egg with attached sperms, only one of which unites with the egg.

FIG. 281. Brown seaweeds, principally species of *Fucus*, *Ascophyllum*, and *Laminaria*, on the coast of Nova Scotia at low tide.

swollen ends of the branches. The reproductive organs, oögonia and antheridia, are contained within hollow depressions (conceptacles). The eight egg cells formed within each oögonium are discharged into the sea and are there fertilized by the sperms set free from the antheridia. The oöspores germinate at once, and from them the leathery plants develop. The rockweeds do not produce swimming spores.

In tropical waters species of Sargassum or gulf weed that are related to Fucus are abundant. These forms are remarkable for their resemblance to seed plants with leaves and berries. The berry-like bodies are filled with air and aid in flotation. When torn from their native rocks in the Caribbean, these algæ drift to all parts of the North Atlantic.

The kelps. The third line of development is represented by the kelps, which vary from forms a few feet in length (*Laminaria*), with a root-like holdfast, a stalk, and a large, leaf-like blade, to forms in which the stalk is terminated above by a float and several branches each with one or more large blades. Here belongs the *Nereocystis* of our own northwest coast, and the

Macrocystis, which is best developed on the west coast of South America. The former attains lengths of 10 to 30 feet, and the

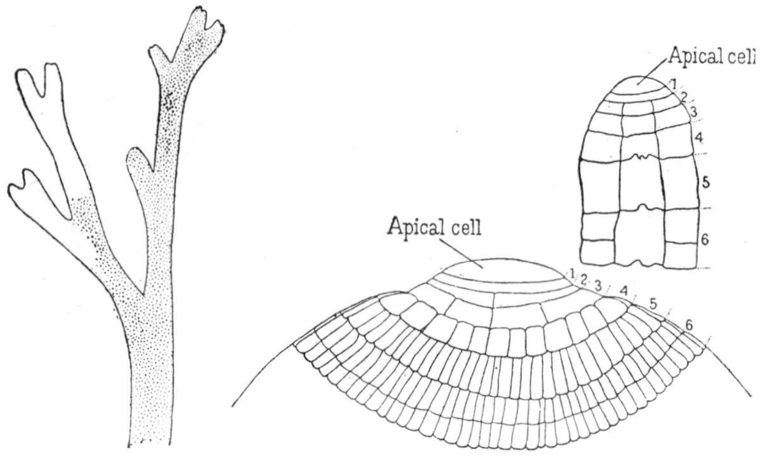

FIG. 282. Thallus of *Dictyota*, a brown alga. At the right is a growing point sectioned parallel and perpendicular to the flat surface to show the regularity of the cell division.

latter, growing in 200 feet of water, may reach a total length of 500 feet. These large, leathery plants produce swimming spores which germinate and produce a small filamentous, or single-celled, generation. These in turn produce antheridia and oögonia, and sperms and eggs. After fertilization the resulting oöspore develops into the large, leathery generation.

Here, then, are two distinct generations — one a large food-manufacturing plant which is called a *sporophyte* (spore-plant), because it develops spores; the other a small, or microscopic, generation, the *gametophyte* (gamete-plant), which ends in the production of gametes.

Economic importance of the brown algæ. In China, Japan, and along the northern coasts various brown algæ are cooked with fish and used as food. Japan exports many tons of dried kelps to China.

The kelp beds of our own western coast have during recent

years been used as a source of potassium salts and of other chemicals. It has been estimated that they are capable of furnishing

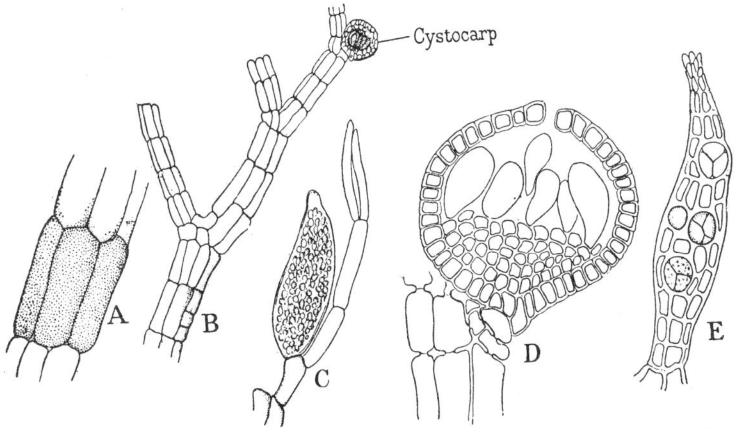

Fig. 283. Parts of a red alga, *Polysiphonia*, showing vegetative branch (*A*, *B*) and reproductive structures; *C*, antheridium; *D*, cystocarp and carpospores; *E*, branch forming tetraspores.

all the potassium needed for agricultural fertilizers in the United States, and more iodine than we now annually use. In Europe and Asia the kelps were formerly the chief source of iodine.

RED ALGÆ OR RHODOPHYCEÆ

The red algæ, noted for their beautiful colorings and graceful forms, reach their greatest development in the warm temperate and tropical seas. Many species occur in shallow water, but some likewise grow at great depths. The red pigment found in the cells with the chlorophyll aids in photosynthesis in deep water. A few genera occur in fresh water. More than 3000 species are known.

The red algæ are usually filamentous and highly branched; sometimes they are irregular blades, or have slender stalks with leaf-like branches. Among the red algæ are a number of families that deposit calcium carbonate about them. These are the

Corallines, which are often associated with the true corals on the coral reefs of the tropics, and a few species of which extend into cold waters.

The spores and sperms of the red algæ are without cilia and are not motile. The egg cell is inclosed and stationary. Their methods of reproduction and their life histories are highly complicated and cannot be detailed here.

Economic importance. The common "dulse" and Irish moss of northern coasts are used in the production of blancmange and jellies. In Asia several species of red algæ, together with a few species of brown algæ, are used in the making of agar, which is in composition a complex of gelatinous carbohydrates. In Japan the red algæ are not only collected, dried, and used as food in enormous quantities, but the algæ are actually cultivated in shallow arms of the sea. The edible bird's-nest of the Orient is constructed of seaweeds.

REFERENCES

COLLINS, F. S. *Green Algæ of North America* (3 Supplements). Tufts College Library, Medford, Massachusetts; 1918.

TILDEN, J. E. *Myxophyceæ of North America.* Minnesota Biological Survey; 1910.

WARD, H. B., and WHIPPLE, G. C. *Freshwater Biology.* John Wiley & Sons, Inc., New York; 1918.

WEST, G. S. *Algæ.* G. P. Putnam's Sons, New York; 1916.

—— *Treatise on British Freshwater Algæ.* G. P. Putnam's Sons, New York; 1904.

CHAPTER FORTY-THREE

BRYOPHYTES: LIVERWORTS AND MOSSES

THE phylum *Bryophyta* includes two diversified kinds of plants commonly known as the *liverworts* and *mosses*. Their structures and life histories are somewhat more complicated than those of the algæ. All together they comprise some 16,000 species, three fourths of which are mosses.

The largest of the mosses and liverworts never attain a height or length of more than a few inches, and they are of very simple structure in comparison with the flowering plants. In contrast with the algæ, which on the whole are water plants, mosses and liverworts for the most part live on land. The passing of plants from a water to a land habitat is one of the notable steps in the evolution of the plant kingdom, and in connection with the study of this group we shall contrast the environments of land and water plants and consider the modifications in structure that accompany the passing of simple plants from the water to a land habitat.

Living conditions of land and water plants contrasted. In the preceding chapter attention was called to the conditions under which the algæ grow. The water environment is most favorable for the growth of simple plants, because of (1) the avoidance of heating and drying effects of intense sunshine, (2) abundant supply of carbon dioxide, oxygen, and mineral salts, (3) more uniform temperature, and (4) longer growing season.

The environment of the land plant, on the other hand, furnishes through wet cell walls a supply of carbon dioxide and oxygen from the atmosphere, and mineral salts may be secured only from the soil water with which the plants are in contact. If the plant grows in full sunlight, it is subjected to much more intense illumination and heating than are water plants, and it must withstand the drying effects of the air. A study of the amphibious liverworts shows that they have become adjusted only to a medium light and a moderate amount of drying. These plants, therefore, grow in moist, shaded situations. During wet periods many individuals start in other places, only to be killed off later by the light and its secondary temperature and drought effects. The shaded situation where the water is near the surface of the soil is evidently the habitat where these plants suffer the least, and this explains why liverworts persist in moist situations and not in the open.

Responses of plants to the aërial environment. In contrast to the algæ the land liverworts show several changes in structure that are of advantage to plants in an aërial habitat. The more important of them are:

(1) *Firmer, and in some cases thicker, cell walls and water-storage tissue.* The firmer cell walls are less permeable to water and reduce the rate of water loss. Furthermore, the plants grow flat on the soil in contact with the water supply, and some of the forms develop layers of water-storage cells and mucilage pockets on the side in contact with the soil. This enables them to withstand short dry periods better than do those forms that have only the usual aquatic type of cell wall.

(2) *The development of rhizoids.* Land plants are favored by being anchored, and by having structures that will bring them into contact with the soil-water supply. In the liverworts, rhizoids anchor the plant and to some extent absorb water and mineral salts from the soil. Rhizoids are elongated cells that develop on the lower side of the plant body and penetrate the soil. They resemble root hairs in form.

(3) *The development of an epidermis.* The land liverworts are covered by an epidermis which decreases the rate of water loss. The liverworts with thicker bodies have pores in the epidermis which afford a ready access to the carbon dioxide and oxygen necessary for photosynthesis and respiration. In the more complex liverworts the epidermis is raised like a transparent roof on ridges of supporting tissues, leaving beneath it a series of small air chambers in which the chlorophyll-bearing cells stand

up in short chains. Each chamber is connected with the air by epidermis, but the epidermal pores in them are chimney-like

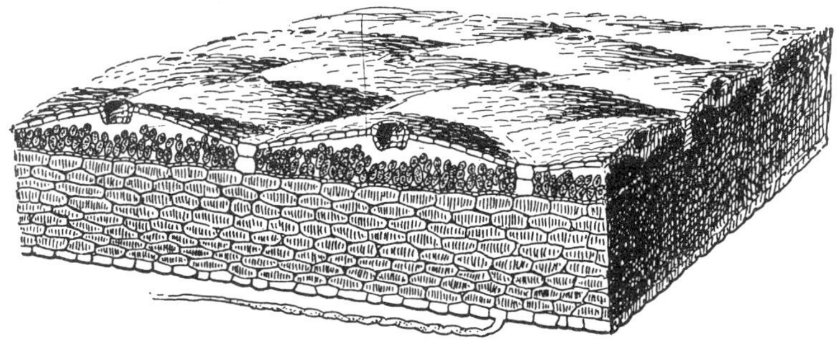

FIG. 284. Diagram of a small portion of a *Marchantia* thallus showing, above, the upper epidermis with chimney-like openings and the air cavities containing the chlorenchyma. Below are the water-containing tissue and the lower epidermis with a single rhizoid.

openings and are incapable of closing as do the stomata of the higher plants. The presence of a distinct epidermis having pores is a third feature of plants which improves their chances of living on land.

(4) *The ability to withstand drying.* When the vegetative cells of water plants are dried, the protoplasm dies at once; but a few of the liverworts, like many mosses and like Protococcus and a few other algæ, do not die when water is lost from the cells. Just what quality the protoplasm possesses that enables it to withstand drying, it is at present impossible to say; but some of the liverworts that grow on trees and rocks possess this quality, and certain mosses have to a remarkable degree the ability to withstand drying. A fourth factor which enables some plants to live in the land environment is the ability to withstand drying.

(5) *The production of light spores.* At some point in the life cycle of most land plants, spores of small size and light weight are produced. The food contained in these spores is largely oil — the lightest form in which a given amount of energy may be stored. Spores of this kind are readily carried scores of miles by

the wind. The development of light-weight spores is a fifth important characteristic of plants fitted to the land environment.

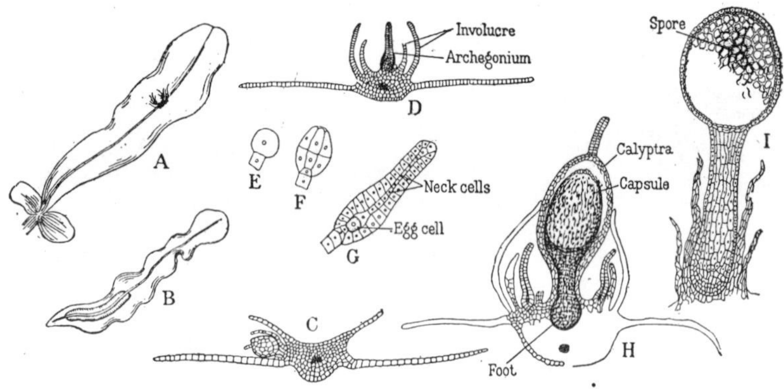

FIG. 285. Figures showing the life history of a liverwort (*Pallavicinia*): *A*, *B*, archegonial and antheridial thalli; *C*, cross-section of thallus, showing antheridium; *D*, cross-section of thallus, showing archegonium; *E*, *F*, *G*, stages in development of archegonium; *H*, embryonic sporophyte within the greatly enlarged archegonium wall (calyptra); *I*, mature sporophyte with spores.

LIVERWORTS

The body of many liverworts is flat and leaf-like, and is called a *thallus* (plural, *thalli*). It may be from one to several cell layers in thickness. Growth takes place by repeated divisions of a single cell at the tip. The thalli branch at intervals by forking. Liverworts do not stand erect, but usually have their thalli in close contact with the substrata on which they grow. In most forms the thallus is a continuous plate of cells, but some forms have prostrate stems with small leaves on either side. Even the thalloid forms like *Marchantia* have scales on the under surface. All the forms have small, hair-like rhizoids that anchor the plant and absorb water and minerals. They reproduce by spores, produced either directly on the thallus or on special reproductive branches. In some liverworts there are produced also special bodies called *gemmæ* (singular, *gemma*), which propagate the plants vegetatively.

The 4000 species of liverworts are widely distributed but are most numerous in the tropics. Liverworts may be found along streams, on overhanging rocks, on shaded moist soil, and on trunks of trees. In the tropics they often occur as epiphytes on the stems and leaves of trees.

The liverworts are probably descended from plants like the green algæ; for it is thought that the simplest plants existed first and that plant life (as well as animal life) had its origin in the water. The liverworts may be considered, therefore, as a group of simple plants that exhibit some of the evolutionary stages through which plants passed in taking up life upon the land. In this respect they can be compared to the amphibious (Greek: *amphi*, double, and *bios*, life) frogs and salamanders of the animal world.

Life history of a liverwort. The most common of the aquatic liverworts is *Ricciocarpus*, a small, heart-shaped thallus which floats on the surface of ponds and lakes. On its lower side are hair-like rhizoids and scales that aid in absorption. On its

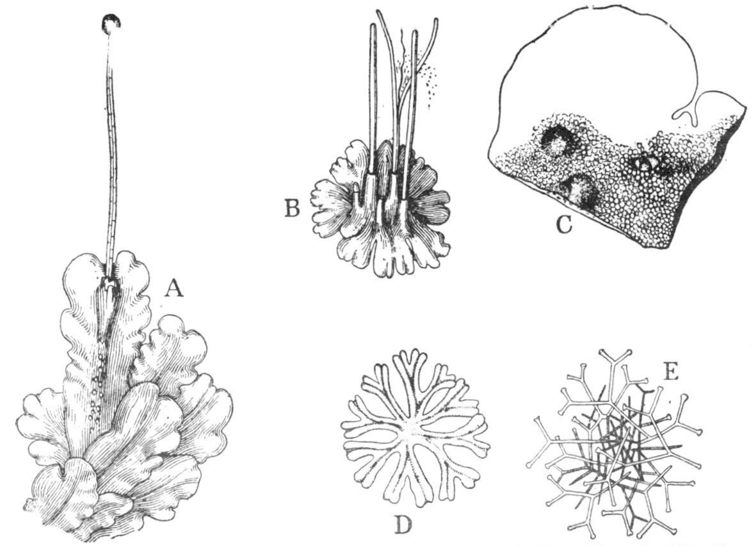

FIG. 286. Some widely distributed liverworts: *A*, *Pellia* thallus with antheridia (dots on surface) and a sporophyte; *B*, archegonial thallus of *Anthoceros* with sporophytes; *C*, antheridial thallus of the same; *D* and *E*, land and water forms of *Riccia*. (*After Velenovsky.*)

upper side are two divergent grooves in which antheridia and *archegonia* (singular, *archegonium*[1]) are formed. The antheridia produce the sperms. The archegonium is a flask-shaped organ which contains the egg. Fertilization is effected by the small sperm swimming to the archegonium when the thallus is wet, passing down the neck of the archegonium and fusing with the egg. The oöspore, or fertilized egg, germinates directly, producing a rounded body of cells. The inner cells of this body divide, each forming four spores, while the outer layer of cells forms the sporangium wall. At maturity the sporangium wall breaks, liberating the spores.

As will become more evident when we study the ferns, this life history is made up of two distinct phases, or generations. The one producing the gametes is called the *gametophyte;* the one ending with the production of spores is the *sporophyte.* The gametophyte of all liverworts and mosses is a food-manufacturing phase, the sporophyte is a parasitic phase.

Other liverworts. *Marchantia* is a common thallose liverwort found on moist rocks and in swamps. It differs from *Ricciocarpus* mainly in having specialized branches (Fig. 287), arising from the thallus, on which the antheridia and archegonia are produced. The sporophyte also has a short stalk below the sporangium, the base (foot) of which grows downward into the tissue of the gametophyte, thus becoming a distinct absorbing organ.

Anthoceros is another thallose form, in which the sporophyte is greatly elongated, growing upward from the thallus in which the archegonium is embedded. The anthoceros sporophyte is of further interest because the sporangium wall is several cell layers in thickness and the cells contain chlorophyll. Moreover, the epidermis has guard cells and stomata.

There are many genera of leafy liverworts, and about 3000 species have been described. These forms are very abundant in

[1] The archegonium is found in the mosses, liverworts, ferns, and in gymnosperms. It is analogous to the oögonium of the algæ.

FIG. 287. The liverwort, *Marchantia*, showing capsules and various stages in the development of the antheridial and archegonial branches.

the tropics. *Porella* is a rather common example found at the bases of trees, or on rocks in moist ravines. The life histories of the leafy liverworts are very similar to that of the thallose forms.

MOSSES

The mosses form a very large group found in all parts of the world. Like the liverworts, they are most abundant in moist, partly shaded habitats. A few, however, grow on rocks and trees where they are exposed to intense light and periodic drought. When dry, they are dormant; and when wet, they carry on the usual life processes.

As a result of their methods of vegetative multiplication, mosses have the habit of growing in compact clusters. This gives them an external means of conserving water and maintaining the water balance. The dense masses of plants take up water from rains and hold it for some time like a sponge.

The plant body. Mosses usually have upright and radially symmetrical stems, though many live close to the substratum and have only horizontal or inclined stems. They possess very simple leaves, frequently only one cell layer in thickness, sometimes thicker toward the midrib. Like the liverworts, mosses have rhizoids. But the rhizoids of the liverworts are one-celled structures, while those of the mosses are branching, many-celled structures which penetrate the soil. These afford a firm anchorage for the plant and absorb a part of the water used by it. The stems of the largest mosses have elongated cells forming the central axis. These cells probably form a primitive conducting tissue.

FIG. 288. Habitat sketch of three common mosses: *Climacium* (at left), *Polytrichum* (above at right), and *Mnium*.

Mosses, therefore, show some advances over the liverworts in their upright radial stems and branching rhizoids, in the regular

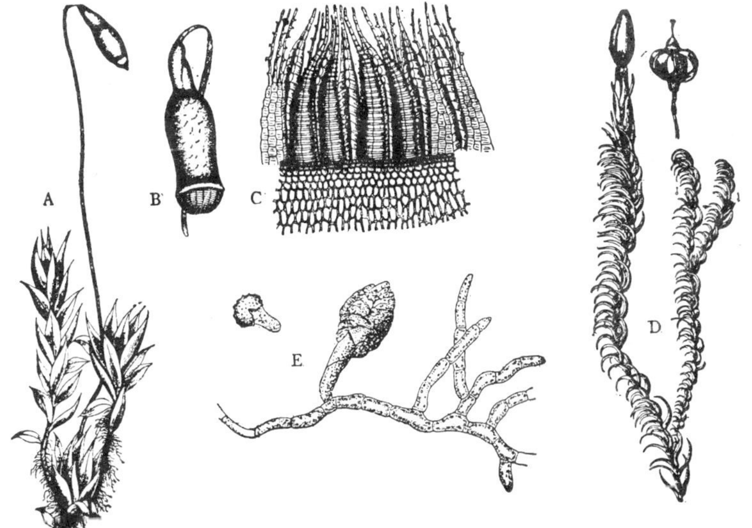

FIG. 289. Mosses: *A*, ***Bryum***, showing leafy gametophyte with attached sporophyte; *B*, sporangium enlarged, showing the peristome teeth; *C*, details of peristome teeth. *D*, *Andræa*, showing leafy gametophyte and sporophyte with two sporangia; the one at the right has shed its spores. *E*, germinating spore and protonema of a moss, showing bud from which a leafy gametophyte develops. (*After Frank.*)

occurrence of simple leaves, and in their ability to grow in drier habitats.

Life history of the moss. Mosses reproduce freely by vegetative propagation and by spores. A study of each of these methods will make clear the somewhat complicated life history of the moss plant.

Vegetative multiplication. When a moss spore germinates on the soil, it produces a branching, filamentous body, the *protonema* (Greek: *protos*, first, and *nema*, thread), which resembles some of the branching forms among the green algæ. The protonema spreads over the soil for some distance and then develops numerous buds (Fig. 289). The buds give rise to the upright leafy branches which we commonly call the moss plant. Because of the numerous buds developed on the protonema, the moss plants stand in thick clusters or masses.

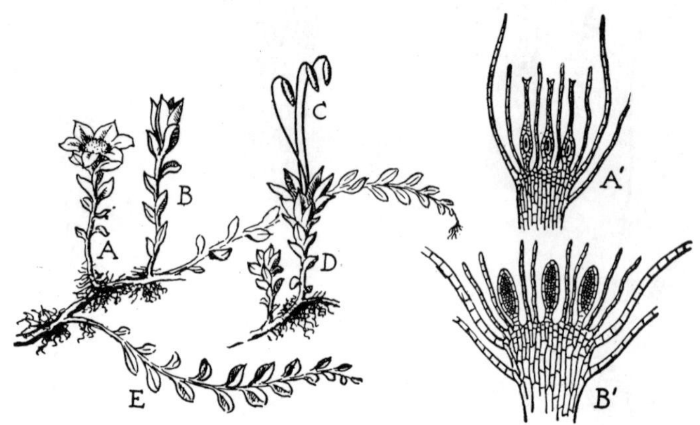

FIG. 290. A moss plant (*Mnium*). *E* is a vegetative branch, *B* a branch that produces eggs, and *A* a branch that produces sperms. After fertilization, an upright stalk bearing a spore case (*C*) develops from the egg. *A'* is a longitudinal section of a female branch, showing three egg cells in the archegonia in which they are produced; *B'* is a section of a male branch, showing three of the antheridia that produce the sperms.

The upright leafy stems of the moss also have the power of producing protonema-like branches which spread still farther over the soil, thus serving to multiply the plants and to make the plant mass denser and larger. In some mosses with horizontal or inclined stems, the stem tips when in contact with the soil develop rhizoids and give rise to new branches, much as the stems of the raspberry develop new plants. These methods of vegetative propagation are common among the mosses, and some mosses are not known to multiply in any other way. Some mosses also produce gemmæ.

Gametophyte and sexual reproduction. Archegonia and antheridia are produced on the mature upright stems of most mosses. The antheridia are many-celled structures, each of the smaller interior cells of which produces a sperm. The archegonium is a multicellular, flask-shaped body in which a single large egg is formed at the base of the neck. These

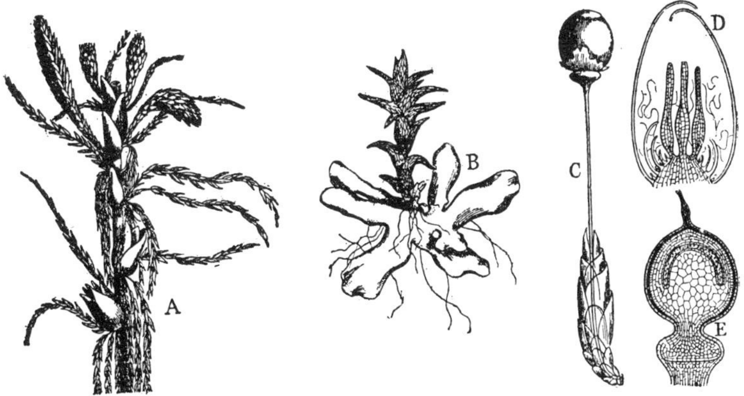

FIG. 291. Sphagnum moss: *A*, upright shoot, with antheridial branches above and two archegonial branches below; *B*, prothallus with young sporophyte; *C*, archegonial branch with mature sporophyte; *D*, archegonia within the scales of the archegonial branch; *E*, tip of archegonial branch and the attached sporophyte seen in section. The old archegonium wall still surrounds the sporangium. (*After Frank.*)

organs may be on the same branch tip or on different branches. The sperms are discharged from the antheridium by the absorption of water and consequent bursting when the moss is wet. The sperms swim about in the film of water on top of the plants. Some reach the archegonia and fertilization follows. The interior row of cells (neck-canal cells) of the archegonium (Fig. 290) disintegrate as the egg matures and form a mucilaginous mass from which sugars diffuse into the water. When this diffusing sugar reaches the swimming sperms, their direction of swimming is changed toward the diffusing sugar and in this way they swim into the archegonium and one finally reaches the egg and fuses with it. When the sperm unites with the egg, it forms an oöspore.

The protonema and the leafy branches that arise from the spore make up the gametophyte generation of the moss. Generally the gametophyte is perennial and gametes are produced each year from new branches.

Sporophyte and asexual reproduction. The oöspore germinates while still within the archegonium on top of the stem, and produces a slender, stalk-like body. The base of this body grows downward into the parent stem and draws water and nourishment from it. At the top of the stalk a sporangium, or capsule, develops which contains spores. The stalk and sporangium live parasitically on the green, leafy moss plant and constitute the sporophyte generation.

Summary. The Bryophytes probably present some of the features that characterized the first land plants. They are comparatively simple in structure, but they are more differentiated than the green algæ. They show (1) tendencies toward the development of distinct absorptive and photosynthetic tissues; (2) the presence of chloroplasts similar to those of the seed plants in both gametophyte and sporophyte; (3) the development of intercellular spaces, air pores, and (in the sporophyte) guard cells and stomata; and (4) a life cycle of two distinct phases, each producing a spore that develops the alternate generation.

The vegetative plant is the gametophyte, and it is among the Bryophytes that the gametophyte attains its greatest size and differentiation among land plants.

REFERENCES

CAMPBELL, D. H. *Structure and Development of Mosses and Ferns.* The Macmillan Company, New York; 1918.

GROUT, A. J. *Mosses with a Hand-lens.* Published by the Author, Brooklyn, New York; 1905.

—— *Mosses with Hand-lens and Microscope.* Published by the Author, Brooklyn, New York.

JENNINGS, O. E. *Manual of the Mosses of Western Pennsylvania.* Published by the Author, Pittsburgh, Pennsylvania; 1913.

CHAPTER FORTY-FOUR

THE PTERIDOPHYTES

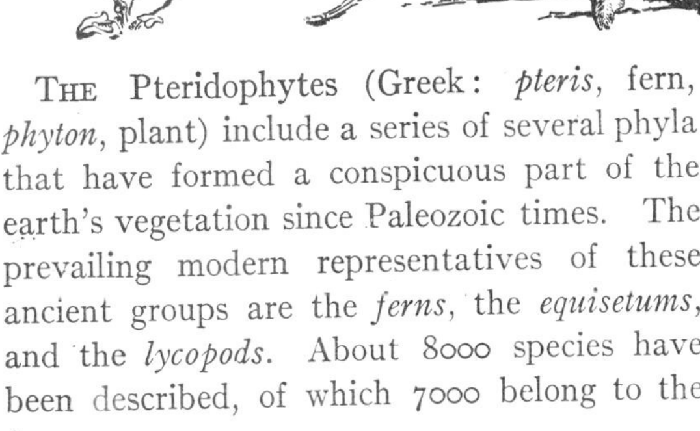

THE Pteridophytes (Greek: *pteris*, fern, *phyton*, plant) include a series of several phyla that have formed a conspicuous part of the earth's vegetation since Paleozoic times. The prevailing modern representatives of these ancient groups are the *ferns*, the *equisetums*, and the *lycopods*. About 8000 species have been described, of which 7000 belong to the ferns.

Like the Bryophytes, the Pteridophytes reproduce by spores, but in contrast to them the two generations of the life cycle are distinct plants, both living on the soil, and the conspicuous generation is the sporophyte. Furthermore, the sporophyte is differentiated into leaves, stems, and roots. Some of these plants attain large size, with stems 10 to 50 feet in height and leaves 10 to 30 feet in length. This remarkable differentiation is made possible by the presence of vascular bundles and mechanical tissues that are not very different from those of the seed plants.

FIG. 292. The walking fern (*Camptosorus rhyzophyllus*). New plants are developed from buds at the ends of the leaves.

H. N. Whitford

FIG. 293. A large tropical fern (*Marattia*), with leaves 15 feet in length.

The appearance of a vascular system in the evolution of the plant kingdom may be compared with the coming in of a backbone in the evolution of animals. There could be no large land plants raised far above the soil without efficient conductive tissues through which water and food may move rapidly. The vascular conductive system is therefore a most important adjustment to land conditions.

The Pteridophytes are at once the simplest of the land plants, with true roots, stems, and leaves, and the most highly organized plants without seeds. The origin of all these phyla is unknown, and although there have been many evolutionary developments since Paleozoic times, the distinctive features of each of the phyla are found in the oldest known fossil forms.

THE FERNS (FILICALES)

The ferns attain their greatest size, number, and variety in the moist tropical and subtropical regions. Some of the smaller epiphytic forms, the filmy ferns, have leaf blades only a few cell layers in thickness and are confined to the dripping forests of the

FIG. 294. A large tree fern in the Philippines.

FIG. 295. A roadside group of the cinnamon fern (*Osmunda cinnamomea*) growing in the shade of red maples, eastern Pennsylvania.

FIG. 296. The shield fern (*Aspidium marginale*).

rainy tropics. Most ferns are mesophytic and attain their best development in partial shade and in rich humus soils.

The ferns are readily distinguished by their divided and compound leaves. The leaves arise near the apex of the stems and uncoil as they develop. The youngest part of the leaf is the apex. The venation is also characteristic, being forked or dichotomous.

The stems of most ferns are horizontal and branched, extending either at or below the surface of the soil. The cinnamon fern of the United States has an erect stem, sometimes rising a

FIG. 297. The sensitive fern (*Onoclea sensibilis*), showing foliage leaf and sporophylls.

foot above the soil. In the tropics woody, erect stems give rise to the tree ferns, which attain an extreme height of 60 feet.

The roots of most ferns are comparatively small and less branched than the roots of seed plants. In the herbaceous ferns they arise irregularly from the sides and under surface of the rhizome. In the tree ferns the root systems are more complex, but they do not attain the size and spread of the root systems of the seed plants. A restricted water-absorbing system is one of the reasons why ferns are uncommon in dry habitats.

Ferns multiply vegetatively by their branching rootstocks. Some, like the walking fern, develop new plants at the tips of the leaves when in contact with the substratum. In some trop-

C. J. Chamberlain

FIG. 298. One of the largest known specimens of the staghorn fern, an epiphyte on trees in the tropics. The upright leaves are the photosynthetic organs; the rounded leaves pressed against the tree cover masses of roots; the pendant leaves produce the reproductive bodies. The photograph was made on a small island near Brisbane, Australia.

ical species new plants develop vegetatively from the swollen leaf bases.

The sporophyte. The familiar fern plant is the sporophyte. In many species the foliage leaves develop groups of fruiting bodies called *sori* (singular, *sorus*), on the under surfaces of the later leaves. Each sorus consists of several or many *sporangia*, and within each sporangium from 32 to 64 spores develop. In the cinnamon fern the fertile leaves differ from the foliage leaves, being reduced in size, without chlorophyll, and with the leaflets (pinnæ) acting merely as supports of sporangia. Special spore-bearing leaves are called *sporophylls*. An average fern plant in this way produces several to many million spores each season.

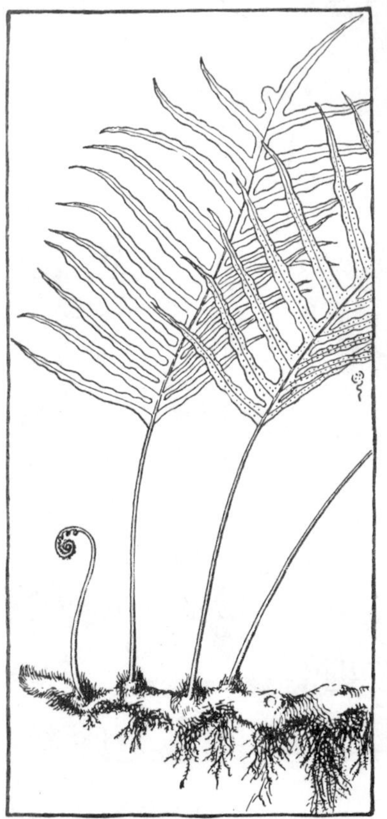

FIG. 299. Underground stem, roots, and leaves of a fern.

The gametophyte. The spores germinate either immediately, or after a dormant period. From them there develops a small, heart-shaped thallus that superficially resembles a liverwort. This is called the *prothallus*, and constitutes the gametophyte generation of the fern. Prothalli may be found commonly on moist rocks, or on the soil near fern plants.

As the flat expanse of cells forming the prothallus develops, rhizoids appear on the lower side; and soon afterward, in the vicinity of the rhizoids, antheridia develop. The antheridia are comparatively simple structures, with a wall composed of several cells, inclosing the sperm mother cells, each of which produces a sperm. The sperms have a spirally twisted body and a beak with forty or fifty long cilia. The archegonia appear as the gametophyte matures, and like the antheridia are located on the under side near the notch, or growing region, of the prothallus. They are simpler in structure than those of Bryophytes; the neck is curved, and the egg cell is embedded in the prothallus.

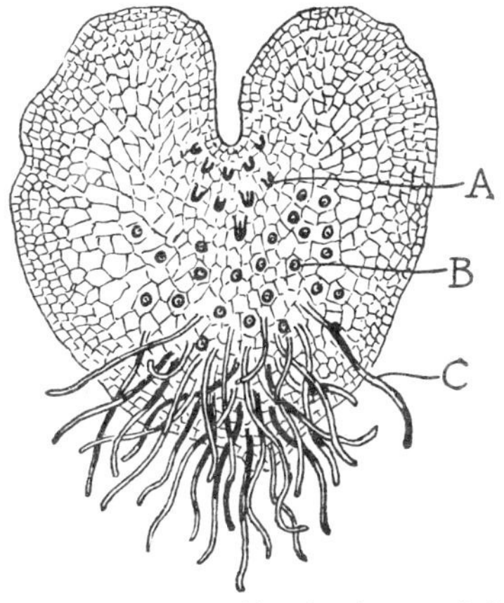

FIG. 300. Under side of a fern prothallus, showing egg-producing organs (archegonia) (*A*), the sperm-producing organs (antheridia) (*B*), and the rhizoids (*C*).

Fertilization. The sperms are released by the swelling and bursting of the antheridium, when water stands under the prothallus. Under similar conditions the archegonium opens and the products of the disintegration of the neck canal cells diffuse into the water. The sperms are directed in their swimming by these substances, and one of the sperms after entering the archegonium fuses with the egg cell, forming an oöspore. When fertilization has taken place in one of the archegonia, the further development of the remaining immature archegonia ceases. For this reason fern prothalli usually produce but a single sporophyte. The same general statement might also be made for the Bryophytes.

Embryo of sporophyte. The oöspore germinates directly after fertilization. Cell division takes place rapidly, and an embryo is soon formed that shows four general regions: (1) the foot, a holdfast and absorbing region by which the embryo is attached for a short time to the prothallus; (2) a root, which rapidly elongates and pushes into the soil; (3) a leaf of very simple structure, which soon rises above the prothallus and forms the first photo-

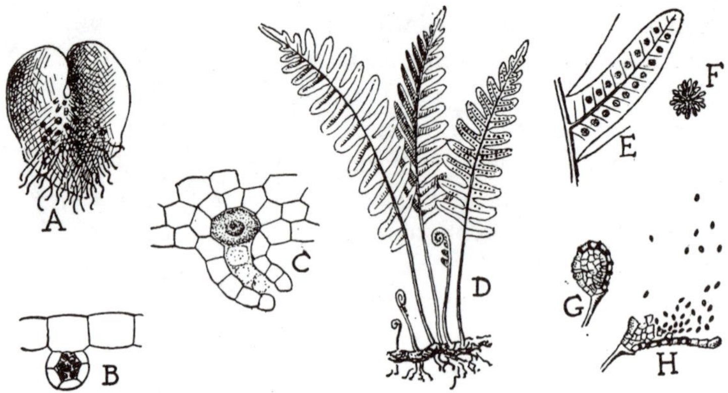

FIG. 301. The life history of a fern. The prothallus (*A*) produces egg cells and sperms in organs on the lower surface. One of the sperms set free from *B* unites with an egg cell (shown in *C*) and produces an oöspore. This germinates and produces the leafy fern plant (*D*), which in turn produces spores in sporangia (*F* and *G*) on the lower side of the leaves. By the bursting of the walls of the sporangium (*H*) the spores are set free. They then germinate on the soil (in some species on rocks or trees) and produce a new generation of prothalli like the one shown in *A*. The prothallus is here shown about four times its natural size.

synthetic organ of the sporophyte; and (4) a stem tip, which extends more slowly and gives rise to the successive leaves and adventitious roots. The sporophyte is thus at first parasitic on the gametophyte, but it soon becomes independent and the prothallus dies and disappears.

The embryo develops into the mature sporophyte, which has already been sufficiently described.

Alternation of generations. Among the algæ both the oöspores and the spores formed from vegetative cells usually reproduce the plant directly. In *Œdogonium* the oöspore, when it germinates, produces an enlarged cell (*sporangium*), in which four swimming spores are formed, and these reproduce the filamentous plant.

Among the Bryophytes the spores, formed asexually, develop

a flat thallus, or a protonema and leafy branched gametophyte. The gametophyte in turn produces oöspores which germinate *in situ*, and a simple parasitic or partially parasitic sporophyte, consisting of a sporangium, foot, and stalk, ensues. Its life terminates with the production of spores.

Among the ferns the asexually formed spores germinate on the soil and produce a prothallus. This is the gametophyte, and it in turn produces an oöspore, from which the large, leafy fern plant develops. Among all plant groups, beginning with the Bryophytes, a corresponding alternation of generations may be discerned.

Chromosome numbers. In Chapter XXXI attention was called to the importance of chromosomes as carriers of heritable qualities. The statement was made that the number of chromosomes is usually definite, and that at one step (formation of pollen and embryo sac) in the life cycle of complex plants there is a reduction division. Following the reduction division the cells have just one half the number of chromosomes. In the life cycle the last of the cells, with the reduced number, are the sperm and egg. When they unite, forming a zygote or oöspore, the number of chromosomes is restored.

The reduction division occurs in the mother cells that produce the spores. The spores, the cells of the subsequent gametophyte,[1] and the sperms and eggs have the reduced number of chromosomes (usually written n number of chromosomes). The oöspore and the sporophyte generation up to and including the spore mother cells have the $2n$ number of chromosomes.[2]

[1] It should be mentioned in this connection that the form of neither the gametophyte nor the sporophyte is determined by the number of chromosomes. These alternate generations are sometimes produced vegetatively in both the mosses and ferns. A gametophyte developed vegetatively from a sporophyte has $2n$ chromosomes; likewise a sporophyte developed vegetatively from a gametophyte has n chromosomes.

[2] In the brown alga Dictyota the chromosome numbers are 16 and 32; in the red alga Polysiphonia, 20 and 40; in the liverwort Pellia, 8 and 16; in Anthoceros, 4 and 8; in the moss Bryum, 10 and 20.

THE EQUISETUMS

The equisetums constitute a small group of about twenty-five living species that superficially bear little resemblance to the ferns. Nevertheless, their life histories are quite similar. Like the ferns they are representative of a very ancient phylum. During the Carboniferous period there were allied plants that formed extensive forests, with trunks 90 feet in height and 3 feet in diameter. The modern species are usually less than 3 feet in height, although there are two tropical species that reach a height of 10 to 15 feet and a South American species that attains a height of 40 feet when partly supported by trees.

The equisetums usually have columnar, upright, jointed stems, externally fluted and internally characterized by long, tubular air cavities. The upright stems arise as branches from underground horizontal rhizomes. The leaves are scales arranged in whorls at the nodes. As the leaves are without chlorophyll, the photosynthetic work is carried on by the chlorenchyma of the stems. In several species the upright stems bear a multitude of slender whorled branches, whose brush-like character suggests one of the common names, "horsetail." Another common name, "scouring rush," is suggested by the fact that the cell walls contain large amounts of silica and that in pioneer days the plants were used to scour metal utensils.

The roots are small and arise along the rhizomes mostly at the nodes. The plants are essentially hydrophytes and are found commonly on stream margins, swamps, and lake shores. A few of the species occur in dry situations, but are there much dwarfed.

The sporophyte. The sporophyte generation is the plant we have just described. It is a perennial. The reproductive structures consist of whorls of peculiar shield-shaped sporophylls, each bearing five to ten sporangia, that together form a terminal cone.

Within the sporangia spores arise which are peculiar in

having four long appendages, that coil around the spore when moist and uncoil when dry. The spores contain chlorophyll and do not withstand drying, and they die unless germinated within a month.

The gametophytes. In most species of equisetum the gametophytes are irregularly lobed, thalloid structures which grow on moist banks of streams. They are bisexual, producing both antheridia and archegonia. In the most-specialized species (*E. arvense*) the gametophytes are usually unisexual; that is, about one half produce antheridia only and the other half archegonia. Fertilization takes place when a swimming sperm fuses with an egg cell. An oöspore results, which germinates at once as in the ferns, and from it the sporophyte develops (Fig. 304, *C*, *D*).

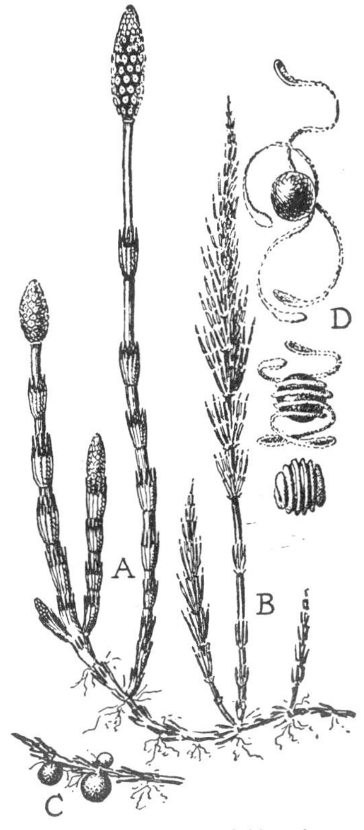

Fig. 302. The common field equisetum (*Equisetum arvense*). Rootstock with sterile branches (*B*), spore-bearing branches (*A*), and tubers (*C*). *D* shows the spores with their appendages.

THE LYCOPODS (LYCOPODIALES)

Another, and perhaps the most ancient, group of Pteridophytes includes the lycopods or club mosses. Only two genera remain, *Lycopodium* (100 species) and *Selaginella* (500 species). Both consist of scale-leafed creeping plants, often several feet in length, with upright or inclined branches. The stems branch by repeated forking, and this is true even of the fossil tree forms. In all cases the leaves are scale-like. In the more primitive species the sporangia develop in the axils of every leaf; in the more specialized types the sporangia occur in cones of modified scales, or sporophylls.

In Lycopodium the spores are all alike, and when they germinate produce green, fleshy thalloid gametophytes, or thick underground tuberous gametophytes. The subterranean gametophytes are saprophytes, and in the other species they are partial saprophytes. Each gametophyte produces both sperms

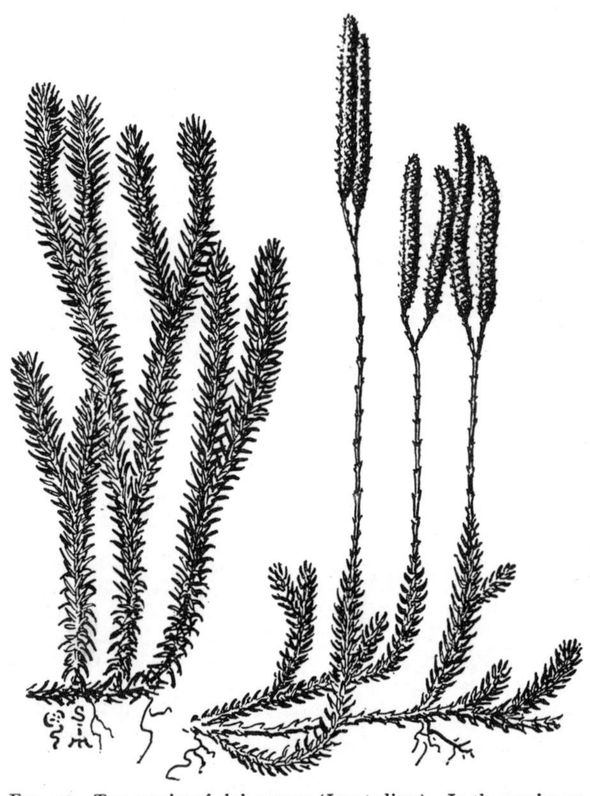

FIG. 303. Two species of club mosses (*Lycopodium*). In the species on the left (*Lycopodium lucidulum*) the sporangia are borne in the axils of the upper leaves; in the other (*Lycopodium clavatum*) they are borne in the terminal cones.

and eggs. The sperms of lycopods differ from those of other pteridophytes in being very small and in having only two cilia.

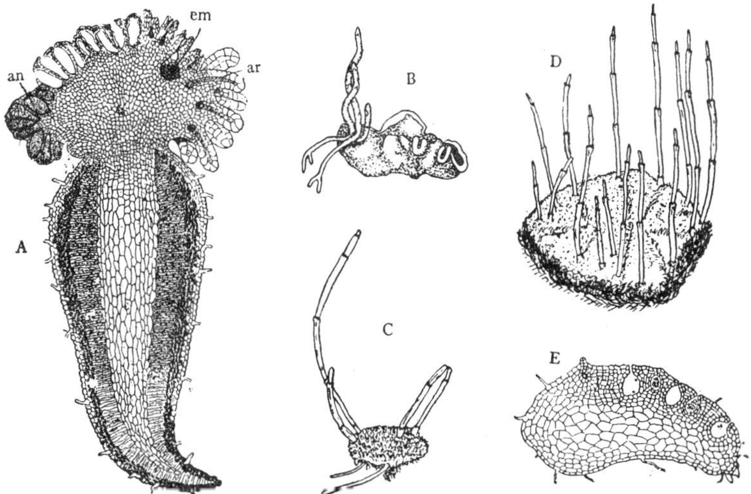

FIG. 304. Gametophytes: *A, Lycopodium complanatum,* longitudinal section showing antheridia, archegonia, and one embryo; *B, Lycopodium annotinum,* with three young sporophytes; *C, Equisetum lævigatum,* with four young sporophytes; *D, Equisetum debile; E,* the fleshy prothallus of a fernwort, *Ophioglossum vulgatum.* (*A, B, after Bruchmann; C, after Walker; D, after Kashyap; E, after Frank.*)

In Selaginella the spores are of two kinds, produced in two different kinds of sporangia. The small spores (microspores) are produced by hundreds in small sporangia (microsporangia); and four large spores (megaspores) develop in each large sporangium (megasporangium). These spores have special interest because from the microspores only antheridial gametophytes develop, and from the megaspores only archegonial gametophytes are formed (Fig. 306).

Heterospory. The occurrence of two kinds of asexually formed spores is known as *heterospory,* in contrast to *homospory,* the formation of only one kind of spores. While we have described heterospory only in the case of Selaginella, it should be mentioned that heterospory occurs among the ferns as well as among the lycopods, and has occurred among the equisetums that are now extinct.

Gametophytes. Both the male and female gametophytes are small or microscopic, being formed partly or wholly inside the spore wall. Here, then, are the most-reduced gametophytes among the Pteridophytes.

Seeds. Before leaving the Pteridophytes, attention should be called to the fact that occasionally the large spores of Selaginella germinate within the sporangium and produce a prothallus. Fertilization may take place, followed by the development of an embryo sporophyte before the megaspore leaves the megasporangium. These are rare and accidental happenings in Selaginella, but when they occur we have the same arrangement of structures that regularly occurs in the formation of the seeds in spermatophytes. A seed is the result of telescoping a gametophyte and a new sporophyte within a sporangium:

FIG. 305. *Selaginella martensii,* showing leaf-like shoots with cones at the ends of the smaller branches.

Megasporangium	+ female prothallus	+ embryo	= a seed
Part of 1st sporophyte	+ (gametophyte)	+ (2d sporophyte)	
Seed coat	+ endosperm (as in conifers)	+ embryo	

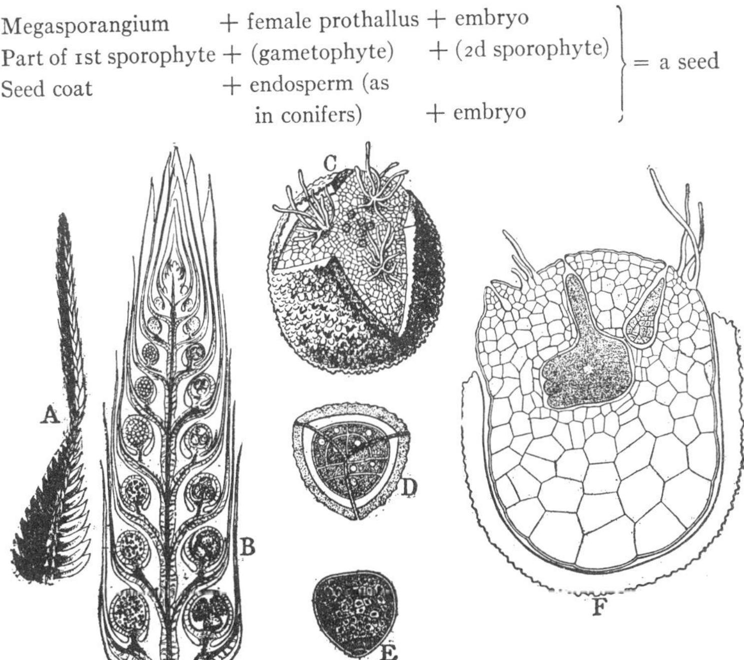

FIG. 306. *Selaginella: A*, vegetative branch with terminal cone; *B*, longitudinal section of cone, showing microsporangia on one side, megasporangia on the other; *C*, female gametophyte protruding from the megaspore wall with several archegonial openings among the rhizoids; *D*, male gametophyte within the microspore wall; *E*, male gametophyte with sperms formed in the cells; *F*, section of female gametophyte, or prothallus, after fertilization, showing two embryos. (*After Frank.*)

Summary. The occurrence of conductive tissues in the sporophyte of the Pteridophytes not only made possible the development of large land plants with roots, stems, and leaves, but it gave the sporophyte generation possibilities of evolution far beyond that of the gametophyte.

The three phyla, ferns, equisetums, and lycopods, all have independent thalloid gametophytes. In the ferns they are autophytic and bisexual; in the equisetums, autophytic and usually bisexual; and in the lycopods, partially saprophytic, sometimes bisexual (Lycopodium) and sometimes unisexual (Selaginella).

In all three phyla the production of two kinds of spores (heterospory) occurs either in the modern representatives or in the fossil forms.

Among the ferns the leaves are often extremely large and are characterized by forked venation. During development the leaves uncoil, as the growing points are in the tips of the leaves and leaflets. The leaves of the lycopods are poorly developed, being only scales, and those of the equisetums are scales devoid of chlorophyll.

The roots of Pteridophytes are usually small and scattered along the horizontal stems. In the large, upright tree types they are generally basal and sometimes of considerable size, but even then they do not compare with those of the seed plants in relative absorbing area.

All of the phyla have forms in which the spores are not produced on foliage leaves. The special spore-bearing leaves are reduced in size and in extreme forms lack chlorophyll. These leaves are termed *sporophylls*. In the equisetums and lycopods the sporophylls are arranged in cones. The extreme forms also have two kinds of sporangia and spores: microsporangia and microspores, and megasporangia and megaspores.

REFERENCES

ATKINSON, G. F. *Biology of Ferns*. The Macmillan Company, New York.

CAMPBELL, D. H. *Structure and Development of Mosses and Ferns*. The Macmillan Company, New York; 1918.

CLUTE, W. N. *Our Ferns in Their Haunts*. F. A. Stokes Company, New York; 1901.

WATERS, C. E. *Ferns*. Henry Holt & Co., New York; 1903.

CHAPTER FORTY-FIVE

FOSSIL PLANTS

When a leaf falls on soft mud, it may become imbedded in it. Later the mud may be covered by other layers of sediment. When the mud dries, a perfect imprint of the outline and veins may be left. As time goes on and the mud becomes more deeply buried, it may harden into rock and retain the imprint of the leaf as a record of a plant that lived when the rock was merely soft mud. In this way leaves, fruits, seeds, stems, and roots have left their imprints to testify, thousands and millions of years afterward, to their former existence.

Plant remains also accumulate in deep water or in water containing large amounts of mineral matter in solution. In such places they may decay very slowly, and the material of which they are composed may be gradually replaced by the mineral substances in the water. Under these favorable conditions the

T. D. A. Cockerell

Fig. 307. Fossil oak leaf from the Tertiary shales at Florissant, Colorado, and a modern oak leaf from the same region.

T. D. A. Cockerell

FIG. 308. Fossil flower from the Tertiary shales at Florissant, Colorado.

internal structures of the plant are preserved. As animal remains are preserved in the same way, we have in the rocks a record of the plants and animals of the past. These petrified plant and animal remains and the plant and animal imprints from former geological ages are called *fossils*.

Fragmentary nature of fossil record. Present-day observations on the fate of fallen leaves and of other plant organs show that they usually decay and disappear within a few months. Fossils are being formed at the present time only in lakes, in bogs, in muddy estuaries, and in a few other exceptional situations. It is only by rare chance that upland plants leave a record. We should therefore expect to find the geological record of plants very fragmentary.

The Pteridophytes and their near relatives, being plants of low grounds, swamps, and bogs, were situated in the most favorable habitats for preservation and their record is more complete, perhaps, than that of any of the groups of land plants.

The record of plant groups below the Pteridophytes is very scant for two evident reasons. The simple thalloid plants lacked hard tissues which would resist bacterial action until prints and casts were made. Furthermore, the rocks of the early Paleozoic and preceding periods have been metamorphosed by being subjected to great pressure by overlying rocks and by heat due to crushing, faulting, and warping of the earth's crust. Even though there had been a fossil record in them, it would have been erased by the changes that have occurred during the millions of years that have elapsed since the rocks were deposited.

Importance of fossils in tracing relationships. In spite of the fragmentary character of the record, hundreds of species have been found and they have been of great importance in establishing the relationships between some of the phyla of plants. Conditions during the Carboniferous period were such that plant remains are very abundant in coal seams and in the shales associated with coal deposits. The rocks of the Carboniferous and succeeding periods have not been so greatly modified, and they have accordingly yielded many fossils. Nevertheless, there is as yet very little geological evidence concerning the origin of the conifers and flowering plants. Until fossils of the ancestors of these groups are discovered, there is no satisfactory basis for explaining their origin from the seedless plants, or their relationships to the known plants of earlier ages.

FIG. 309. Fossil imprints of fern-like leaves in a rock of the Carboniferous period.

The fossil record. The diagram on the next page shows in a general way the occurrence of the larger plant groups and some of the probable relationships. The diagram shows very

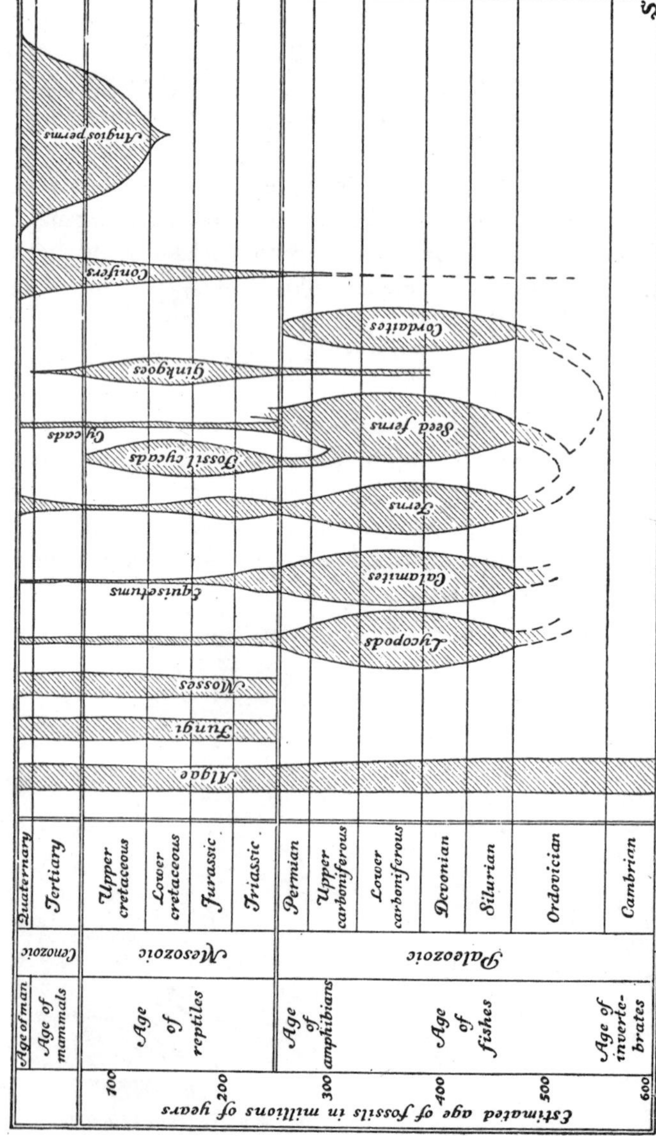

FIG. 310. Diagram of the fossil record of plants, showing the relative importance of the several groups in the successive geological periods. Read the diagram from below upward.

clearly that the origin of most of the groups is unknown. It also shows that most of the great groups have had a very long history on the earth. Furthermore, there is evidence of progressive changes in structure as we follow the plants in any one group from the earliest records to the present time. One of the large plant groups of the Carboniferous, the cordaites, became extinct. This also has been the fate of many smaller groups not shown on the diagram. These groups disappeared because their structures were not suited to the changed environments of later times.

The dominant animal groups of the several periods are indicated and also the time roughly estimated to have been necessary for the deposition and consolidation of the rocks belonging to each period.

Later Paleozoic forests. During the latter half of the Paleozoic there were five great groups of plants that dominated the vegetation. The lycopods were represented by large tree forms with stems that showed secondary thickening, with scale, or lance-shaped leaves, and spores produced on large sporophylls arranged in cones. The ancestral forms of the equisetums are the calamites (*Calamariales*), with tall, straight, hollow-jointed stems with whorls of branches bearing slender simple or forked leaves, and cones. While abundant in the coal measures, they contributed little material to the coal itself. The calamites seem to have attained in some instances a height of 90 feet, but most of the forms were smaller and with the ferns formed a conspicuous part of the vegetation of open places in the lycopod forests.

The seed-ferns (*Pteridospermophyta* or *Cycadofilicales*) include several families of plants with fern-like leaves and stems but which produced a simple type of seed. Many of the leaf imprints of the coal measures that were formerly classed as ferns belong to this group. From the early Paleozoic to Permian there is a gradual increase in the complexity of the stem structure, in the direction of the cycads. This group became extinct about the close of the Paleozoic.

A fifth group of plants that contributed to the forests of the Carboniferous is the family of cordaites (*Cordaitales*). These were much-branched trees, sometimes a hundred feet in height, with dense foliage of parallel-veined narrow, simple leaves. They had cones of two types in which the small and large spores were developed, much as they are formed in the cycads. In the large cones nut-like seeds were produced. The wood of cordaites bears a striking resemblance to the wood of some of the living conifers.

In the vegetation of the later Paleozoic, then, there were forms that combined in various ways the special characteristics of the ferns, the lycopods, the equisetums, and such seed plants as the cycads and conifers. The development of complex vascular systems, of stems with wood and cambiums, and the development of seeds were the great advances made during the Paleozoic.

The Paleozoic closed with an uplift of the continents, and consequent increase of land areas, and increased drought. There is also evidence of glaciation during the Permian. These were doubtless important factors in the extinction of many forms of Paleozoic plants.

The vegetation of the Mesozoic. The Mesozoic era was marked by the extinction of the cordaites and seed ferns and the reduction of the lycopods and equisetums to herbaceous remnants that are of slight importance in the vegetation. The ferns continued their existence as forest undergrowth, but were early forced to compete with a new group of seed plants, the "fossil cycads" (*Bennettitales*). These plants had compound, fern-like leaves and usually short, thick, woody trunks like those of modern cycads. In most forms the reproductive structures consist of a whorl of microsporophylls surrounding a central cone-shaped body bearing the megasporangia. In the extreme forms the inflorescence is highly suggestive of certain angiosperm flowers. The forests of the Mesozoic were dominated by the

FIG. 311. Map showing the glaciation of the Wisconsin epoch, and the probable distribution of the forests when the ice extended farthest south.

ancestors of cur modern conifers. The Cretaceous period is second only to the Carboniferous as a coal-making period.

The great event of the later Mesozoic era was the appearance of many types of angiosperms. The rocks of the Upper Cretaceous contain an abundance of fossils of broad-leafed plants such as oak, willow, beech, maple, tulip, sassafras, and palm. The sudden appearance of so great a diversity of forms shows that as a group they must have diverged from the other fossil groups a long time previously. Thus far very few of the pre-Cretaceous ancestors of the angiosperms have been discovered.

The Tertiary vegetation. During the Tertiary the forests were dominated by angiosperms and conifers, much like the forests of today. It was during the Tertiary that there came a gradual lowering of temperature on the earth and the differentiation of distinct torrid, temperate, and frigid zones, replacing the previous uniformly mild temperatures of the Cretaceous. This lowering of temperatures culminated in the Glacial period, which closed the Tertiary.

REFERENCE

Seward, A. C. *Fossil Plants* (4 vols.). G. P. Putnam's Sons, New York; 1917.

CHAPTER FORTY-SIX

GYMNOSPERMS: THE CYCADS

THE term *gymnosperm* (naked seed) is applied to those plants whose seeds are attached to a sporophyll, but are not inclosed in an ovulary. Attention was called in an earlier chapter (page 492) to the fact that a Selaginella may occasionally form a structure which could not be excluded from any definition of a seed. Furthermore, it is definitely known that some of the ancestral lycopods attained the seed habit in Paleozoic times. The cordaites formed a second group of Paleozoic seed plants. At the same time there were plants, Pteridosperms, so closely resembling ferns that, until the rather recent discovery of seeds attached to their leaves, they were classified as ferns. From the seed ferns came the Mesozoic cycads (*Bennettitales*), having thick tuberous stems with a crown of foliage leaves superficially like some of the modern cycads. All these forms were gymnosperms, and attention is directed to them again merely to emphasize the fact that the record of the transition from Pteridophytes to gymnosperms is remarkably complete, and that seeds arose in several quite independent phyla of plants.

The cycads. Of the living gymnosperms the most primitive are the cycads. This interesting group of seed plants with fern-like leaves and stems and many other characteristics reminiscent of their fern-like ancestors is practically confined to tropical and subtropical regions. There are about 100 species belonging to nine genera, of which five occur only in the eastern hemisphere and four only in the western. Specimens of several species are common in conservatories, among them the "sago palm" (*Cycas revoluta*). The graceful, rigid leaves of this species are frequently seen in floral decorations and on Palm Sunday.

The cycad sporophyte. The cycad stem is either an underground erect tuberous body, or a columnar trunk 5 to 60 feet in height. The columnar stems are covered with an armor of old leaf bases like that of certain tree ferns and of the fossil cycads.

C. J. Chamberlain

FIG. 312. South African cycads (*Encephalartos*), showing characteristic leaning trunks. These specimens are probably 500 years old.

The root system consists of a long tap root and usually some basal adventitious roots. This root system is a distinct advance over that of the ferns. The stem is surmounted by a crown of leaves that is renewed by the growth of the terminal bud at intervals of from 1 to 3 years. The pinnate leaves, like those of the ferns, uncoil during their development.

Cycads produce the two kinds of spores on separate plants. The sporangia are borne on spirally arranged sporophylls that are aggregated into cones. The microspores or, as we are accustomed to call them in the seed plants, pollen grains are produced in large numbers, in sporangia scattered over the under surface of the microsporophylls or stamens. We may therefore call this aggregate of microsporophylls the *staminate cone.*

The *ovulate cones* consist of aggregates of megasporophylls, each of which bears from two to eight megasporangia (ovules) on its lower margins. In the more primitive species (*Cycas*) the ovulate sporophylls are divided, resembling greatly reduced leaves; in the most specialized genus (*Zamia*) the sporophylls are scale-like. The ovulate cone thus varies from loosely aggregated, leaf-like sporophylls each with several ovules, to tight cones of scale leaves each bearing two ovules.

The gametophyte generation. The mature microspore, or pollen grain, consists of three cells, one of which forms the sperms. The pollen is carried by the wind, and by chance some reaches the open end and pollen chamber of the ovule. There the pollen germinates, developing pollen tubes that grow into the nucellus and absorb food from the adjacent cells (Fig. 314). After several months the pollen chamber has been enlarged by the breaking down of the cells of the nucellus, and the

C. J. Chamberlain

FIG. 313. Large cycads (*Dioön*) in southern Mexico, showing staminate cones on plants at left and carpellate cone at right.

pollen tubes have enlarged downward into this cavity. Finally two free-swimming sperms are liberated from each tube into the

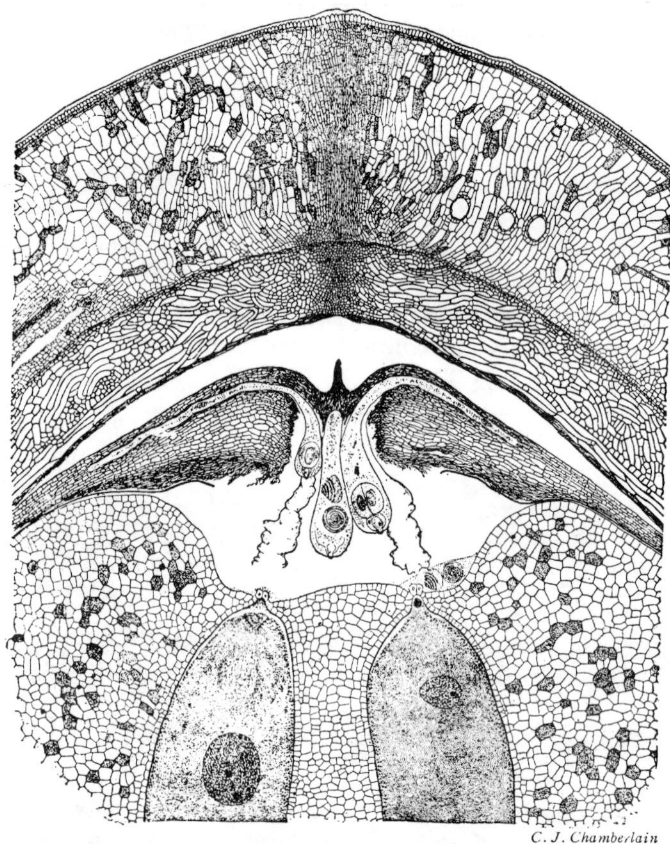

FIG. 314. Fertilization in a cycad (*Dioön*), showing pollen tubes in the nucellus; the spirally ciliated sperms; and two archegonia, one (right) with egg, the other (left) after union of sperm and egg nucleus. Surrounding the archegonia is the female prothallus.

pollen chamber. The sperms are relatively very large (.2 to .3 mm.) and are propelled by a much-coiled spiral line of cilia (Fig. 314). The pollen chamber at this time contains the liquid discharged from the pollen tubes, and in this liquid the sperms swim about.

In the cycads, then, the male gametophyte is a pollen tube. As in the flowering plants, it is parasitic. It is unique, however, in its manner of growth and in producing motile sperms — a habit that has been carried along through all the Bryophytes and Pteridophytes from the swimming sperms of the algæ.

The female gametophyte. A single megaspore is formed within each ovule or megasporangium. The megaspore germinates inside the ovule, and utilizing the food in the inner soft tissue (nucellus) of the ovule ultimately fills most of the space inside the hard wall of the ovule. This is the female gametophyte. Like the female gametophyte of Selaginella, it develops within the megasporangium, but in this case the gametophyte is entirely shut away from the light. Like the pollen tube, it is wholly parasitic.

The female gametophyte at maturity organizes several archegonia that consist merely of two small neck cells and the very large egg cell. The neck cells open into the pollen chamber at the time of the liberation of the sperms from the pollen tube.

Fertilization. A sperm moves down between the neck cells and enters the egg. The sperm nucleus slips out of its covering of cytoplasm and cilia and unites with the egg nucleus. The fertilized egg is the first cell of the new sporophyte generation. It soon begins to divide and ultimately forms the embryo. The embryo pushes back into the gametophyte tissue until it occupies the whole longitudinal axis of the seed. The embryo has two cotyledons and develops the growing points of the stem and root.

The seed. The seed at maturity has an outer soft fleshy layer surrounding a stony layer. Within these seed coats is a membranous tissue, the remains of the nucellus and inner fleshy wall of the sporangium. Next inside is food-containing tissue (usually termed *endosperm*), the remnant of the female gametophyte. The gametophyte incloses the embryo or young sporophyte.

REFERENCE

CHAMBERLAIN, C. J. *The Living Cycads.* University of Chicago Press, Chicago

CHAPTER FORTY-SEVEN

GYMNOSPERMS: THE CONIFERS

THE most important of living gymnosperms are the conifers. They comprise about 350 species generally distributed from the subtropics to the polar limits of tree growth. In North America and Eurasia, pine, spruce, fir, hemlock, cypress, larch, juniper, cedar, and sequoia cover the larger part of the forested areas. In the southern hemisphere the araucarians and podocarps also form extensive conifer forests.

Among the conifers are the largest and oldest of living plants. They have a deep and wide-spreading root system, an efficient water- and food-conductive system, much-branched stems, and a larger leaf display than the Pteridophytes and cycads. Consequently they grow far more rapidly, and are less restricted to particular habitats. Many of the conifers are traversed throughout by resin ducts.

The conifers have scale-covered buds, and are able to withstand droughts and the low temperatures of winter. With the exception of the larch and bald cypress, the leaves remain on the trees from 3 to 10 years. Because of the strong terminal buds, they usually form a large excurrent trunk with many small horizontal branches, and the trees become conical in form. Some species, however, after attaining their height growth, become ovoid through the lengthening of their upper lateral branches.

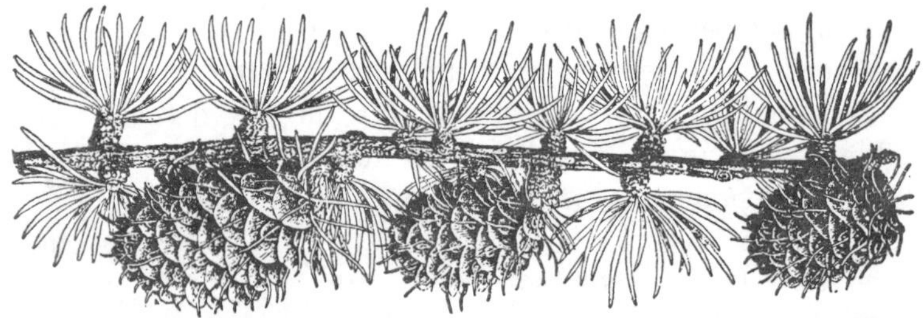

U. S. Forest Service

FIG. 315. Long branch of Western larch (*Larix occidentalis*), showing lateral dwarf branches with clusters of leaves and mature cones.

There are two distinct types of branches: those which increase from year to year, the *long branches*, and those which grow only a

U. S. Forest Service

FIG. 316. Branch of Douglas fir, showing ovulate cone and seeds.

fraction of an inch, the *dwarf branches*. From the latter the foliage leaves develop. A young growing stem of the pine is at first clothed with spirally arranged brown scales. These fall off as the stem elongates, and from the axil of each scale a spur branch develops crowned by one or several needle leaves.

The leaves of most conifers are needle-shaped, but in some families the leaves are reduced to scales and in others they are quite broad. The life history of the pine will be used to exemplify that of the group.

The pine sporophyte. The vegetative sporophyte consists of the root system, the stem and its branches, and the scales and needle leaves. The stem, in cross-section, is quite similar to that of a woody dicot. It differs chiefly in the absence of true vessels or tracheæ.

Two kinds of *cones*, the *staminate* and *ovulate*, are produced on the same tree. In them are formed the microspores and megaspores, within which the male and female gametophytes develop.

The staminate cone. A cluster of staminate cones develops in the spring at the *base* of the new stem segment. These cones are small, short-lived structures, falling from the tree as soon as the pollen is shed. Each cone is made up of yellow, membranous sporophylls, each bearing two microsporangia (pollen sacs) on its lower face. The pollen, or microspore, at first consists of a single cell, but before it is shed cell division occurs and the mature pollen grain consists of four cells. Two of these cells soon degenerate. The third cell is called the *generative cell;* and the fourth, which occupies most of the pollen grain, is called the *tube cell.* The outer wall of the pollen grain also enlarges and separates from the inner wall, forming on either side of the living cells two miniature balloons which help support the grain in the air. The four cells within the pollen represent the remnants of a male gametophyte.

In late spring the pollen sacs break open and the pollen is blown about by the wind. The amount of pollen produced by a pine forest is enormous, and when scattered may give the soil and all near-by objects a yellow tinge as though powdered sulfur had been sprinkled about.

The ovulate cone. The megasporophylls are at first small, green, fleshy scales, but ultimately they enlarge and become woody. They develop on small lateral branches near the *upper* end of the year's growth segment. There are usually two or three of these ovulate cones formed near each other.

Each sporophyll has two megasporangia or ovules on its upper

FIG. 317. Shoot of *Pinus densiflora*, with one-, two-, and three-year carpellate cones and a few of the staminate cones.

surface (Fig. 321). The ovule consists of an outer *integument* (sporangium wall), inclosing an oval body of tissue, the *nucellus*. At the inner end there is an opening in the integument, the *mi-*

FIGS. 318 and 319. At the left, staminate cones of *Pinus rigida* clustered about the bases of the new shoots. At the right, one-, two-, and three-year-old cones of *Pinus pungens.*

cropyle. Within the nucellus four megaspores are formed. Three of these degenerate as the fourth enlarges. The production of the megaspore ends the sporophyte generation.

The megaspore germinates the following spring and forms within the nucellus a mass of tissue, the female gametophyte, at the expense of the food contained in the surrounding cells. The female gametophyte grows during the spring and by June has organized several archegonia just beneath the micropyle. Each archegonium consists of a large *egg cell* and several very small *neck cells.*

Pollination. At the time the pollen is shed the axis of the ovulate cone elongates, separating the sporophylls. Some of the pollen grains drift in between the sporophylls and become lodged near the micropyle. These are filled with a sticky fluid at the time, and the pollen grains adhere to it. As the fluid subsequently dries, the pollen grains are drawn within the micropyle.

Pollination occurs in May or June, about the same time that the megaspore is being organized within the nucellus. The pollen grains, now in contact with the nucellus, begin developing pollen tubes into the nucellus. Elongation is very slow, and it is not until the following June, or early July, that the tubes pass entirely through the nucellus. Meanwhile two sperm nuclei have been formed from the generative cell. This occurs about the time that the archegonia are formed by the female gametophyte.

Fertilization and growth of the embryo. When a pollen tube passes between the neck cells and reaches the egg, the sperm

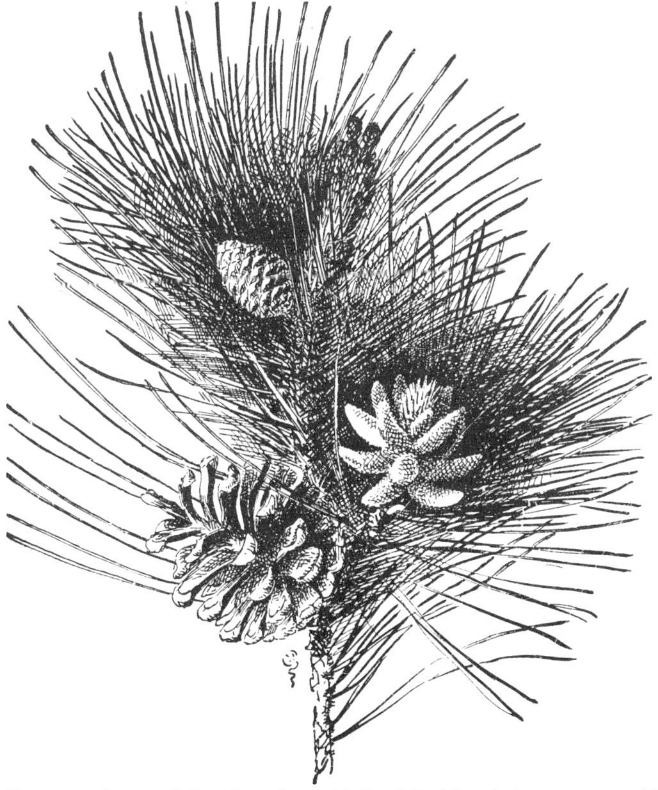

FIG. 320. Spray of Austrian pine. At the left (above) is a one-year-old ovulate cone and (below) a two-year-old ovulate cone. On the right is a cluster of staminate cones.

nuclei are discharged into the egg. One of the nuclei unites with the egg nucleus; the other breaks down and disappears.

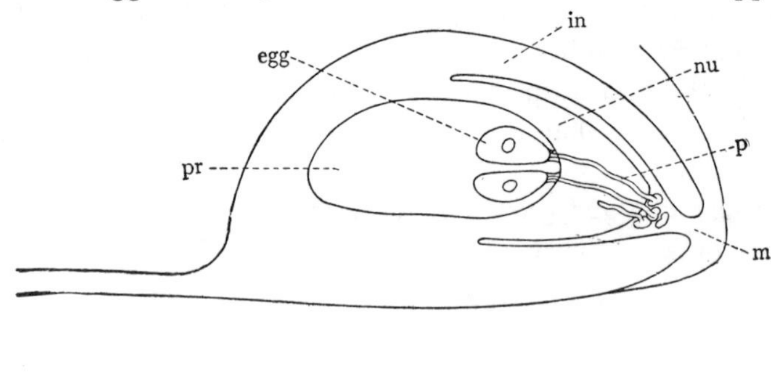

FIG. 321. Sketch of a vertical section of a pine ovule and the scale to which it is attached, showing male and female gametophytes at the time of fertilization: *pr* is the prothallus with two archegonia; *in* is the integument; *nu* is the nucellus; *m* is the micropyle; and *p* is a pollen tube, two of which have reached the neck cells of the archegonia. (*Redrawn from Strasburger.*)

The fertilized egg is the beginning of a new sporophyte generation, and its growth and development take place within and at the expense of the food accumulated in the female gametophyte. Two months after fertilization the young sporophyte, or embryo, is fully formed and occupies the axis of the ovule.

The seed. During the two years following pollination the whole ovulate cone has been enlarging, and the ovules have greatly increased in size. The integument has hardened into a seed coat and the nucellus has been reduced to a membranous layer inside it. Food has accumulated within the remaining portion of the female gametophyte, usually called the *endosperm*. The embryo has several cotyledons and a stem with growing points at either end. The growing point which will ultimately form the root is inclosed in a long sheath.

In late autumn, or winter, the ovulate cone dies and its tissues dry out; the sporophylls curl outward and the seeds are liberated. As the seed separates from the sporophyll, a thin blade of sporophyll tissue goes with it, forming the wing of the seed.

Summary of the conifers. In comparison with the Pteridophytes the conifers show a greatly enlarged root system, more massive trunk, with numerous branches, and an internal structure approaching that of the dicots. The leaves have the form of needles or scales which are sometimes broad, and they are peculiar in being produced, not on the elongating branches, but on small dwarf branches. The rate of growth of the conifers far exceeds that of the cycads and ferns. The conifers likewise are able to grow in almost all land habitats.

The conifers show another step in the simplification of the gametophyte generation. The cycad male gametophyte retains the habit of producing swimming sperms. In the conifers this habit is gone and the male nuclei pass directly from the pollen

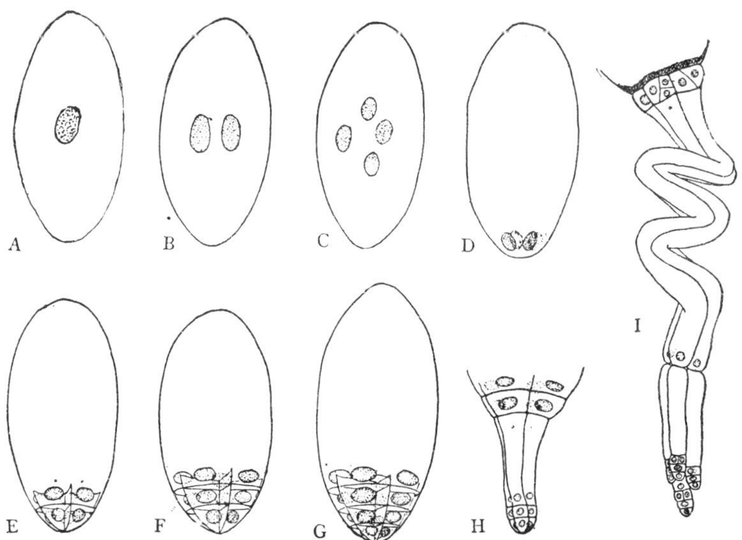

FIG. 322. The formation of the embryo of a pine from the fertilized egg (*A*) to the development of four rudimentary embryos. Only one of the embryos will survive in the mature seed. (*After Buchholz.*)

tube to the egg. The female gametophyte retains its habit of producing a prothallus and archegonia, though the latter are greatly simplified and the prothallus is strictly a parasite.

REFERENCE

COULTER, J. M., and CHAMBERLAIN, C. J. *Morphology of Gymnosperms*. D. Appleton & Co., New York.

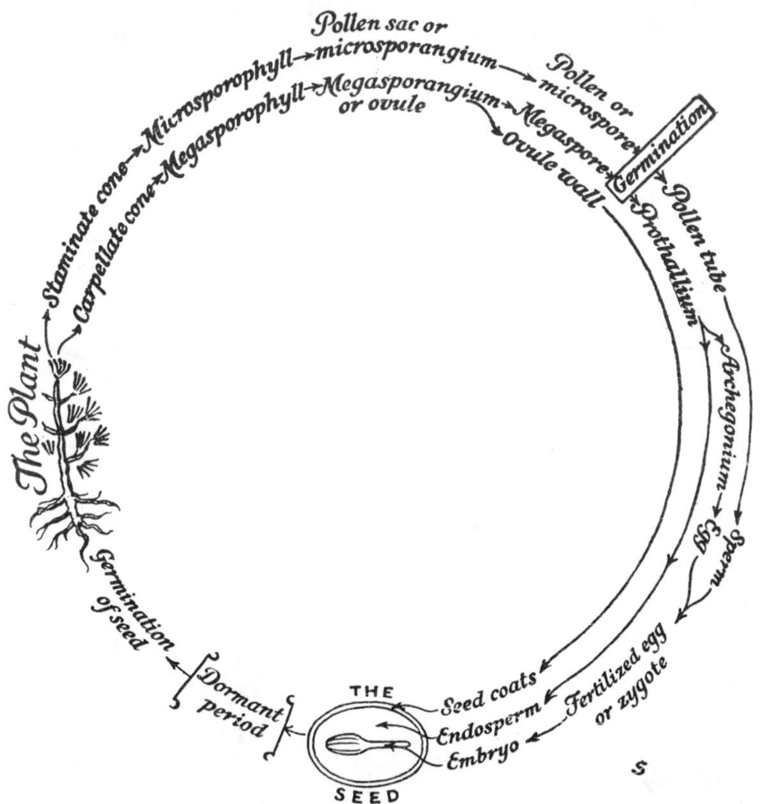

FIG. 323. Diagram of life cycle of a conifer.

CHAPTER FORTY-EIGHT

THE ANGIOSPERMS OR FLOWERING PLANTS

THE flowering plants, or angiosperms, have the shortest geological history of all the vascular plants, though they include the largest number of species and the greatest diversity of vegetative forms and reproductive structures. Since Cretaceous times they have been gradually replacing the gymnosperms, until today they are the dominating plants of the earth.

There are at least 140,000 angiosperms, or about 40,000 more than of all other known species of plants combined. They include herbs, shrubs, vines, and trees, and they vary in size from the duckweed, *Wolffia* (half the size of a pin head), to the Australian *Eucalyptus*, 340 feet high. To this variety of forms is added structural, physiological, and chemical diversity that enables the angiosperms to live in almost every habitat on the earth. Consequently we find the angiosperms in control of most tropical forests, the temperate deciduous forests, the prairies, the plains, and the deserts, and forming the undergrowth of nearly all conifer forests.

With the angiosperms came not only the habit of forming seeds in closed ovularies, but also the production of variously colored bracts and floral leaves surrounding the spore-bearing stamens and carpels. The primitive angiosperm flowers, represented today by the magnolias and the tulip tree (" yellow poplar "), had the sporophylls spirally arranged like the scales on the cones of conifers. Another primitive type, represented by willows, have the very simple flowers spirally arranged in spikes and catkins.

Among the floral organs nectaries appeared, usually near the base of the sporophylls. The Paleozoic landscapes may have rivaled our own, with their varied textures and shades of green, but they were devoid of that color interest which flowers added to the Tertiary and more recent landscapes.

Insects and flowers. The increasing importance of the angiosperms among plants of the Tertiary was paralleled among

Fig. 324. Black willow (*Salix nigra*): *C*, vegetative branch; *A*, branch with spike of pistillate flowers; *B*, spike of staminate flowers.

animals by the increasing number and diversity of insects. This is of botanical interest because flowers and insects seem to have reacted on each other; and there have come to be numbers of insects that are dependent upon certain plants, and likewise many plants whose pollination is effected only by certain insects. Like the gymnosperms, the early angiosperms seem to have been mostly wind pollinated, while the later and more specialized angiosperms are largely pollinated by insects.

The life history. The details of the life history of the angiosperms are somewhat variable in different orders and families. We have already given a general account of floral structures, pollination, fertilization, and seeds in Chapters XXV to XXVIII (pages 232–271), and the first half of the book is concerned with the vegetative structures and processes of the angiosperms. It may be well, however, to repeat the life history in the same terms that we have used to describe those of the preceding plant groups, in order that the student may see that the sequence of events in the life cycle of an angiosperm is very similar to that of a fern, a cycad, or a conifer.

The sporophyte. In addition to the root, stem, scale leaves, foliage leaves, and sporophylls displayed by the gymnosperms, the angiosperms have *bracts* and *floral leaves.*

The bracts are small leaves that occur at the base of flowers and flower clusters. Usually they are green, but in the flowering dogwood they are white or pink and in poinsettia are scarlet.

The floral leaves form a perianth about the *microsporophylls* (stamens) and the *megasporophylls* (carpels). In the simplest flowers they are green; in more advanced flower types the inner cycle forming the *corolla* is white, or variously colored.

FIG. 325. Flowers of *Magnolia conspicua,* showing cone-like arrangement of stamens and carpels. The magnolia is a representative of the order Ranales, believed to be the most primitive group of the angiosperms.

The members of the outer cycle, or *calyx*, are usually green, but in the lily family they are often similar to the corolla in form and color.

The microspores are produced in four sporangia which make up the *anther*. At maturity the sporangia open in pairs so that they appear as two pollen sacs. The young pollen grain, or microspore, consists of a single cell; when shed, the microspore has divided internally and the mature pollen contains two or three cells.

The megaspores arise within megasporangia, which are usually called *ovules*. These are variously arranged inside the carpels (megasporophylls). There may be one or many hundred ovules within each ovulary. Each ovule consists of a *nucellus* inclosed by two integuments, except for one small opening, the *micropyle*. Within the nucellus four megaspores form, but only one of them matures.

The female gametophyte. The megaspore germinates, or continues to enlarge within the nucellus, and ultimately forms the embryo sac or female gametophyte. This consists of seven cells, one of which is the egg nucleus, and near the center of the embryo sac is the fusion nucleus. The female gametophyte, then, has been greatly simplified in comparison with that of gymnosperms.

The male gametophyte. Previous to the time of pollination the microspore has divided internally, forming two nuclei — one, the *tube nucleus;* the other, the *generative nucleus*. The latter may have divided a second time, forming the two *sperms*. The male gametophyte is, therefore, reduced to its simplest form.

Fertilization. After the pollen has reached the stigma and germinated, the pollen tube penetrates the tissues of the style and enters the nucellus of the ovule through the micropyle. As the pollen tube elongates, the tube nucleus maintains a position near the tip and the two sperms follow just behind it.

When the pollen tube reaches the embryo sac, it discharges

the sperms into it. One sperm unites with the egg, forming the zygote or *fertilized egg*; the other unites with the fusion nucleus, forming the *endosperm nucleus*.

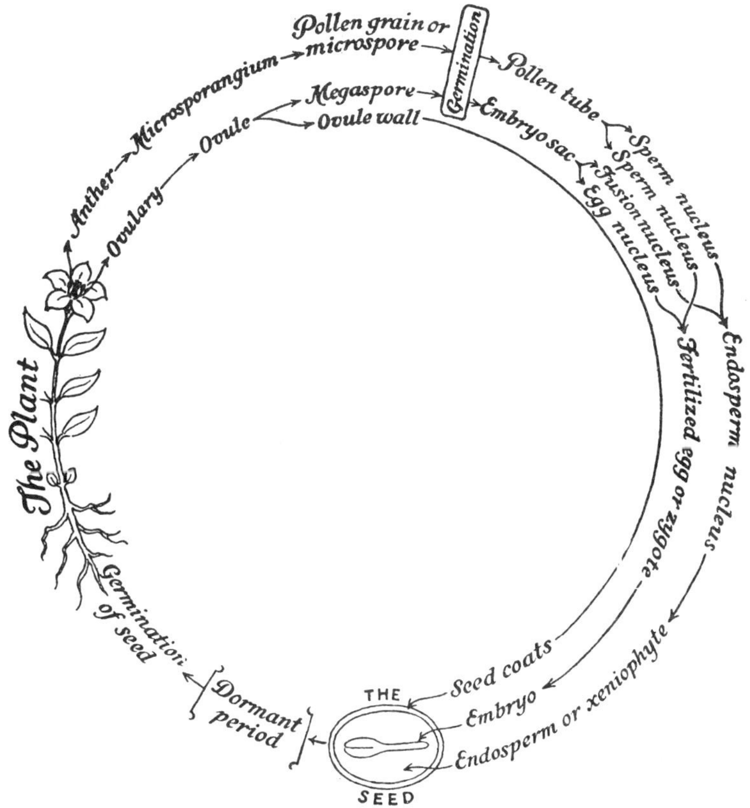

FIG. 326. The life history of an angiosperm.

The embryo. The zygote germinates at once and by cell division forms the embryo, which pushes backward into the developing endosperm. The embryo may consist of only a few cells, as in orchids, or it may grow to considerable size, using up the entire contents of the endosperm, the nucellus, and in rare instances, as in the skunk cabbage, even the integuments.

The embryo consists of a hypocotyl, a plumule, and one (monocots) or two (dicots) cotyledons. This is the new sporophyte generation. Since the sperm and the egg each contain n chromosomes, the cells of the sporophyte contain the double number ($2n$) of chromosomes.

The xeniophyte. The germination and growth of the endosperm nucleus leads to the formation of a tissue surrounding the embryo, called the *endosperm*. In some seeds (e.g., castor bean and corn), this tissue persists and becomes very large, and accumulates starch and other nutritive materials. In other seeds (e.g., common bean) the endosperm is a temporary tissue, that is consumed by the growing embryo.

The endosperm of the angiosperms is formed by successive divisions of a nucleus that resulted from the union of three nuclei. Its cells, therefore, contain three sets ($3n$) of chromosomes. To distinguish this kind of an endosperm from the very different one of the gymnosperms (female gametophyte), it is sometimes called the *xeniophyte*.

The seed. In the seed of an angiosperm, therefore, there are (or for a time were) three distinct generations represented: (1) the integuments, a part of the first sporophyte producing the seeds; (2) the endosperm, or xeniophyte; and (3) the embryo, or second sporophyte.

The seed coats. Strictly speaking, the seed coats are the mature integuments of the ovule. In many plants having simple pistils that contain a single ovule, the seed coat at maturity includes the pistil wall. This is the case in grasses and in the buttercup, rose, carrot, and sunflower families. Such a seed is termed an *akene*, or, in the grasses, a *grain*.

Divisions of the angiosperms. As noted earlier in the book, the angiosperms consist of two great classes of plants, the dicotyledons and the monocotyledons. The former seem to be the older and to have given rise to the monocots in comparatively recent geological time.

Distinguishing features of the dicotyledons. The embryo has two cotyledons, except in certain parasites and saprophytes. The stems are either herbaceous or woody, with open vascular bundles arranged in a circle, and usually with a cambium intersecting the bundles between xylem and phloem tissues. The stems of most dicots are profusely branched. The primary root is retained until maturity, and not infrequently develops into a long tap root. The leaves are usually net-veined, simple or compound, entire or variously lobed and divided. The flowers are usually made up of five cycles of floral parts: the sepals, petals, two whorls of stamens, and the carpels. The number of members in each cycle varies from two to six, but is usually five or four. In some of the simplest flowers the corolla is wanting, and in others there is but a single whorl of stamens or pistils. In some families the number of stamens and pistils is very large.

Distinguishing features of the monocotyledons. The monocots are herbs and woody plants, with closed and scattered bundles. The embryo has usually only one cotyledon, sometimes one large one and a very small one. A cambium is usually absent, but when secondary cambiums are present they arise outside the vascular bundles. The stems of monocots are not as highly branched as those of the dicots. The primary root is usually short-lived, and is replaced by adventitious roots, which are in turn succeeded by new adventitious roots that develop from points higher and higher up the stem.

The leaves are devoid of stipules and commonly lack a petiole, though they frequently have a sheathing leaf base. The flowers consist of five cycles of floral organs, as in the dicots, but the second whorl of stamens is not infrequently wanting. The number of members in each cycle is usually three, but in the grasses this may be reduced to two.

REFERENCE

COULTER, J. M., and CHAMBERLAIN, C. J. *Morphology of the Angiosperms.* D. Appleton & Co., New York; 1903.

CHAPTER FORTY-NINE

SOME FAMILIES OF ANGIOSPERMS

CERTAIN families of the angiosperms are of special interest because of their economic importance, their common occurrence, or their peculiarities. The following families have been selected from among several hundred to exemplify the variety of plants included among the angiosperms, and to show some further characteristics of the monocots and dicots.

MONOCOTYLEDONS

The grass family (***Gramineæ***). The grass family includes the most important food plants in the world. Wheat, corn, rice, barley, rye, oats, and sugar cane furnish the bulk of human food. Corn, wheat, oats, and a variety of wild and cultivated grasses furnish most of the food of grazing animals.

FIG. 327. Flower cluster and spikelet of orchard grass (*Dactylis*), showing glumes, pistils, and stamens.

FIG. 328. Bamboo thicket on an island in the Caribbean. The bamboo is the largest of the grass family and one of the most useful plants in the tropics and subtropics.

There are 350 genera and about 5000 species of grasses. Mostly they are mesophytes, but there are grasses in nearly every habitat where plants grow. In size they vary from grasses of 2 or 3 inches high to the woody bamboos 60 to 100 feet in height and a foot in diameter. The larger grasses are an important source of building materials, fiber, paper pulp, and in the Oriental tropics of innumerable household articles.

The spikelet is the unit flower cluster of the grasses. Spikelets may be variously arranged on a single axis, or on highly branched axes, forming spikes, racemes, or panicles. Each spikelet consists of two or more flowers inclosed by bracts, called *empty glumes*. Each flower is also enveloped by two *flowering glumes*. Next above are two (rarely three) very small bracts called *lodicules*, which represent the remnants of a perianth. Next above are three stamens; and in the center of the flower is the pistil, made up of three carpels forming a single ovulary with one ovule. It is very evident that the grass flower and inflorescence are highly specialized structures.

Bureau of Agriculture, P. I.

FIG. 329. Japanese cane, a near relative of the sugar cane. It is grown as a fodder crop in the Philippines, and is one of the many useful members of the grass family.

Grass stems are usually round, and the internodes are hollow. The leaves are arranged alternately in two ranks.

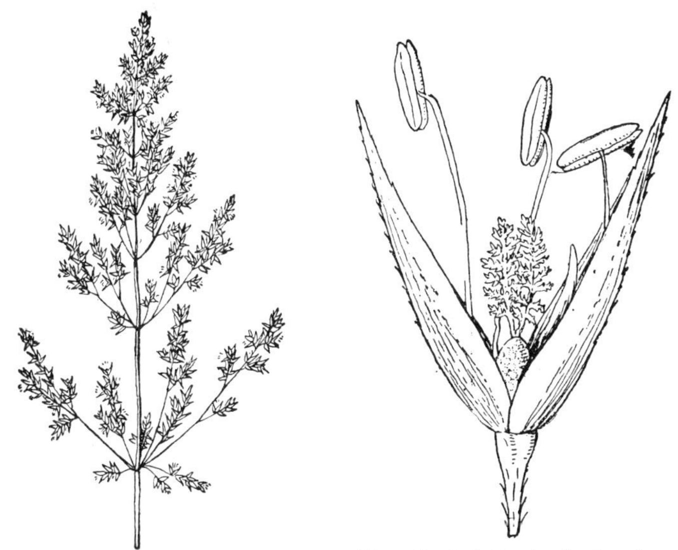

FIG. 330. Panicle and flower of bent grass (*Agrostis*), the latter showing two glumes and the much shorter lemma adjoining the pistil and stamens.

Closely related to the grasses are the sedges (*Cyperaceæ*), with triangular, solid, unjointed stems, and leaves in three ranks. The sedges are common in swamps, marshes, and low grounds. The papyrus, a sedge of the Nile Valley, was one of the first plants used in paper making. In fact, our word "paper" comes from the name of this sedge.

The aroids (***Araceæ***). The calla lily, jack-in-the-pulpit, skunk cabbage, and caladiums are examples of a large family of tropical and temperate herbs, embracing not less than 1000 species. The flower cluster consists of a spike (*spadix*) of very simple flowers, inclosed by a large, sometimes highly colored bract (*spathe*). Many of these flowers are noted for their disagreeable odors.

The rootstocks of some species of caladium, colocasia, and arum accumulate starch and are a source of food in the tropics.

FIG. 331. A tropical climbing aroid (*Monstera*) in bloom. Note the white spathe which incloses the spike (spadix) of flowers. Behind the flower can be seen a ripe fruit. Most members of this family are tropical, but jack-in-the-pulpit and skunk cabbage are familiar representatives.

The palm family (*Palmaceæ*). The most familiar mark of a tropical landscape is the presence of palms, with their tall unbranched stems, topped with a rosette of large divided leaves. Some palms attain a height of 150 feet, others have only a short, upright rootstock, while a few, like the rattans, are climbing vines several hundred feet in length. They are for the most part intolerant of shade, and consequently occur along streams, in clearings, on forest borders, and in oases in deserts. In addition to their edible fruits — dates, coconuts, palm nuts — they are important sources of fibers, oils, wax, starch, sugar, and alcohol. The leaves are used in thatching, in basket making, and in weaving mats and hats. Vegetable ivory is derived from a palm nut. The woody stems do not yield plank timber, because of the scattered bundles. The palms include 130 genera and about 1200 species.

W. S. Cooper

FIG. 332. A yucca (*Yucca whipplei*) in bloom at Cajon Pass, California.

The lily family (*Liliaceæ*). Tulips, hyacinths, lilies, lily-of-the-valley, and trillium are familiar plants of this family. Here also belong the edible onions, leeks, garlic, and asparagus. The yuccas and aloes are the desert representatives of the family. Some of these plants form woody stems and attain the size of trees.

The flowers are characterized by a perianth of six white or colored parts surrounding six stamens, and a three-carpel pistil containing numerous ovules. Two hundred genera and about 2700 species have been described.

The amaryllis family (*Amaryllidaceæ*). Closely related to the lily family, and likewise noted for its large number of ornamental

FIG. 333. Adder's tongue (*Erythronium americanum*), an early spring member of the lily family.

FIG. 334. Flowers of amaryllis

plants, is the amaryllis family. They are characteristic of dry climates and many have leaves only during the rains.

Here belong the amaryllis, narcissus, and tuberose. In the American desert the agaves are widely distributed, and some of the species are important sources of fibers for binder twine and other coarse cordage. Seventy-five genera and 700 species are known, mostly from the tropics and subtropics.

The pineapple family (*Bromeliaceæ*). Found only in the American tropics is the pineapple family, with 40 genera and 1000 species. Some of the species, like the pineapple, are terrestrial plants, but most are epiphytes and form a characteristic feature of the jungle growths from southern Florida and Mexico to southern Brazil. Many of the epiphytes have rosettes of leaves with the bases pressed tightly together, forming water pockets from which the plants secure most of their water. The " Spanish moss " is very different and consists of festoons of leafy branches without even holdfast roots. It is an extreme xerophyte, living not only in the moist subtropics but some species occurring even in semi-desert regions.

The orchid family (***Orchidaceæ***). The orchids form the culminating family of the monocots. They are all perennial herbs, noted for their beautiful, highly specialized and diversified flowers. In temperate regions the species are mostly terrestrial, but in the tropics they are largely epiphytes. During the dry season they drop their leaves, and only the thickened stems, tubers, or fleshy roots pass the dormant period. The fruit is a capsule containing a vast number of minute seeds. In many of the tropical epiphytes the outer layers of the root consist of dead perforated cells, which form a water-holding tissue (the velamen). The great variety of orchids is indicated by the occurrence of 750 genera and 7500 species.

FIG. 335. A tropical orchid (*Cypripedium callosum*).

Other monocot families. The banana family includes many gigantic herbs of the tropics, noted not only for their fruit, but in some species for the fibers obtained from their leaf stalks (Manila fiber).

The yam family includes many climbing herbaceous plants, with thick underground tubers that in tropical countries are eaten like potatoes. The leaves of this family resemble those of dicots.

To the monocots also belong many of the families which include our commonest water plants, like pondweeds, pickerel weed, cat-tails, rushes, eel grass, and water hyacinths.

Without further examples it is evident that most of the monocot families attain their best development in the tropical and subtropical countries.

FIG. 336. Leaves and flower of the tulip tree (*Liriodendron tulipifera*). This magnificent tree of the deciduous forest is closely related to the magnolias.

DICOTYLEDONS

Willow family (***Salicaceæ***). Among the most widely distributed trees and shrubs of the northern hemisphere are the willows and poplars. Both have simple naked flowers in catkins, and the staminate flowers and carpellate flowers occur on different individuals. Both willows and poplars reproduce freely from cuttings. The poplars, especially the cottonwood, is used for paper pulp. Some species of willow are grown for their sprouts, which are used in making baskets and furniture.

Beech family (***Fagaceæ***). The beech family includes the beeches, oaks, and chestnuts. The fruit consists of a cup-like structure inclosing one, two, or three nuts. They are chiefly valuable for their timber products.

Closely related to the beech family is another family of trees and shrubs, the birch family (*Betulaceæ*), which includes the birches, hornbeams, hazels, and alders. All these plants have very simple flowers in spikes or catkins, and most of them are wind-pollinated.

Buttercup family (***Ranunculaceæ***). This family is typical of a large order, known as the *Ranales*, which includes many common herbs, trees, and shrubs: the buttercups, water lilies, anemones, columbine, May apple, larkspur, sassafras, tulip tree,

FIG. 337. Flowers of the Japanese anemone (*Anemone japonica*). It belongs to the buttercup family.

and magnolias. The flowers are solitary and conspicuous. The receptacle is usually elongated, and the parts of the flower are arranged spirally about it. The calyx and corolla are not distinct in shape or color, and the sporophylls are indefinite in number.

The flowers of this order are generally regarded as primitive, and it has been suggested that the monocots were derived from the ancestral forms of the order *Ranales*.

The mustard family (***Cruciferæ***). The scientific name of the family is derived from the cross-like arrangement of the four petals. In this family the four sepals are green and quite distinct from the petals. The stamens are six in number, four long and two short. The single ovulary is divided by a membrane into two compartments, each of which contains a row of ovules.

To the family belong many troublesome weeds; a variety of edible herbs like cabbage, cauliflower, turnip, radish, caper, and cress; and the ornamental wallflowers and stocks.

The pitcher-plant family (***Sarraceniaceæ***). This small family is one of three families belonging to the order *Sarraceniales*.

They are mentioned here merely because of the fact that all three families are made up of insectivorous plants. In the pitcher plants the insects die by drowning in the water contained in the pitcher-like leaves. In the closely related sundews, the insects are caught by the sticky secretion from glandular hairs on the upper surface of the leaves. In the Venus' flytrap the blade of the leaf consists of two halves which fold together. On the upper surface of each half are three hairs and numerous small reddish glands. The margins of the blades have tooth-like projections. When the hairs are touched the two halves of the blade suddenly close, the marginal teeth interlock, and small insects may be caught. In all these plants the insects are subsequently digested and the products absorbed by the plants. It is rather remarkable that three such unusual habits should have arisen within a single order of plants. It should be stated that all these plants may be grown in conservatories without feeding them insects.

FIG. 338. Pitcher-like ends of the leaves of *Nepenthes*, one of a group of tropical epiphytes. These pitchers contain water and are provided with glands that secrete enzymes and absorb the products resulting from the digestion of insects that drown in them.

FIG. 339. A wild rose (*Rosa lucida*) of the Northeastern states, that occurs on swamp margins and rocky shores.

FIG. 340. The flowering raspberry (*Rubus odoratus*), a member of the rose family.

Rose family (***Rosaceæ***). This is a cosmopolitan family of 100 genera and more than 2000 species. It is notable because of the large number of useful plants that are cultivated for their flowers or fruits. Here belong the roses, spiræas, cinquefoils, strawberries, raspberries, blackberries, pears, apples, cherries, and plums. The receptacle of the flower is usually hollowed, so that the five sepals and petals surround a cup in which the numerous stamens and the five to many carpels are borne. In the strawberry the fruit is the enlarged fleshy receptacle.

The legume family (***Leguminosæ***). This is the second largest family of flowering plants, and in the importance of its food products is second only to the grasses. It includes 500 genera and not less that 12,000 species, some of which grow in every climate and habitat. Most of the plants have tubercles on their roots and are hosts to nitrogen-fixing bacteria. Here are included the sensitive plants (*Mimosa*), the acacias, red buds, locusts, peanuts, lupines, clovers, beans, peas, and soy beans.

FIG. 341. Portion of a plant of hairy vetch (*Vicia*). The papilionaceous flowers identify it as a member of the legume family. The outer branches of the compound leaves are tendrils, as in the pea.

The acacias and mimosas are largely tropical and subtropical trees and shrubs, with regular flowers. The genera common to

FIG. 342. Flowering branch of *Acacia senegal*, one of the many leguminous plants with radial flowers widely distributed in the tropics. In the southern United States there are several common species of *Acacia* and the closely related *Mimosa*. *Acacia senegal* is the source of gum arabic. (*After Strasburger*.)

FIG. 343. Peppermint (*Mentha piperita*). The square stems and opposite leaves are characteristic of the mint family.

temperate regions have irregular flowers like the sweet pea. All produce the pod, opening by two sutures called a *legume*.

The cactus family (***Cactaceæ***). This group of succulent desert plants seems to have originated in tropical America and to have spread sparingly into the dry temperate regions both north and south. There are about 25 genera and 1500 species. For the most part they lack leaves and the stems are covered

FIG. 344. The cranberry (*Vaccinium macrocarpon*), one of the low heaths that grows naturally in bogs. This plant was brought into cultivation many years ago. The size of the berries has been doubled by selection of large-berried mutants.

FIG. 345. The great laurel (*Rhododendron maximum*), an evergreen shrub of the Alleghenies. Like the mountain laurel (*Kalmia*) and the azalea, it belongs to the heath family.

with spines. They vary from small perennial herbs to large, much-branched, tree-like forms.

The carrot family (*Umbelliferæ*). The scientific name of the family comes from the umbrella-shaped inflorescence. They are mostly herbs with stout stems, hollow internodes, and divided leaves. The carrot, parsnip, celery, fennel, coriander, and water hemlock are familiar examples of the family. Some of these plants are poisonous when eaten, and many are noted for their peculiar flavors. They are chiefly found in the north temperate zone and include 200 genera and 2700 species.

The heath family (*Ericaceæ*). The family is distributed throughout the world, except in deserts and the moist tropics. Most of the plants have simple, evergreen, entire leaves which

FIG. 346. *Andromeda floribunda*, one of the heaths common on moist hillsides in the southern Alleghenies.

tend to be grouped at the ends of the branches. They are confined to acid soils, and many are found in bogs. The azaleas,

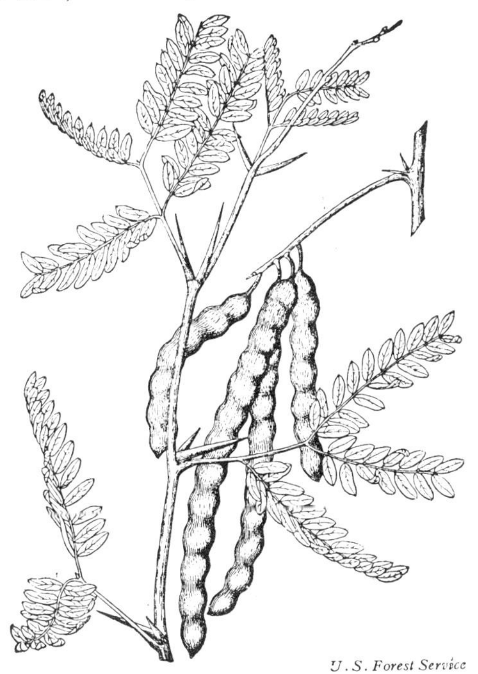

U. S. Forest Service

FIG. 347. Mesquite (*Prosopis juliflora*), a widely distributed shrub in the Southwestern states, belonging to the legume family. (See Figure 240.)

rhododendrons, laurels, arbutus, heather, huckleberries, blueberries, cranberries, and wintergreen are examples of both the ornamental and fruit-producing members of this family.

The flowers usually have five parts in each whorl, and the corolla differs from that of all the preceding families in showing a tendency to have the petals united. In the succeeding families this tendency culminates in the production of tube-like corollas.

The mint family (***Labiatæ***). The corolla of the mints is tubular, and frequently two-lipped, which suggested the technical name. They are world-wide in their distribution, and they num-

FIG. 348. Salvia, a large-flowered member of the mint family, with the characteristic square stem and two-lipped tubular corolla and calyx.

ber 200 genera and 3000 species. They are mostly herbs, with square stems and simple leaves, and with epidermal glands secreting volatile oils that give the characteristic odors to many of the species. The floral whorls, except the pistil, each consist of five members. The pistil is composed of two carpels, each of which is two-lobed, so that the fruit consists of four nutlets. The oils of peppermint, spearmint, thyme, lavender, rosemary, and horehound are of commercial importance in the manufacture of flavoring extracts, perfumes, and medicines.

FIG. 349. Climbing nightshade (*Solanum dulcamara*), one of the wild species belonging to the potato family.

The potato family (*Solanaceæ*). This group of 75 genera and 1500 tropical and temperate species is best developed in Central and South America. It includes herbs, shrubs, and small trees with petals united into a disk, or forming a tube with flaring end. The fruit is a berry or a capsule. Many of the species are poisonous. The potatoes, tomatoes, and peppers are familiar garden species. Equally important commercially is the tobacco plant. The "deadly nightshade" is the source of the drugs atropine and belladonna.

The sunflower family (*Compositæ*). This is the largest family of flowering plants, comprising about 900 genera and more than 13,000 species. Most of the species are herbs, though in the tropics there are a few shrubs and trees. The flowers are usually small, with tubular or strap-shaped corollas. The flowers, however, are grouped in heads, with an outer circle of green bracts so that the flower cluster is frequently mistaken for a single flower. The fruits are achenes, and in many species the fruits have a ring of bristles which lead to their distribution by the wind. In this, the culminating family of the dicotyledons, the production of a multitude of seeds is accomplished by the occurrence of many small flowers in heads. This is in striking contrast to the orchids, which represent the most specialized family of the monocotyledons, in which the flowers are

FIG. 350. Flower clusters of dahlia, sunflower, and thistle, members of the composite family. The small flowers are collected in heads, surrounded by bracts.

few and the seeds are produced in enormous numbers in each capsule.

The composites include the various species of chicory, dandelion, lettuce, ragweed, cocklebur, aster, sunflower, ironweed, goldenrod, fleabane, everlasting, rosinweed, coneflower, Spanish needle, chrysanthemum, and thistle. Many of these plants are weeds, some are cultivated as ornamentals, and a few are of economic importance as food.

REFERENCES

In addition to the well-known manuals:

HARSHBERGER, J. W. *Pastoral and Agricultural Botany.* P. Blakiston's Son & Co., Philadelphia; 1920.

ROBBINS, W. W. *The Botany of Crop Plants.* P. Blakiston's Son & Co., Philadelphia; 1917.

SAUNDERS, C. F. *Useful Wild Plants of the United States and Canada.* Robert M. McBride & Co., New York; 1920.

WILLIS, J. C. *Manual and Dictionary of the Flowering Plants and Ferns.* The Macmillan Company, New York; 1919.

CHAPTER FIFTY

EVOLUTION OF PLANTS

THOSE who have studied plants most have been led to the conclusion that simple plants lived first on the earth, and that from these simple forms all the varied and highly complex plants of today have been derived; that is, that the present-day plants were *evolved* from simpler plants that existed on the earth in former times. Some of the simple plants of the past still persist, and many plants of intermediate degrees of complexity survive; but during the long period of geological time, new and increasingly *complex* plant forms have been produced, and these higher forms now dominate the vegetation of the earth. The process by which the plants of today have come from the plants of the past is called *evolution* (Latin: *evolutio*, an unrolling). Evolution, with regard to plants, implies (1) that the plants of today are the modified descendants of earlier forms, (2) that modifications are going on now as in the past, and (3) that there will be new plants in the future, evolved from plants now living through modification of present plant forms.

The proofs of evolution in plants have been gathered from many sources by many different students. These proofs include the evidence furnished (1) by plant remains found in rocks and coal, (2) by the distribution of plants on the earth's surface, (3) by the remarkable similarity of organs, tissues, and cells among the thousands of plants now in existence, (4) by the similarity in the life histories of all plants, (5) by intergrading species, (6) by the experience of plant breeders and the history of our cultivated plants, and (7) by the discovery of new mutants from time to time.

The geological record. The earliest rocks (Precambrian) contain few recognizable plant fossils, not because plants were rare when these rocks were laid down, but because the rocks during the long subsequent history of the earth were acted upon by water, high temperature, and the enormous pressure of overlying later strata. The occurrence of carbon in these rocks is presumptive evidence that plant remains were present when they were originally laid down.

FIG. 351. Fossil imprint of a leaf of a species of sassafras in rock of the Cretaceous period. But few fossil angiosperms have as yet been found in rocks formed earlier than the Cretaceous period.

The known fossil record shows that during successive periods of the earth's history plant groups succeeded one another and that there was a gradual increase in the diversity of plant forms, accompanied by progressive changes in both the vegetative and reproductive structures of the plants. Modern plant structures are clearly derived by further development and modification of the structures of plants of former geological periods.

That the geological history of each of the plant phyla provides positive evidence for the evolving of new and more complex forms from previously existing forms is clear and unmistakable. In the phyla Cordaites and Pteridosperms, we have the record of the evolution, the world-wide dispersal, and the decline and extinction of two great plant groups.

The trend of evolution. Not only does the geological record furnish abundant proofs of evolution, but it shows the course of the evolution of plants. The series of reproductive structures, for example, beginning with simple sporangia on foliage leaves, may be traced upward through the development of sporophylls and finally to the production of flowers and seeds. The vascular systems of plant stems show a progressive series of changes from the primitive ferns to the modern flowering plants. The development of large and effective root systems may be traced in the same way.

It is quite impossible to account for these gradual and progressive changes in plants except on the basis of evolution. When we understand that the geological record of evolution covers a period of an estimated length of several hundred million years, we should not become impatient at failing to see new genera and families of plants arising during our own brief period of observation. The time that has elapsed since critical observations upon the evolution of living plants have been made, when compared with the time represented by the geological record, is like one second for the observation of the events of a year.

Plant geography. Closely related to the fossil record is the evidence of evolution that is derived from the present distribution of plant groups. Closely related species of plants are not scattered haphazard over the earth. Many families bear evidence of having originated on some particular continent, or part of a continent, and of having spread from the center of origin as new species appeared. Some species have not spread far from their point of origin, while others have moved far from the place of their first appearance because of characteristics which enabled them to live in a variety of conditions. Families which, because of their structures and the absence of a fossil record, are believed to be very modern are usually restricted in their distribution. Ancient families, on the other hand, often have species scattered over several of the continents.

The cactus family, represented by about 1500 species, is native in North and South America only. In North America the family is best developed in Mexico, but it has spread northward and eastward into the United States and to the islands of the West Indies. The geographic distribution of all the North American species points to a common origin in the Mexican plateau. The yucca family and the agave family also appear to have originated there and to have spread in a similar way to the

United States and the West Indies. All these families are comparatively modern.

The laurel family (*Lauraceæ*), to which the European laurel, sassafras, cinnamon, and spice bush belong, is a very ancient family. Its fossil record extends back to the lower Cretaceous. Today its members are scattered widely over the earth, with numerous species in Brazil and southeastern Asia.

Hundreds of examples of this kind might be cited, and all would afford evidence that related plants are distributed over the earth's surface as though they had originated in some one locality and had then spread to other regions. Sometimes they became diversified chiefly at their center of origin; sometimes as they spread they formed secondary centers of diversification. But in all cases the species that occur along a given line of migration are closely related.

The geography of plants, therefore, furnishes a second line of evidence that existing plant species have been derived from preëxisting species.

Comparative anatomy and physiology. One of the most striking proofs of evolution is the remarkable similarity of the cells, tissues, and organs that make up plants belonging to diverse groups. However much they may differ in superficial appearance and in detail, they all have a common plan and organization. The diversity has been brought about through modification in one direction or another. Even more remarkable is the similarity of the physiological processes underlying life, not only in all plants, but in animals also.

Life histories. Except on the basis of evolution it would be impossible to account for the fact that throughout the whole plant kingdom the life histories are so strikingly similar. As we pass from simple plants to the flowering plants, the life histories become more and more complex. Attention has already been called to these facts, and it is only necessary here to repeat that the changes in life history have been made by comparatively small steps. There has been an occasional addition of a new tissue or organ to the life cycle, or the replacement of one structure by another.

The life histories of related groups are similar in essentials and differ only in details. This repetition of the stages in the life cycles of the plants of different groups, when viewed along with other facts of evolution, indicate that the plants with the more complex life histories have evolved from those with less complex life histories. Increase in complexity is one of the general tendencies of evolution. The order in which we should arrange plants on the basis of the geological record is the same as the order suggested by their life histories and structures.

Intergrading species. All who have attempted to classify plants — that is, to determine the species to which individual specimens belong — have been impressed by the intergrading of related species. The existence of individuals intermediate between species long ago suggested that one species may have arisen from another. For example, the common asters, violets, hawthorns, evening primroses, and willows are highly variable; and in any of these genera it is frequently impossible definitely to classify a particular specimen and to say that it belongs to this or that species. If forms intermediate between species were rare, they would only suggest the possibility of evolution; but they are numerous, occurring in hundreds of genera throughout the plant kingdom. These intergrades make it impossible for us to think of the plant kingdom as being made up of distinct and unrelated species, and so they must be regarded as evidences of evolution.

Plant breeding and evolution. Our cultivated plants are the modified descendants of wild species. Many of them, perhaps most of them, were brought into cultivation by wild tribes of men long before the dawn of written history. In many instances the plants have been so greatly modified that it is difficult or impossible to trace their origin to any known wild species. Thus corn was cultivated by the earliest races of men on the American continents. When first found by the early explorers the Indians not only had corn, but they were growing all of the subspecies that we now distinguish as starchy, sweet, soft, waxy, flint, pod, and pop corns. The wild species from which corn was derived is unknown. Since the discovery of America some of these subspecies have been greatly improved by crossing and selection.

All the modern varieties of cultivated plants which supply our fruits, flowers, roots, tubers, and fibers have resulted from the activities of plant breeders. They are mutants selected either from former wild plants or from previously grown varieties. In many plants which cross-pollinate readily, the selection was preceded by hybridization, which often produces plants with new combinations of desirable characters.

The experience of plant breeders furnishes abundant evidence that plants produce mutants which differ in one or more characters from their parents, and that these new characters are heritable. These are the starting points of new varieties and species.

Plant breeding, then, has afforded us an opportunity to see new varieties of plants evolve from older ones. The evolution of many cultivated plants is a matter of historic record. There is evidence that wild plants also produce mutants, and there is every reason to believe that they have evolved in the same way.

The fact of evolution conceded. The time has long since passed when botanists have asked for further proofs of evolution. Nevertheless, new evidences of evolution are appearing from day to day in every field of botany, for the discovery of new facts about the structure, physiology, or chemistry of plants frequently furnishes important new proofs of evolution. We may say, then, that the evolution of plants is a fact, not a theory.

The method of evolution. While evolution may be considered a fact, the methods and causes of evolution are still problems upon which field observations, intensive laboratory study, and extensive field experimentation are going on. The experimental study of evolution is a comparatively new field, and the accumulated data are not yet sufficiently numerous to do more than suggest some of the factors which cause evolution to occur. Among these are *variation*, *heredity*, and *natural selection*.

Variation. Mutants seem to be the chief sources of new varieties of plants. These have been discussed in Chapter XXX. There is another type of variation which is also important in evolution, and that is the variations that result from hybridization. Mutants result apparently from some change in the germ cells, due to unknown causes. Hybrid variants are due to new combinations of characters, derived in part from the pollen parent and in part from the ovule parent. Among evening primroses, oaks, and hawthorns, hybrid variants are very common. Mutants are known among evening primroses, sunflowers, grasses, hemp, flax, and many other wild and cultivated species.

Heredity. The tendency of heredity is to make the offspring like the parent. When a mutation has occurred, heredity becomes an important factor in evolution, since only through heredity can the new variety be maintained. In hybrid variants heredity can maintain a new variety only when both the pollen and ovule parents have the same constitution.

Natural selection. Most plants produce offspring by the hundreds, thousands, or even millions, and there is room for only a small part of the offspring to live. It is said that those plants survive that are more vigorous, that are better adjusted to their environments, or that happen to start in favorable places; the weak and the unfortunate perish. Certain variations or mutations may fit plants the better to survive, and the persistence of the forms showing these changes may lead to the formation of new varieties and species. The wholesale destruction of individual plants in nature, with the survival of a few, is called *natural selection*, and it has been thought to resemble in some respects the selection made by the plant breeder. It is unquestionably true that most of the plants that start life in nature die before reaching maturity; but there are differences of opinion as to whether or not the plants that do survive can through repeated selections *in nature* develop into new species. Man can pick out new forms that originate among the plants that he cultivates and by breeding from them secure new varieties, and it is believed by some that in nature certain advantageous mutations are selected or preserved in a similar way.

The agencies that are supposed to do the selecting in nature are the factors of the environment. Almost any environmental factor may become a limiting factor for the growth of some particular variety of plant. If mutants occur which are not limited in the same way or to the same degree, such mutants survive. During geological time the great changes in the elevation of continents, in connections between continents and islands, in the climate, and in the habitats available have been major factors in determining the changes in the kinds of plants that survived.

Summary. In this final chapter an attempt is made to define evolution, to show the sources from which the proofs of evolution are obtained, and to distinguish between the fact of evolution and the tentative explanations which have been offered to account for evolution. Botanists are now generally agreed (1) that variations are the possible sources of evolution, (2) that those variations which are inherited, particularly mutants, are the only ones which lead to new varieties and species, and (3) that from among these mutants some survive and some perish, according as they fit into or fail to fit into the environment.

INDEX